Applications of Artificial Intelligence in Forestry

Applications of Artificial Intelligence in Forestry

Guest Editors

Guojie Wang
Zengxin Zhang

MDPI

Basel • Beijing • Wuhan • Barcelona • Belgrade • Novi Sad • Cluj • Manchester

Guest Editors
Guojie Wang
School of Remote Sensing and Geomatics Engineering
Nanjing University of Information Science & Technology
Nanjing
China

Zengxin Zhang
College of Forestry
Nanjing Forestry University
Nanjing
China

Editorial Office
MDPI AG
Grosspeteranlage 5
4052 Basel, Switzerland

This is a reprint of the Special Issue, published open access by the journal *Forests* (ISSN 1999-4907), freely accessible at: https://www.mdpi.com/journal/forests/special_issues/T4SHY8C92E.

For citation purposes, cite each article independently as indicated on the article page online and as indicated below:

Lastname, A.A.; Lastname, B.B. Article Title. *Journal Name* **Year**, *Volume Number*, Page Range.

ISBN 978-3-7258-3455-6 (Hbk)
ISBN 978-3-7258-3456-3 (PDF)
https://doi.org/10.3390/books978-3-7258-3456-3

Contents

About the Editors

Guojie Wang

Prof. Guojie Wang is a Professor in the School of Remote Sensing and Geomatics Engineering at the Nanjing University of Information Science and Technology. He is interested in the assessment of forest functions and forest–climate interactions in the warming climate using a variety of deep learning techniques. The main goals of this current research are to assess forest functions and their contributions to energy and water fluxes between land and atmosphere to understand the forests' response to and, conversely, impact on climate change in the recent decades.

Zengxin Zhang

Prof. Zengxin Zhang is a Research Fellow in the College of Grasa and Forestry at Nanjing Forestry University. Prof. Zhang is interested in species' responses to climatic and environmental changes, and his current areas of research mainly focus on foresty ecology, landscape ecology, vulnerability, and climate change. The main goals of his current research are to assess the factors that contribute to the success or failure of ecological restoration and the additional impacts of vegetation greening on the WUE and water resources in Northwestern China.

forests

MDPI

Article

Tree Recognition and Crown Width Extraction Based on Novel Faster-RCNN in a Dense Loblolly Pine Environment

Chongyuan Cai [1,2,3], Hao Xu [4], Sheng Chen [5], Laibang Yang [6], Yuhui Weng [7], Siqi Huang [8], Chen Dong [1,2,3,*] and Xiongwei Lou [1,2,3,*]

1 College of Mathematics and Computer Science, Zhejiang A&F University, Hangzhou 311300, China
2 Key Laboratory of State Forestry and Grassland Administration on Forestry Sensing Technology and Intelligent Equipment, Zhejiang A&F University, Hangzhou 311300, China
3 Key Laboratory of Forestry Intelligent Monitoring and Information Technology Research of Zhejiang Province, Zhejiang A&F University, Hangzhou 311300, China
4 Zhejiang Forestry Bureau, Hangzhou 310000, China
5 Center for Forest Resource Monitoring of Zhejiang Province, Hangzhou 310000, China
6 Hangzhou Ganzhi Technology Co., Ltd., Lin'an 311300, China
7 College of Forestry and Agriculture, Stephen F. Austin State University, Nacogdoches, TX 75962, USA
8 Longquan Urban Forestry Workstation, Longquan 323700, China
* Correspondence: dongchen@zafu.edu.cn (C.D.); lxw@zafu.edu.cn (X.L.); Tel.: +86-158-6718-0919 (C.D.); +86-135-8822-8755 (X.L.)

Abstract: Tree crown width relates directly to wood quality and tree growth. The traditional method used to measure crown width is labor-intensive and time-consuming. Pairing imagery taken by an unmanned aerial vehicle (UAV) with a deep learning algorithm such as a faster region-based convolutional neural network (Faster-RCNN) has the potential to be an alternative to the traditional method. In this study, Faster-RCNN outperformed single-shot multibox detector (SSD) for crown detection in a young loblolly pine stand but performed poorly in a dense, mature loblolly pine stand. This paper proposes a novel Faster-RCNN algorithm for tree crown identification and crown width extraction in a forest stand environment with high-density loblolly pine forests. The new algorithm uses Residual Network 101 (ResNet101) and a feature pyramid network (FPN) to build an FPN_ResNet101 structure, improving the capability to model shallow location feature extraction. The algorithm was applied to images from a mature loblolly pine plot in eastern Texas, USA. The results show that the accuracy of crown recognition and crown width measurement using the FPN_ResNet101 structure as the backbone network in Faster-RCNN (FPN_Faster-RCNN_ResNet101) was high, being 95.26% and 0.95, respectively, which was 4.90% and 0.27 higher than when using Faster-RCNN with ResNet101 as the backbone network (Faster-RCNN_ResNet101). The results fully confirm the effectiveness of the proposed algorithm.

Keywords: ResNet101; FPN; UAV; deep learning; loblolly pine

Citation: Cai, C.; Xu, H.; Chen, S.; Yang, L.; Weng, Y.; Huang, S.; Dong, C.; Lou, X. Tree Recognition and Crown Width Extraction Based on Novel Faster-RCNN in a Dense Loblolly Pine Environment. *Forests* **2023**, *14*, 863. https://doi.org/10.3390/f14050863

Academic Editor: Helmi Zulhaidi Mohd Shafri

Received: 22 February 2023
Revised: 1 April 2023
Accepted: 21 April 2023
Published: 23 April 2023

1. Introduction

A tree crown comprises the part of the tree bearing live branches and foliage. Photosynthesis occurs in leaves, and its resulting products are translocated to other tree parts via branches. Therefore, foresters always use the tree crown's characteristics, particularly the crown width, to describe a tree's growth potential. Previous studies have confirmed strong, positive relationships between crown width, tree growth, and carbon sequestration [1]. Hao et al. studied the relationship between teak growth factor and crown width, and established a crown growth prediction model, providing theoretical support for the management of teak plantations [2]. In a 10-year comparative study, Jones et al. demonstrated relationships between crown damage and survival, diameter growth, and tree height growth in Douglas firs [3]. Putney and Maguire studied nitrogen use efficiency in

Douglas fir plantations in western Oregon, where tree growth was measured by changes in crown shape and vertical leaf distribution [4]. Feng et al. argued that the vertical functional variation in leaf traits might indicate niche partitioning within forests [5]. It has been noted that crown width information is also vital in forest modeling, especially for models that include competition indices [6–8]. Therefore, it is of great interest to foresters to develop methods that can accurately measure crown characteristics such as crown width and height.

Tree crown width is often defined as the average width of a tree crown in the north–south and east–west directions [9]. Despite the wide use of tree crown width data in managing forests, accurately measuring crown width is always challenging. Conventional crown width measurement methods include the vertical sighting method [10] and the projection method [11]. The vertical sighting method is quick but less accurate than the projection method. The projection method takes a long time and has low measurement efficiency [12]. However, trees often grow in rows, with tree crowns of varying shapes overlapping, and there is also incompleteness caused by occlusion, making individual tree crown extraction a challenging problem [13]. The use of new techniques to measure crown width has become a hot topic in recent years. With the popularity of smart mobile devices, some scholars have used smartphones to identify and measure tree crowns. For example, Xinmei et al. proposed a passive method for the measurement of tree height and crown diameter based on a smartphone monocular camera [14]. With the development of artificial intelligence and unmanned aerial vehicle (UAV) technology, interest in using UAVs equipped with laser radar and high-definition cameras to measure the crown width of trees is increasing. For example, Ahmadi et al. proposed segmenting early Ganoderma-infected oil palms based on UAV images and artificial neural networks [15]. Safonova et al. proposed a method for extracting tree crowns from UAV images for species classification and stand assessment [16]. Kolanuvada et al. used a UAV paired with a multispectral camera to obtain photos of multiple frequency bands of a forest, employed a simple deep learning convolutional neural network (CNN) to train the images, and developed a linear clustering algorithm to optimize the crown extraction and obtain the crown measurements [17]. Guerra-Hernández et al. used a UAV equipped with an aerial camera and a laser scanner to obtain the high-density 3D point cloud of a eucalyptus plantation, and then conducted 3D modeling to obtain the 3D canopy structure of the eucalyptus forest, which was then incorporated into a prediction of the volume of eucalyptus plantations [18]. Gurumurthy et al. proposed a method for the semantic segmentation of mango trees in high-resolution aerial images and a new method for single crown detection using the segmentation output [19]. To mitigate the impacts of the great homogeneity of neighboring trees and the interlaced crown, Li et al. proposed a crown width estimation method based on an adaptive neuro-fuzzy inference system to improve the intelligence level of crown width estimation [20]. Ritter and Nothdurft proposed a multi-layer seeded region growing-based approach for automatically assessing crown projection areas (CPAs) based on 3D point clouds derived from terrestrial laser scanning (TLS) [21]. In a study based on larch plantations with different stem densities, a two-stage individual tree crown (ITC) segmentation method using airborne light detection and ranging (LiDAR) point clouds was presented [22]. Quan et al. (2019, 2020) evaluated the ability of a UAV laser scanning (UAVLS) system to extract crown structure information from larch plantations [23,24]. They also compared the accuracy of the UAVLS system and airborne laser scanning (ALS) in extracting crown feature attributes. Currently, most crown extraction methods are based on laser scanning and semantic segmentation techniques. Laser scanning technology and segmentation technology can extract more information about the crown, but laser scanning equipment is expensive, and segmentation technology is complicated to use in terms of dataset establishment and it requires outlining along the crown edge. Therefore, it is necessary to find a low-cost, hardware-intensive crown extraction method. The dataset construction of the object-detection model is highly convenient for rapid crown detection and crown width measurement.

Loblolly pine is the second most widely distributed tree species in the United States, and it is the most important commercially in the southeastern United States. Therefore, monitoring the growth of these loblolly pine stands is vital to efficiently manage the stands. In 2021, Lou et al. applied object-detection technology to the measurement of loblolly pine crowns [25]. A UAV was used to obtain the orthophoto images of young and mature stands of *Pinus taeda* in eastern Texas, USA, and three advanced object-detection methods were used to identify the crown and extract the crown width. The faster region-based convolutional neural network (Faster-RCNN) method performed significantly better than the single-shot multibox detector (SSD) on sparse young loblolly pine forests, but on the mature loblolly pine stand, the Faster-RCNN model performed poorly in recognizing the crown and measuring the crown width. The poor performance of Faster-RCNN in the mature stand was unexpected since, in theory, Faster-RCNN is a second-order detector, while you-only-look-once (YOLO) and SSD are single-order detectors. Compared with single-order networks, second-order networks are often more accurate with advantages in multi-scale, high-precision, and small-object detection [26]. Faster-RCNN also outperforms the other two methods in handling the spatial constraints of the algorithm. The main Faster-RCNN improvement is to enhance the adaptability of Faster-RCNN for the crown detection and measurement of both sparse young stands and dense mature stands of loblolly pine. In order to enhance the performance of Faster-RCNN in dense loblolly pine forest sample sites, this study proposes two new Faster-RCNN algorithms, which are then applied to a mature stand to evaluate their accuracy in recognizing tree crowns and measuring tree crown widths.

2. Materials and Methods

2.1. Materials

2.1.1. Image Acquisition

Dataset creation is a critical step in object detection using deep learning models. The study area was located in east Texas, which has a subtropical climate, with heavy rain during the summer.

The study area was located in Rusk, Cherokee County (31°45′31.3″ N, 95°02′318″ W). The site was originally an old field on flat terrain. The site was planted with loblolly pine seedlings in 2001, and in September 2019, when the photos were taken, it had become a mature pine stand with a closed canopy and a high density. The trees averaged 22.2 cm in diameter at breast height (DBH), 16.9 m in total height, and the stand had 35.2 m^2 in basal area per hectare. Figure 1 shows a global orthophoto image of the study area.

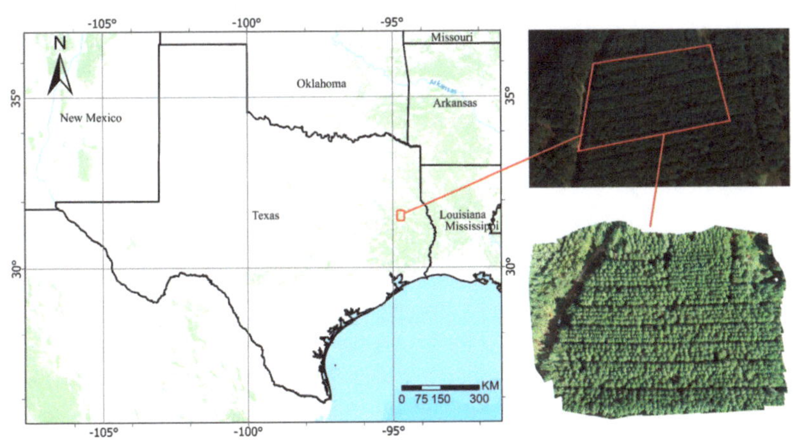

Figure 1. Orthophoto image of loblolly pine plot.

The UAV model used in this study was the DJI Phantom 4 Pro, manufactured by Shenzhen DJI Technology Co., Ltd. This UAV is equipped with a 1-inch 20-million-pixel image sensor with a maximum ascent speed of 6 m/s, a maximum descent speed of 4 m/s, and a maximum horizontal flight speed of 58 km/h in attitude mode. We used Pix4Dcapture (PIX4D) software to control the flight and PhotoScan (v1.2.5) to generate the orthophoto images. Pix4D capture is a mobile flight planning app that allowed us to set flight heights, camera angles, image overlaps, and flight speeds. PhotoScan is an excellent real-world modeling software that automatically generates high-quality 3D models based on images without setting initial values or camera-check calibration. It can process photos according to multi-view 3D reconstruction technology and generate 3D models with real coordinates through control points. In order to maintain sufficient light and to reduce the influence of clouds and ground shadows, the photos were taken during calm periods with stable light intensity. In Pix4Dcapture, we selected the rectangular simple grid route planning mode to instruct the UAV to collect images automatically. The UAV flight parameters were set as follows: an altitude of 46 m, a camera angle of 90° vertically downward, an overlap rate of 90%, and a flight speed of 27 km/h. The original image was in the JPEG format, and the image data included position and orientation system (POS) data, along with precise GPS coordinates. The main orthophoto production steps were as follows: (1) PhotoScan quickly found matching points between all overlapping images, estimated the camera position for each image, and built a sparse point cloud (the processing time depends on the number of photos and the image resolution). (2) A dense point cloud was built. Based on the estimated camera position, the software calculated its depth information and merged it into a dense point cloud model. (3) A grid was generated. After the dense point cloud was reconstructed, a polygon network model was generated based on the dense point cloud data. (4) The DEM model was constructed based on the grid model, and then the high-resolution orthophoto image was generated according to the DEM model. Figure 2 shows the UAV flight routes and the real-time images taken in the study area.

(a) (b)

Figure 2. (**a**) UAV (unmanned aerial vehicle) flight path planning; (**b**) UAV real-time image.

2.1.2. Image Annotation and Development of the Dataset

In this study, the orthographic images were cut into several 500 × 500-pixel images, each of which contained several loblolly pine tree crowns. LabelImg is a commonly used dataset annotation tool for deep neural network training that is written in Python and uses Qt (a cross-platform C++ graphical-user-interface application-development framework) as its graphical interface. It was used to manually annotate the obtained samples, and the rectangular boxes marked with this tool are shown in Figure 3.

A total of 207 samples were randomly selected from the whole orthoimage as datasets, and the 207 samples were also cut into 500 × 500-pixel datasets. During the training process, the datasets were further divided into training sets and validation sets according to a 9:1

ratio. In the model test section, 185 trees were selected as the test set independent of the training samples. Figure 3 shows the annotated crowns.

Figure 3. The crowns of each image in the training set were labeled using LabelImg software. The red box is the location of the crown marked by LabelImg.

Each annotated image was saved in PASCAL VOC format as XML files [27]. The file content included the image's path, name, size, and annotated border coordinate.

2.1.3. Image Augmentation

The deep convolution neural network is ideal for many tasks in the field of computer vision. However, using a neural network for object detection generally relies on thousands of pictures for training. Therefore, it is necessary to fine-tune and optimize the model parameters for distinct objects to facilitate the convergence of the model's loss function to its global minimum and enhance its efficacy in detecting diverse objects. However, in the process of real data collection, it is not easy to collect such a huge amount of data; for the model to achieve a better detection result in practical scenarios, and to improve the robustness and generalization ability of the model, data augmentation on the existing dataset is needed [28].

Common augmentation techniques include flipping the image, moving the object position in the image, adding Gaussian noise, improving image contrast, and exposing the image. The crowns in this study exhibited similarities in spectral features. To capitalize on these features, we augmented the dataset by applying operations that manipulate the brightness levels and add Gaussian noise, enhancing the crown's color characteristics. Figure 4 shows a set of data enhancement samples.

(a) (b) (c) (d)

Figure 4. (**a**) Original sample; (**b**) darkened sample; (**c**) variable sample; (**d**) added Gaussian noise sample.

2.2. Methods

To improve the crown recognition and crown width extraction results of different models, the orthophoto map of the whole sample plot was first input into the model for recognition and extraction. However, the orthophoto map was too large so it was cut into several small images using the cropped part of the picture in the crown width extraction program. The size of each small graph was 900 × 900 pixels. To avoid missed detection in the process of object detection, the two connected images maintained a 50% coincidence degree for traversal identification. The prediction box was then scaled and offset. Finally, the crown coordinates were identified. A red detection box was used to mark each identified crown position, and the model could automatically extract the position coordinates of the detection box, such as (990, 1245, 560, 962). Using the position coordinates of the detection box, we computed the number of pixels corresponding to the length and width of the detection box. We then computed the predicted length and width according to the actual length corresponding to a single pixel. Finally, the predicted crown width was calculated by averaging. To measure the actual size of the crown width, we used LabelImg to frame the border of the tree crown, resulting in an XML file with generated position coordinates. The program was then used to extract the coordinates of the framed border to compute the number of pixels, along with the length and width. After extracting the number of pixels for length and width, the real length and width were calculated according to the actual length corresponding to a single pixel. The real crown width was obtained by averaging.

2.2.1. Crown Detection Using Faster-RCNN

As mentioned earlier, object detection in complex environments remains a challenge in machine vision and deep learning. In the field of object detection, RCNN is a classic method. Compared with the traditional method of extracting the target position by traversing images with candidate boxes of different sizes, RCNN introduces the convolutional neural network to extract the depth features, and then maps the extracted features to the classifier, which determines whether the target is contained in the search area and calculates its confidence, obtaining more accurate results.

Ren et al. proposed Faster-RCNN [29], which is based on RCNN and Fast-RCNN [30]. Compared with RCNN and Fast-RCNN, Faster-RCNN has dramatically improved detection accuracy and efficiency. The notable improvement of Faster-RCNN over Fast-RCNN is that it does not use a selective search to create region proposals. However, it introduces a region proposal network (RPN) to extract candidate regions to realize the sharing of convolution features between region proposal and object detection. It can conduct end-to-end training for generating candidate regions, which saves training time.

Faster-RCNN is composed of two parts: Fast-RCNN and RPN. The primary function of RPN is to filter out the high-quality regional proposal boxes in the feature map. Then, the sliding window traverses each point in the feature map and configures k anchor boxes of different sizes on each point. The anchor box is used to extract features, and the softmax is used to determine whether the anchors extract objects that are positive or negative. The bounding box regression is then used to correct them to obtain a more accurate regional proposal. Subsequently, the proposal is input into the region of interest (ROI) pooling layer. This layer mainly transforms the features corresponding to the candidate regions in feature maps and proposals to a fixed size. It inputs the next whole connection layer (classifier) for category judgment and object localization. Figure 5 shows the structure of Faster-RCNN.

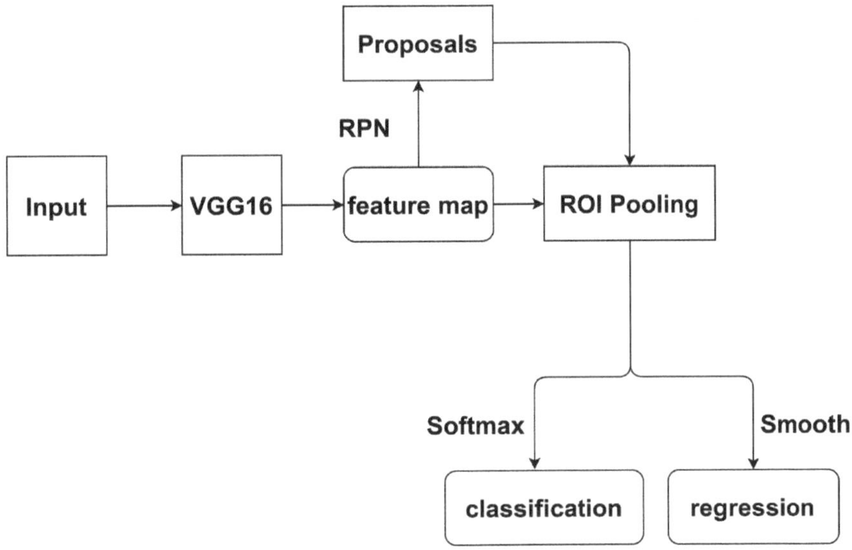

Figure 5. Flow chart of the Faster-RCNN algorithm based on VGG16 backbone network.

2.2.2. Proposed Algorithm: Faster-RCNN with ResNet101

The backbone network of Faster-RCNN is visual geometry group 16 (VGG16) [31], composed of thirteen 3 × 3 convolution layers, three fully connected layers, and several pooling layers. This improves the accuracy of classification results by increasing the number of small convolution kernels and increasing the depth of the network. The network structure is simple and uses the superposition of small convolution kernels instead of large ones, with more nonlinear transformations than a single convolution layer. To further optimize the model recognition effect, this study first adopted the method of deepening the backbone network depth. However, with the deepening of the network, the model may produce gradient disappearance in the training process.

Based on the above premise, this study used ResNet101 [32] to replace VGG16 as the backbone network for feature extraction. Based on the ConvNet model, ResNet introduces numerous identical mappings of $y = x$ across the convolutional layers. Here, x and y represent tensors within the input and output feature maps, respectively. Its main function is to increase the network with depth change without producing the phenomenon of gradient disappearance or weight attenuation. The residual block structure is shown in Figure 6. $F(x)$ and $G(x)$ represent residuals, and $G(x) + x$ is the mapping output; thus, the final network output is $H(x) = G(x) + x$. Since there are three relu functions and three convolution layers in the residual block of the instance, the final framework output results can be expressed as follows:

$$F(x) = relu_1(w_1 \times x) \tag{1}$$

$$G(x) = relu_2(w_2 \times F(x)) \tag{2}$$

$$H(x) = G(x) + x \tag{3}$$

Figure 6 shows the specific structure of the residual block.

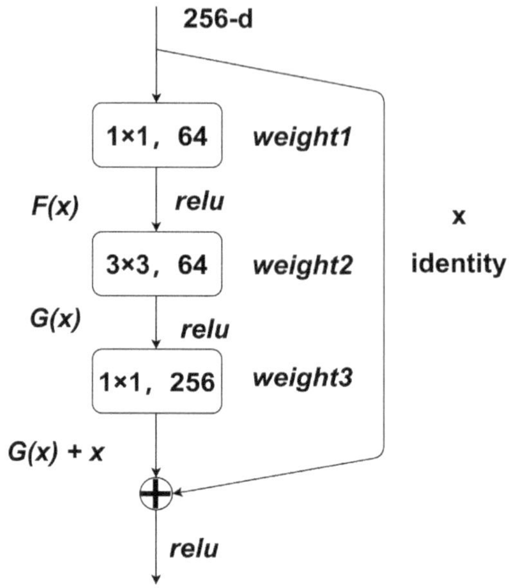

Figure 6. ResNet101 second-layer residual block structure.

The specific network structure of ResNet101 used in this experiment is shown in Figure 7.

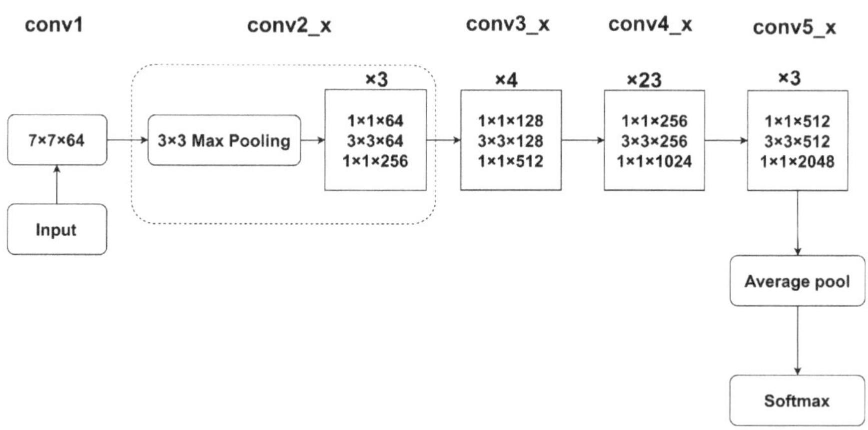

Figure 7. ResNet101 structure chart.

2.2.3. Proposed Algorithm: Faster-RCNN with ResNet101 and FPN

To solve the problem of deep information loss that may occur when ResNet101 replaces VGG16 as the backbone network, this study proposed a combination of ResNet101 and a feature pyramid network to create the FPN_ResNet101 structure. The feature pyramid network (FPN), proposed by Lin et al. [33], is a top-down feature fusion method with horizontal connection. Common object-detection algorithms only use top-level features to predict, while shallow location information is lost. Figure 8 shows the structure of FPN fusing high-level and shallow features for prediction.

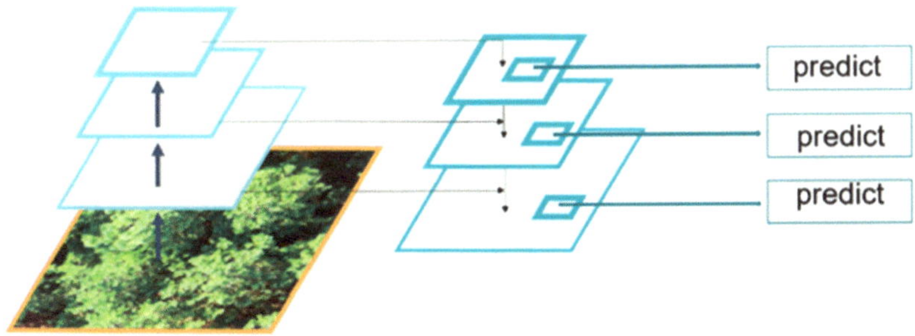

Figure 8. FPN structure.

Finally, the Faster-RCNN model based on FPN and ResNet101 was improved in this study. Since the canopy occupies most of the area of each image in the dataset, while the background area occupies only a tiny portion, to output the canopy color features with greater weight during the training process, the RGB averaging module was added before the base FPN_ResNet1010 structure. An image-averaging operation was performed before inputting each dataset into the model. The resulting values were input to the model as part of the parameters to facilitate more targeted canopy color features trained in the model. The FPN_ResNet101 structure replaced the VGG16, and the Region Proposal Network (RPN) in the Faster-RCNN was scale-separated. The FPN can fuse different scales for detection, and it comprises a three-stage architecture that involves bottom-up feature map generation at multiple scales, top-down feature enhancement, and lateral connections. Given the convolutional outputs at different levels, denoted by Cx, the intermediate feature maps represented by Mx, and the ultimately fused feature map illustrated by Px, the three components are mutually aligned. In the five feature layers of FPN, anchors with different sizes were defined, which were 32×32, 128×128, 256×256, and 512×512. There were three ratios of 1:1, 1:2, and 2:1. Therefore, there were 15 anchors. The improved model structure of FPN_Faster-RCNN_ResNet101 is shown in Figure 9.

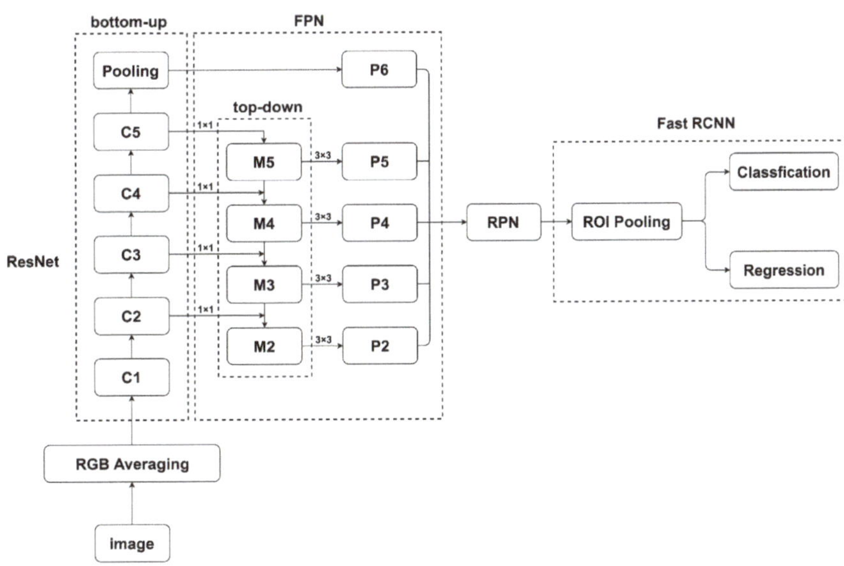

Figure 9. Structure diagram of FPN_Faster-RCNN_ResNet101, where C represents convolutional outputs, M denotes features maps, and P is fused feature maps.

3. Results and Discussion

3.1. Experimental Procedures and Metrics

3.1.1. Experimental Configuration and Dataset

The network training configuration environment was Windows 11, Intel$^{(R)}$ Core^{TM} i7-10750H CPU@2.60 GHz processor, 16 GB memory, and NVIDIA GeForce GTX 1650Ti with 4 GB of video memory as the GPU. Microsoft headquarters in Redmond, Washington, USA. Intel's headquarters and NVIDIA's headquarters are both located in Santa Clara, California, USA. The experimental environment was Python 3.6, TensorFlow-GPU1.12, CUDA9.0, and CUDNN7.3.

Since the Faster-RCNN model requires a large amount of data training to improve its robustness, but the number of existing datasets is limited, migration learning helps to improve this situation. Specifically, it trains on a large dataset and then takes the obtained weight as the training initialization parameter. This study used the initial weights of ResNet101 network model weights from pre-training on the ImageNet dataset. The total number of iterations was 20,000, and the model was saved every 5000 times. The learning rate was set to 0.001, and the batch_size was set to 256. The FPN_Faster-RCNN_ResNet101 model selected the ResNet101 network model, which was pre-trained on the ImageNet dataset for initialization training. The format of the dataset was VOC, and the input image size was set to 512 × 512.

3.1.2. Evaluation Index

For the model evaluation, it is necessary to evaluate the crown recognition and crown width extraction of the model, respectively.

The crown recognition was evaluated by calculating the accuracy, precision, recall, and F1-score:

$$\text{Accuracy} = \frac{TP}{TP + FP + FN} \tag{4}$$

$$\text{Precision} = \frac{TP}{TP + FP} \tag{5}$$

$$\text{Recall} = \frac{TP}{TP + FN} \tag{6}$$

$$F1\text{-Score} = 2 \times \frac{Precision \times Recall}{Precision + Recall} \tag{7}$$

where TP is the number of correctly divided positive cases (i.e., the number of correctly identified crowns); FP represents the number of incorrectly divided positive cases (i.e., the number of incorrectly identified crowns); and FN denotes the number of incorrectly divided negative cases (in this paper, the number of unidentified crowns).

Among the four indexes of crown recognition (Equations (4)–(7)), the accuracy is used to reflect the ability of the model to predict the whole sample, the precision is used to reflect the proportion of the real target in the model prediction, the recall rate is used to reflect the proportion of the model prediction positive cases to the number of real positive cases, and the F1-score, also called the balanced F score, is defined as the harmonic average of the precision and the recall rate.

In the crown width extraction part, the following three indicators were calculated to evaluate the accuracy of the crown width model (Equations (8)–(10)). Bias represents the deviation between the estimated value and the actual value. The accuracy of the crown width model is demonstrated by calculating the root mean square error ($RMSE$) and the coefficient of determination (R^2):

$$Bias = \frac{1}{N} \sum\nolimits_{i=1}^{N} |\hat{y}_i - y_i| \tag{8}$$

$$RMSE = \sqrt{\frac{\sum_{i=1}^{N}(y_i - \hat{y}_i)^2}{N}} \tag{9}$$

$$R^2 = 1 - \frac{\sum_{i=1}^{N}(y_i - \hat{y}_i)^2}{\sum_{i=1}^{N}(y_i - \overline{y})^2} \tag{10}$$

where $\hat{y}_i$ represents the estimated value; y_i denotes the actual value; N is the number of samples; and $\overline{y} = \frac{1}{N}\sum_{i=1}^{N} y_i$.

3.2. *Results and Discussion*

3.2.1. Identify Impressions

In our previous study, the Faster-RCNN, YOLO, and SSD models achieved good results in young forests, for which Faster-RCNN had the highest recognition accuracy. Figure 10 shows the crown-detection effect of the three models on young forests, and Table 1 presents their respective detection results. The data comes from "Measuring loblolly pine crowns with drone imagery through deep learning" [25]. Faster-RCNN outperformed the other two methods in young forest detection.

(a) (b) (c)

Figure 10. (**a**) Faster-RCNN detection effect image; (**b**) YOLO detection effect image; (**c**) SSD detection effect image. The red square is the crown detected by the model.

Table 1. Classification detection results of Faster-RCNN, YOLO, and SSD [25].

Index	Faster-RCNN	YOLO	SSD
TP	128	126	125
FP	1	6	1
FN	0	2	3
Precision (%)	99.22	95.45	99.21
Recall (%)	100.00	98.44	97.66
Accuracy (%)	99.22	94.03	96.90
*F*1-score (%)	99.61	96.92	98.43

However, in the mature forest, the original Faster-RCNN model performed poorly. The objective of this research was to enhance the performance of Faster-RCNN and enhance its versatility when operating in mature forest environments. Figure 11 shows the crown-detection effect of the two models in orthophoto images.

(a) (b)

Figure 11. (**a**) The Faster-RCNN_ResNet101 model with ResNet101 as the backbone network; (**b**) the FPN_Faster-RCNN_ResNet101 model with FPN_ResNet101 as the backbone network. The red square is the crown detected by the model.

3.2.2. Crown Identification

We use Method 1 to represent YOLO, Method 2 to represent SSD, Method 3 to represent Faster-RCNN_ResNet101, and Method 4 to represent FPN_Faster-RCNN_ResNet101.

Due to the slow growth of trees, in crown identification, the accurate detection of each canopy is more critical than rapid crown detection, so models with higher accuracy are more suitable for this task. In this experiment, we used the computer mentioned in Section 3.1.1 as the experimental equipment, and we selected the single crown recognition time to measure the model detection speed. Two-stage detector recognition speed is slower than a one-stage detector, but the accuracy is higher. It can be seen from Table 2 that Method 1 and Method 2 were faster than Method 3 and Method 4. However, in this task, FPN_Faster-RCNN_ResNet101 (Method 4) was better than SSD in recall, accuracy, and F1-score, but slightly worse than SSD in precision. The two-stage detector, FPN_Faster-RCNN_ResNet101 (Method 4), gave the best overall results, achieving better accuracy than Method 2, even at a similar speed.

Moreover, Method 4 improved the accuracy by 4.9% over Method 3, which also proved the feasibility of the improved method.

Table 2. Classification detection results of YOLO, SSD [25], and two improved Faster-RCNN algorithms.

Index	(1) YOLO	(2) SSD	(3) Faster-RCNN_ResNet101	(4) FPN_Faster-RCNN_ResNet101
Time (ms)	55	57	72	69
TP	175	180	170	181
FP	3	4	13	5
FN	10	5	6	4
Precision (%)	98.31	97.83	93.19	97.31
Recall (%)	94.59	97.30	96.74	97.84
Accuracy (%)	93.09	95.24	90.36	95.26
*F*1-Score (%)	96.42	97.56	94.93	97.58

In the actual training process, with the deepening of the network depth, the gradient is backward propagation. After increasing the network depth, the forward gradient will be minimal, while the model also has problems such as learning stagnation and gradient disappearance. Table 1 presents the models' performance in crown recognition based on the independent test dataset. After replacing VGG16 with ResNet101, Method 3 improved the efficiency of crown recognition, with the precision, recall, accuracy, and F1-score reaching 93.19%, 96.74%, 90.36%, and 94.93%, respectively. The accuracy and precision were comparable, although slightly weaker, than those for Method 1 and Method 2. After fusing FPN and ResNet101, VGG16 was replaced by the FPN_ResNet101 structure. In crown recognition, the four indexes of Method 4 were improved to varying degrees, of which accuracy was the most improved, reaching 95.26%. Compared with Method 3, the four indexes increased by 4.12%, 1.10%, 4.90%, and 2.65%, respectively. Using the FPN to help detect objects at different scales can theoretically improve the small-target-detection effect of the model. The experimental results in Table 1 also prove this. It was verified that the improved method helps to enhance the canopy detection performance of Faster-RCNN in dense loblolly pine forests, and the feasibility of the improved means was well illustrated.

3.2.3. Extraction of Crown Width

Table 3 and Figure 12 present the results of the models estimating crown width using the independent test dataset. Overall, the application of Method 3 did not achieve the same accuracy and precision as Method 1 and Method 2. Through the study of the residual block structure, it was found that the ResNet101 network has a deep information loss problem. In ResNet101, identity mapping must be used when the size of the building block does not match the size of the next building block. According to Figure 7, in the four mapping stages of ResNet101, there are only four continuous 1×1 convolutions, but there is no linear relationship between the two, which limits its learning ability and eventually leads to the loss of deep information.

Table 3. The mature loblolly pine crown-width-measurement effect index of YOLO, SSD [25], and two improved Faster-RCNN algorithms.

Index	(1) YOLO	(2) SSD	(3) Faster-RCNN_ResNet101	(4) FPN_Faster-RCNN_ResNet101
Bias (m)	0.92	0.99	1.15	0.98
RMSE (m)	0.66	0.31	1.06	0.45
R^2	0.69	0.94	0.68	0.95

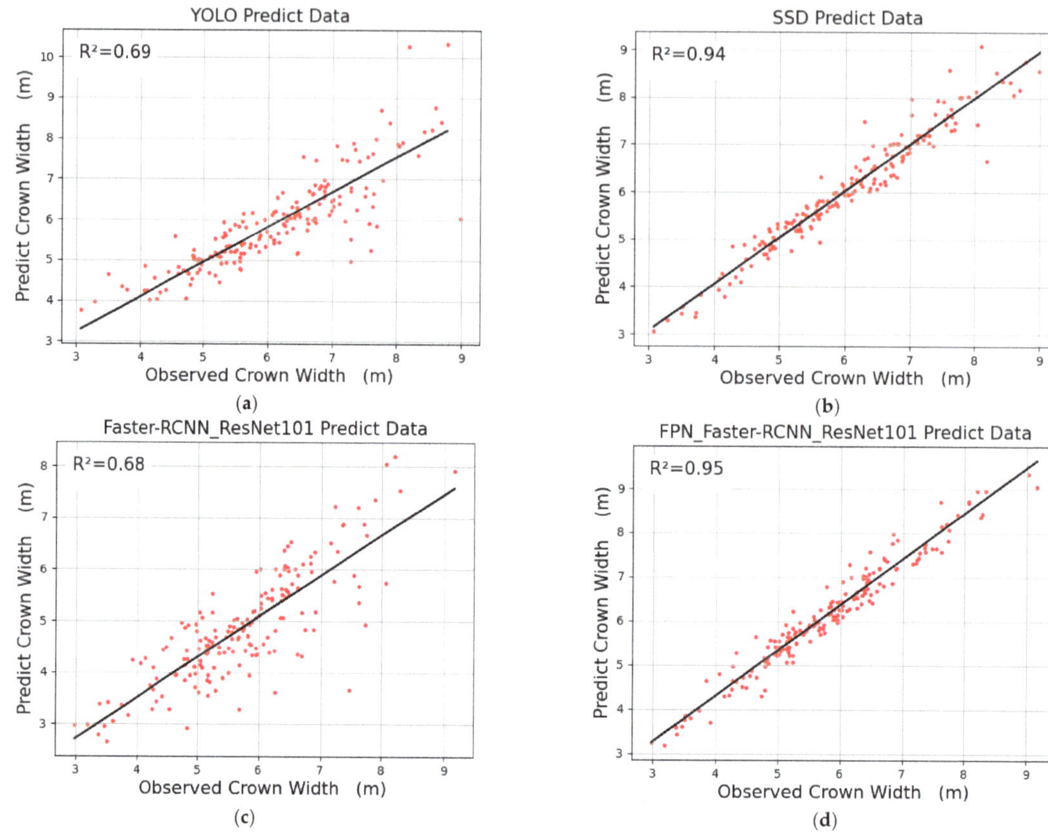

Figure 12. (**a**) Linear regression graphs of (1) the YOLO model; (**b**) linear regression graphs of (2) the SSD model; (**c**) linear regression graphs of (3) the Faster-RCNN_ResNet101 model; (**d**) linear regression graphs of (4) the FPN_Faster-RCNN_ResNet101 model.

The *FPN* is a way to fuse low-level and high-level features. The shallow feature map has a small receptive field and less semantic information, but the spatial location information is accurate. After the fusion of ResNet101 using the *FPN*, we created a new network structure, named FPN_ResNet101, and applied this structure to Faster-RCNN. Method 4 measured crown width very accurately and precisely, resulting in a bias of 0.98, an *RMSE* of 0.45, and an R^2 of 0.95. These estimates were comparable to those of Method 2, but more improved than those of Method 1. Compared with Method 3, the *RMSE* decreased by 0.61 and the R^2 increased by 0.27.

FPN_Faster-RCNN_ResNet101 offers a huge improvement in crown width measurement, with a higher R^2 than all the other methods. The feasibility of using *FPN* and ResNet101 to improve the original model is illustrated.

4. Conclusions

In this study, high-resolution orthophotos, obtained by UAVs shooting a mature loblolly pine forest in eastern Texas, were used as the data source. ResNet101 and FPN_ResNet101 replaced the backbone network VGG16 of the original Faster-RCNN model. Using FPN_ResNet101, the crown recognition accuracy rate of Method 4 reached 95.26%, and the crown width extraction R^2 reached 0.95. Compared with Method 3, the two indexes had increased by 4.90% and 0.27, respectively, which proves the feasibility of improving the original model using FPN_ResNet101 network architecture and the superiority of the improved model in this research field. At the same time, with regard to recognition speed, the improved Method 4 (FPN_Faster-RCNN_ResNet101) was also enhanced to a certain extent in comparison with Method 3 (Faster-RCNN_ResNet101). The speed of the two-stage detector was improved to a level similar to that of the single-stage detector, and some progress was made in comparison with Method 2 (SSD) in terms of crown detection and crown width extraction. However, due to the similarity of trees, the accurate identification and classification of tree crowns in mixed forests remains a significant challenge. In terrain such as hills, accurate canopy width measurement is impossible due to the change in relative distance between the UAV and the ground, which is a crucial direction for future research. Nonetheless, the excellent accuracy of Faster-RCNN suggests the model's applicability in dense loblolly pine forests, providing an alternative for forestry practitioners in tree mensuration.

Author Contributions: Conceptualization, X.L. and C.D.; formal analysis, X.L., C.D., C.C. and L.Y.; funding acquisition, Y.W. and C.D.; methodology, H.X., S.C. and S.H.; resources, X.L. and Y.W.; writing—original draft, C.C. All authors have read and agreed to the published version of the manuscript.

Funding: This research was funded by the Zhejiang Natural Science Foundation Project, grant number LQ21C160018, the ETPPRP, and the McIntire Stennis program.

Data Availability Statement: Not applicable.

Conflicts of Interest: The authors declare no conflict of interest.

References

1. Gratani, L.; Varone, L. Plant Crown Traits and Carbon Sequestration Capability by *Platanus hybrida* Brot. in Rome. *Landsc. Urban Plan.* **2007**, *81*, 282–286. [CrossRef]
2. Hao, J.; Jia, H.; Yang, B.; Huang, X.; Huang, G. Regession analysis of Teak Crown Growth with Tree Height and DBH. *J. Northwest For. Univ.* **2019**, *34*, 144–148.
3. Jones, D.A.; Harrington, C.A.; Marshall, D. Survival, and Growth Response of Douglas-Fir Trees to Increasing Levels of Bole, Root, and Crown Damage. *For. Sci.* **2019**, *65*, 143–155. [CrossRef]
4. Putney, J.D.; Maguire, D.A. Shifts in Foliage Biomass and Its Vertical Distribution in Response to Operational Nitrogen Fertilization of Douglas-Fir in Western Oregon. *Forests* **2020**, *11*, 511. [CrossRef]
5. Feng, J.; Lian, J.; Mei, Q.; Cao, H.; Ye, W. Vertical Variation in Leaf Traits and Crown Structure Promote the Coexistence of Forest Tree Species. *Forests* **2022**, *13*, 1548. [CrossRef]
6. Bella, I.E. A New Competition Model for Individual Trees. *For. Sci.* **1971**, *17*, 364–372.
7. Röhle, H. Vergleichende Untersuchungen zur Ermittlung der Genauigkeit bei der Ablotung von Kronenradien. *Forstarchiv* **1986**, *57*, 67–71.
8. Thurnher, C.; Klopf, M.; Hasenauer, H. MOSES—A Tree Growth Simulator for Modelling Stand Response in Central Europe. *Ecol. Model.* **2017**, *352*, 58–76. [CrossRef]
9. Fu, L.; Sharma, R.P.; Wang, G.; Tang, S. Modelling a System of Nonlinear Additive Crown Width Models Applying Seemingly Unrelated Regression for Prince Rupprecht Larch in Northern China. *For. Ecol. Manag.* **2017**, *386*, 71–80. [CrossRef]
10. Preuhsler, T. Ertragskundliehe Merkmale oberbayerlscher Bergmischwald-Verjüngungsbestände auf kalkalpinen Standorten im Forstamt Kreuth. *Forstwiss. Cent.* **1981**, *100*, 313–345. [CrossRef]
11. Fleck, S.; Mölder, I.; Jacob, M.; Gebauer, T.; Jungkunst, H.F.; Leuschner, C. Comparison of Conventional Eight-Point Crown Projections with LIDAR-Based Virtual Crown Projections in a Temperate Old-Growth Forest. *Ann. For. Sci.* **2011**, *68*, 1173–1185. [CrossRef]
12. Pretzsch, H.; Biber, P.; Uhl, E.; Dahlhausen, J.; Rötzer, T.; Caldentey, J.; Koike, T.; van Con, T.; Chavanne, A.; Seifert, T.; et al. Crown Size and Growing Space Requirement of Common Tree Species in Urban Centres, Parks, and Forests. *Urban For. Urban Green.* **2015**, *14*, 466–479. [CrossRef]
13. Ning, X.; Ma, Y.; Hou, Y.; Lv, Z.; Jin, H.; Wang, Z.; Wang, Y. Trunk-Constrained and Tree Structure Analysis Method for Individual Tree Extraction from Scanned Outdoor Scenes. *Remote Sens.* **2023**, *15*, 1567. [CrossRef]
14. Wu, X.; Xu, A.; Yang, T. Passive Measurement Method of Tree Height and Crown Diameter Using a Smartphone. *IEEE Access* **2020**, *8*, 11669–11678.
15. Ahmadi, P.; Mansor, S.; Farjad, B.; Ghaderpour, E. Unmanned Aerial Vehicle (UAV)-Based Remote Sensing for Early-Stage Detection of Ganoderma. *Remote Sens.* **2022**, *14*, 1239. [CrossRef]

16. Safonova, A.; Hamad, Y.; Dmitriev, E.; Georgiev, G.; Trenkin, V.; Georgieva, M.; Dimitrov, S.; Iliev, M. Individual Tree Crown Delineation for the Species Classification and Assessment of Vital Status of Forest Stands from UAV Images. *Drones* **2021**, *5*, 77. [CrossRef]
17. Kolanuvada, S.R.; Ilango, K.K. Automatic Extraction of Tree Crown for the Estimation of Biomass from UAV Imagery Using Neural Networks. *J. Indian Soc. Remote Sens.* **2021**, *49*, 651–658. [CrossRef]
18. Guerra-Hernández, J.; Cosenza, D.N.; Cardil, A.; Silva, C.A.; Botequim, B.; Soares, P.; Silva, M.; González-Ferreiro, E.; Díaz-Varela, R.A. Predicting Growing Stock Volume of Eucalyptus Plantations Using 3-D Point Clouds Derived from UAV Imagery and ALS Data. *Forests* **2019**, *10*, 905. [CrossRef]
19. Gurumurthy, V.A.; Kestur, R.; Narasipura, O. Mango Tree Net—A Fully Convolutional Network for Semantic Segmentation and Individual Crown Detection of Mango Trees. *arXiv* **2019**, arXiv:1907.06915.
20. Li, Y.; Zhang, H.; Yang, T.; Ma, Z.; Li, S.; Shen, K. A Method of Estimating Chinese Fir Crown Width Based on Adaptive Neuro-Fuzzy Inference System. *Sci. Silvae Sin.* **2019**, *55*, 45–51.
21. Ritter, T.; Nothdurft, A. Automatic Assessment of Crown Projection Area on Single Trees and Stand-Level, Based on Three-Dimensional Point Clouds Derived from Terrestrial Laser-Scanning. *Forests* **2018**, *9*, 237. [CrossRef]
22. Ma, Z.; Pang, Y.; Wang, D.; Liang, X.; Chen, B.; Lu, H.; Weinacker, H.; Koch, B. Individual Tree Crown Segmentation of a Larch Plantation Using Airborne Laser Scanning Data Based on Region Growing and Canopy Morphology Features. *Remote Sens.* **2020**, *12*, 1078. [CrossRef]
23. Quan, Y.; Li, M.; Zhen, Z.; Yuanshuo, H. Modeling Crown Characteristic Attributes and Profile of Larix Olgensis Using UAV-Borne Lidar. *J. Northeast Univ.* **2019**, *47*, 52–28.
24. Quan, Y.; Li, M.; Zhen, Z.; Hao, Y.; Wang, B. The Feasibility of Modelling the Crown Profile of Larix Olgensis Using Unmanned Aerial Vehicle Laser Scanning Data. *Sensors* **2020**, *20*, 5555. [CrossRef] [PubMed]
25. Lou, X.; Huang, Y.; Fang, L.; Huang, S.; Gao, H.; Yang, L.; Weng, Y.; Hung, I.K. Measuring Loblolly Pine Crowns with Drone Imagery through Deep Learning. *J. For. Res.* **2022**, *33*, 227–238. [CrossRef]
26. Soviany, P.; Ionescu, R.T. Optimizing the Trade-Off between Single-Stage and Two-Stage Deep Object Detectors Using Image Difficulty Prediction. In Proceedings of the 2018 20th International Symposium on Symbolic and Numeric Algorithms for Scientific Computing (SYNASC), Timisoara, Romania, 20–23 September 2018; pp. 209–214.
27. Everingham, M.; Van Gool, L.; Williams, C.K.; Winn, J.; Zisserman, A. The Pascal Visual Object Classes (Voc) Challenge. *Int. J. Comput. Vis.* **2010**, *88*, 303–338. [CrossRef]
28. Perez, L.; Wang, J. The Effectiveness of Data Augmentation in Image Classification Using Deep Learning. *arXiv* **2017**, arXiv:1712.04621.
29. Ren, S.; He, K.; Girshick, R.; Sun, J. Faster R-CNN: Towards Real-Time Object Detection with Region Proposal Networks. In *Advances in Neural Information Processing Systems*; Curran Associates, Inc.: Red Hook, NY, USA, 2015; Volume 28.
30. Girshick, R. Fast R-CNN. In Proceedings of the IEEE International Conference on Computer Vision, Santiago, Chile, 7–13 December 2015; pp. 1440–1448.
31. Simonyan, K.; Zisserman, A. Very Deep Convolutional Networks for Large-Scale Image Recognition. *arXiv* **2015**, arXiv:1409.1556.
32. He, K.; Zhang, X.; Ren, S.; Sun, J. Deep residual learning for image recognition. In Proceedings of the IEEE Computer Society Conference on Computer Vision and Pattern Recognition (CVPR), Las Vegas, NV, USA, 27–30 June 2016; pp. 770–778.
33. Lin, T.Y.; Dollár, P.; Girshick, R.; He, K.; Hariharan, B.; Belongie, S. Feature pyramid networks for object detection. In Proceedings of the IEEE Conference on Computer Vision and Pattern Recognition, Honolulu, HI, USA, 21–26 July 2017.

forests

MDPI

Article

Improving the Gross Primary Production Estimate by Merging and Downscaling Based on Deep Learning

Jiao Lu [1], Guofu Wang [2,*], Donghan Feng [3] and Isaac Kwesi Nooni [1]

1 Key Laboratory of Ecosystem Carbon Source and Sink, School of Atmospheric Science & Remote Sensing, Wuxi University, Wuxi 214105, China; jiao_lu@cwxu.edu.cn (J.L.); nooni25593@alumni.itc.nl (I.K.N.)
2 National Climate Center, China Meteorological Administration, Beijing 100081, China
3 Shandong Climate Center, Jinan 250031, China; dorisfdh@163.com
* Correspondence: wanggf@cma.gov.cn

Abstract: A reliable estimate of the gross primary productivity (GPP) is crucial for understanding the global carbon balance and accurately assessing the ability of terrestrial ecosystems to support the sustainable development of human society. However, there are inconsistencies in variations and trends in current GPP products. To improve the estimation accuracy of GPP, a deep learning method has been adopted to merge 23 CMIP6 data to generate a monthly GPP merged product with high precision and a spatial resolution of 0.25°, covering a time range of 1850–2100 under four climate scenarios. Multi-model ensemble mean and the merged GPP ($CMIP6_{DL}$ GPP) have been compared, taking GLASS GPP as the benchmark. Compared with the multi-model ensemble mean, the coefficient of determination between $CMIP6_{DL}$ GPP and GLASS GPP was increased from 0.66 to 0.86, with the RMSD being reduced from 1.77 $gCm^{-2}d^{-1}$ to 0.77 $gCm^{-2}d^{-1}$, which significantly reduced the random error. Merged GPP can better capture long-term trends, especially in regions with dense vegetation along the southeast coast. Under the climate change scenarios, the regional average annual GPP shows an upward trend over China, and the variation trend intensifies with the increase in radiation forcing levels. The results contribute to a scientific understanding of the potential impact of climate change on GPP in China.

Keywords: GPP; deep learning; data merging; image super resolution

Citation: Lu, J.; Wang, G.; Feng, D.; Nooni, I.K. Improving the Gross Primary Production Estimate by Merging and Downscaling Based on Deep Learning. *Forests* **2023**, *14*, 1201. https://doi.org/10.3390/f14061201

Academic Editor: Helmi Zulhaidi Mohd Shafri

Received: 24 April 2023
Revised: 25 May 2023
Accepted: 8 June 2023
Published: 9 June 2023

1. Introduction

Global climate change, especially the impact of climate warming on the human living environment, has attracted more and more attention [1]. With the continuous development of the global industry, the use of fossil fuels is increasing. This process releases large amounts of carbon dioxide and changes the proportions of greenhouse gases, causing global climate problems such as the intensification of the greenhouse effect. The gross primary productivity (GPP), formed by terrestrial vegetation through photosynthesis to fix CO2 and energy in the biosphere, is the initial stage of entering the carbon cycle process and the basis of the ecosystem carbon cycle [2]. Terrestrial ecosystems can influence global climate change through the carbon cycle [3]. Therefore, the research on the carbon cycle of global terrestrial ecosystems has received widespread attention and has become a hot spot in global change research [4].

An essential indicator of terrestrial ecosystem carbon cycle research is GPP. The gross primary productivity, that is, the total photosynthetic uptake or carbon assimilation by plants, is a key component of terrestrial carbon balance [5]. It is the main driving factor of the terrestrial carbon sink, which is responsible for absorbing about 30% of anthropogenic carbon dioxide emissions [6]. In general, it can be said that GPP plays a vital role in the global carbon cycle [7–10]. Characterizing the spatiotemporal changes and trends of GPP is essential for a deeper understanding of the carbon cycle between terrestrial ecosystems and the atmosphere [11]. Therefore, quantifying the GPP is crucial to understand the impact of climate change and atmospheric carbon dioxide concentration changes on the terrestrial carbon cycle [12,13]. The accurate estimation of GPP at regional and even global scales contributes to a more comprehensive and in-depth understanding of ecosystem functions and the global carbon cycle [14]. However, with the intensification of climate change, there is significant uncertainty in estimating GPP [15].

In recent decades, significant progress has been made in estimating GPP, and many global GPP datasets have been published [16]. These products differ in GPP estimates in different ecosystems and external environments, with advantages and disadvantages in various studies [17–19]. Data-driven GPP estimation is the primary tool for GPP research and model evaluation. It mainly includes the following three types. The first type uses eddy covariance (EC) technology to obtain flux tower observation and upscaling [20]. Process-based models such as the Community Land Model (CLM) are commonly considered in the revision of GPP products [21]. However, the distribution of flux towers is sparse and uneven, and the coverage period is different and limited [22], which makes the derivation of global GPP complicated. When estimating GPP products by scaling up, there will be a more significant deviation near the sparse flux tower [22]. The second type is the machine learning (ML) method that uses satellites and other meteorological data as input [23–28]. ML products are widely used for benchmarking and depend more on the spatial representation of flux sites [15,24]. The third type is based on the principle of light use efficiency (LUE), which is used to derive the GPP based on optical remote sensing variables [22,27,29–31]. LUE products are good at detecting the spatial distribution pattern of GPP; however, they usually perform poorly in seasonal estimation and overestimate GPP under dry and cold conditions [28,32–35]. At the regional and even global scale, the land surface model (LSM) and ecosystem model [36–38] are also effective methods to simulate GPP. The process models consider plant physiological and ecological processes, which have solid theoretical significance and can simulate future productivity changes. LSM products are more consistent in spatiotemporal distribution and better in response to climate change, but various model parameters, input data, and model structures cause more significant uncertainty [15,39,40]. Especially in large-scale applications, the accuracy of the model is affected to some extent [41].

However, the current GPP products rely on assumptions. They are based on different estimation models so that no GPP product group can perform consistently better in diverse ecosystems and external conditions [42]. In particular, there is significant uncertainty in CMIP6 datasets under future climate change scenarios. The uncertainties in climate prediction are relevant to the modelling of GPP, although they are independent of the estimation method of GPP. As a result, the changes and trends obtained from different GPP products are often various [43], leading to unconvincing research results of a single GPP product [44]. Therefore, to overcome the limitations of a single dataset of GPP products, the question of how to effectively merge multiple datasets has been raised. Studies have shown that the multi-model ensemble mean is better than the single model in estimating GPP [17,45]. Traditionally, the IPCC report shows that the average value of multiple models is the best. Empirical evidence from various modelling fields indicates that the multi-model ensemble mean eliminates at least some deviations from the single model, which can produce better predictions or values closer to observation [46–49]. However, the simple multi-model ensemble mean depends on the assumption that the uncertainty of each dataset is the same [50]. However, this assumption is not reasonable from the differences between datasets.

On the other hand, the spatial resolutions between prediction CMIP6 GPP datasets are generally low and mismatched. At present, bilinear interpolation is widely used for downscaling. Super-Resolution Convolution Neural Network [51] is a pioneering work of image super-resolution reconstruction. It makes people pay attention to the deep learning convolution neural network, the performance of which is superior to the traditional super-resolution. A neural network called Ynet [52] combines image super-resolution and data merging technology, which has been further used for soil moisture data merging with excellent results [53]. It establishes the relationship between reference data and multi-model datasets in the historical period to generate a merged GPP product with the distribution characteristics of reference data. Furthermore, GPP-merged data in the projection period under multiple climate scenarios can be obtained based on 23 CMIP6 model projections if the relationship built during the historical period is applied to the projection period. The evaluation in the historical period shows that the quality of the merged data is significantly improved compared with the multi-model ensemble mean, indicating that the merged data are reliable. Therefore, the variation trends in GPP under multiple climate scenarios in the future were further analyzed based on the merged data.

This study aims to produce an improved high-resolution merged GPP dataset by adopting a new method to combine the advantages of 23 CMIP6 GPP datasets, without relying on any prior knowledge, to analyze the long-term variation trends in GPP under multiple climate scenarios in the future based on the improved merged data. Based on the historical CMIP6 GPP data, the deep learning method called Ynet [52] neural network is used to train the deep learning model with merging and downscaling functions. The 23 groups of monthly GPP datasets are integrated into this high-resolution long-series dataset called $CMIP6_{DL}$ GPP. The period and the spatial resolution are 1850–2100 and 0.25 degrees. The future projection period includes four climate change scenarios, such as SSP1-2.6, SSP2-4.5, SSP3-7.0, and SSP5-8.5. The performance of the merged product has been evaluated based on eddy covariance tower GPP data, and the effectiveness of the merging method has been discussed. Finally, the variation trends of GPP over China under future climate change scenarios have been analyzed based on $CMIP6_{DL}$ GPP.

2. Materials and Methods

2.1. Study Area

This study takes China as the research region. The climate characteristics of China are complex and various, with a monsoon climate in the east, a temperate continental climate in the northwest, and an alpine climate in the Qinghai–Tibet Plateau. In general, the climate in China is characterized by a monsoon climate with a high temperature, a rainy summer and cold winter with little rain. The high-temperature period is consistent with the rainy period in the study region. There is apparent spatial heterogeneity in GPP due to the differences in topography, vegetation types, and hydrothermal conditions in the different areas [54]. The growth of vegetation is closely related to the climate of the region where it is located. Therefore, according to the Coben climate division and the classic climate division of China, China is divided into four climatic parts for regional analysis: the arid, semi-arid, semi-humid, humid, and Qinghai–Tibet Plateau regions.

2.2. GLASS GPP Data

Global Land Surface Satellite (GLASS) gross primary production (GPP) is a global surface remote sensing product with a long time series and high precision based on multi-source remote sensing data and ground observation [45]. The Bayesian multi-algorithm integration method integrates eight widely used light use efficiency (LUE) models, including CASA, CFix, CFlux, EC-LUE, MODIS, VPM, VPRM, and two-leaf. The spatial and temporal resolution of GLASS GPP are 5 km and eight days, from 1982 to 2018. The unit of the data is $gCm^{-2}d^{-1}$. The development and validation of the algorithm are based on the data from global fluxnet, which contains nine types of terrestrial ecosystems, such as evergreen broad-leaved forest, evergreen coniferous forest, deciduous broad-leaved forest, mixed forest, temperate grassland, tropical savanna, shrub, farmland, and tundra. GLASS GPP can be downloaded from http://www.geodata.cn (accessed on 23 April 2023).

2.3. CMIP6 GPP Data

The monthly GPP data from 23 available CMIP6 models are used in the study, with the unit of $kgCm^{-2}s^{-1}$. The period is over 1850–2100, including the historical and future projection period. More specifically, the projection period of 2015–2100 contains the four most interesting scenarios: SSP1-2.6, SSP2-4.5, SSP3-7.0, and SSP5-8.5. The model and GLASS

GPP data are first uniformly interpolated to 2° due to the inconsistent spatial resolution. Meanwhile, it is necessary to unify the unit of model simulation data and GLASS GPP to $gCm^{-2}d^{-1}$. The first member, r1i1p1f1, is selected if possible (r1i1p1f2 is used if unavailable). CMIP6 GPP can be downloaded from https://esgf-node.llnl.gov/projects/cmip6/ (accessed on 9 June 2023). The information on the 23 models used is listed in the Table 1 below.

Table 1. Detailed information on 23 models used in this study.

Model Name	Institution	Variant Label	Resolution (Longitude × Latitude)
ACCESS-ESM1-5	CSIRO, Australia	r1i1p1f1	1.875° × 1.25°
BCC-CSM2-MR	BCC, China	r1i1p1f1	1.125° × 1.125°
CAS-ESM2-0	CAS, China	r1i1p1f1	1.4062° × 1.4062°
CESM2	NCAR, USA	r1i1p1f1	1.25° × 0.9424°
CESM2-WACCM	NCAR, USA	r1i1p1f1	1.25° × 0.9424°
CMCC-CM2-SR5	CMCC, Italy	r1i1p1f1	1.25° × 0.9424°
CMCC-ESM2	CMCC, Italy	r1i1p1f1	1.25° × 0.9424°
CNRM-CM6-1	CNRM, France	r1i1p1f2	1.4° × 1.4°
CNRM-CM6-1-HR	CNRM, France	r1i1p1f2	1.4° × 1.4°
CNRM-ESM2-1	CNRM, France	r1i1p1f1	0.5° × 0.5°
CanESM5-CanOE	CCCMA, Canada	r1i1p2f1	2.8125° × 2.8125°
CanESM5	CCCMA, Canada	r1i1p1f1	2.8125° × 2.8125°
EC-Earth3-Veg-LR	EC-Earth-Consortium, EU	r1i1p1f1	0.7031° × 0.7031°
EC-Earth3-Veg	EC-Earth-Consortium, EU	r1i1p1f1	0.7031° × 0.7031°
GISS-E2-1-G	NASA, USA	r1i1p1f2	2.5° × 2°
GISS-E2-1-H	NASA, USA	r1i1p1f2	2.5° × 2°
INM-CM4-8	INM, Russia	r1i1p1f1	2° × 1.5°
INM-CM5-0	INM, Russia	r1i1p1f1	2° × 1.5°
IPSL-CM6A-LR	IPSL, France	r1i1p1f1	2.5° × 1.2676°
MIROC-ES2L	MIROC, Japan	r1i1p1f2	2.8125° × 2.8125°
MPI-ESM1-2-HR	MPI-M, Germany	r1i1p1f1	0.9375° × 0.9375°
MPI-ESM1-2-LR	MPI-M, Germany	r1i1p1f1	1.875° × 1.875°
UKESM1-0-LL	MOHC, UK	r1i1p1f2	1.875° × 1.25°

2.4. Deep Learning Network

In this study, a GPP-merged product conforming to the distribution characteristics of the GLASS GPP dataset is developed by merging the simulated data of CMIP6 models. A neural network named Ynet [51] is adopted in the data merging part, which combines data merging and image super-resolution technology. The deep learning model establishes relationships between the reference and the CMIP6 GPP data in the historical period and is then applied in future projections. Compared with the multi-model ensemble mean, the merged data have higher accuracy and spatial resolution over 1850–2100 in China. The multi-model ensemble mean is a simple average of the simulation results of the 23 CMIP6 models. It uses the simplest strategy of assuming that all 23 model data have the same uncertainty; thus, the multi-model ensemble mean is a simple average of each model. The merged data in this study are generated based on the deep learning method, which compensates for this deficiency. The novelty of the deep learning method used in this study is that it combines data merging and image super-resolution technology, which combines the advantages of 23 CMIP6 GPP datasets without relying on any prior knowledge.

The model architecture is shown in Figure 1. This structure consists of three parts. The first part is a symmetric encoder-decoder structure with a skip connection. Similar to the residual encoder-decoder network, this structure solves the problem of gradient disappearance when the error calculated by the loss function in the model training process is propagated back. This structure captures the abstract features of low-resolution images with noise and outputs a "cleaner" image, which can be regarded as a feature extractor. In addition to the 30 convolutional layers, this part also contains 15 deconvolution layers. It may introduce "checkerboard artefacts", resulting in lower image output quality [55]. Therefore, to reduce the problem, a convolutional layer is added after each deconvolution layer to compensate for upsampling. Specifically, this part's input to the model is low-resolution CMIP6 GPP. The second part of the model is upsampling, which is mainly used for downscaling. It consists of one upper sampling layer using bilinear interpolation and two convolution layers with the same feature depth as the input channel. As in the first part, the last two convolution layers were added to eliminate the checkerboard effect.

The third part is data merging, which consists of three input datasets, including upsampling output, auxiliary data, and unsampled CMIP6 GPP. The three datasets calculated by the two convolution layers are joined to obtain high-resolution data. Among them, auxiliary data are used to help improve the results, which remained constant throughout different months over the training and testing periods. The loss function can be calculated according to the following formula:

$$L(\theta) = \frac{1}{N} \sum_{i=1}^{N} \left|\left|f(X_i, \theta) - Y_i\right|\right|^2, \tag{1}$$

Among them, θ is the network parameter that needs to be optimized. $f(X_i, \theta)$ represents the learned function. In addition, n is the total number of training samples. X and Y represent the input data and the target at position i.

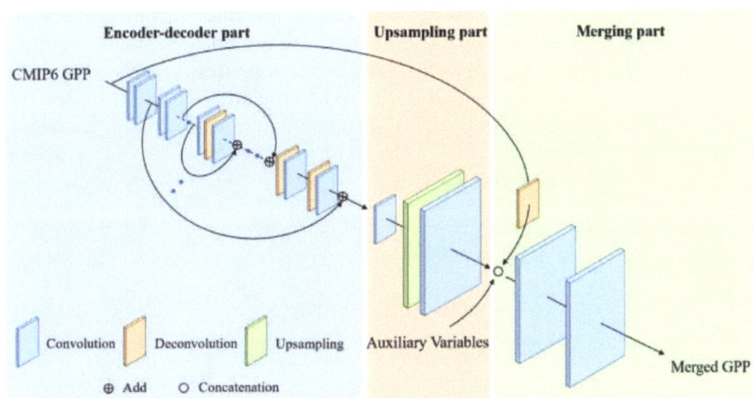

Figure 1. Deep-learning model structure [52].

2.5. Evaluation Indicators

The evaluation indicators, including mean absolute error (MAE), root mean square deviation (RMSD), unbiased root mean square deviation (ubRMSD), and coefficient of determination (r^2), have been used to assess the performance of the merged data objectively and find out if it captures the distribution characteristics of the reference data. The calculation formulae are as follows:

$$MAE = \frac{1}{n} \sum_{i=1}^{n} |R_i - M_i|, \tag{2}$$

$$RMSD = \sqrt{\frac{1}{n} \sum_{i=1}^{n} (R_i - M_i)^2}, \tag{3}$$

$$ubRMSD = \sqrt{\frac{1}{n} \sum_{i=1}^{n} \left[(R_i - \overline{R}) - (M_i - \overline{M})\right]^2}, \tag{4}$$

$$r^2 = 1 - \frac{\sum_{i=1}^{n} (M_i - R_i)^2}{\sum_{i=1}^{n} (\overline{R} - R_i)^2}, \tag{5}$$

where n represents the total number of samples. M_i and R_i represent the merged and reference data at i, respectively. $\overline{M}$ and $\overline{R}$ represent the average of the merged and reference data.

2.6. Deep Learning Merging Model Training

Statistical analysis (Figure S1) shows that GPP data belong to heavy-tail distribution. The characteristic of the skewed distribution data is that there are more samples with little influence and fewer samples with substantial impact. The concrete manifestation in GPP data is that the number of pixels near the value of 0 is far more than that of other values. In addition, the larger the value, the smaller the number of pixels. Such datasets will make the deep learning network perform well in the low-value range, while the efficiency is relatively low in the high-value range, resulting in overall reduced accuracy. Therefore, the log1p function converts GPP data to make it more obedient to Gaussian distribution during data preprocessing.

The period from 1982 to 2014, when CMIP6 GPP data and GLASS GPP data coincide, was selected to establish the relationship between input data and reference data using the deep learning network. The GPP data simulated by 23 CMIP6 models and the reference data in the same month were matched as a group, totaling 396 samples. All samples were divided into three datasets according to time: training (1982–2010), validation (2011–2012), and test dataset (2013–2014). The number of iterations and the initial learning rate were 150 times and 1×10^{-4} during the training stage, respectively.

Figure 2 shows loss function curves for training and validation datasets. As shown in the figure, the loss reduced sharply in the first ten iterations of the training process. Subsequently, it shows a fluctuant reduction. The model is stable after 77 iterations, with a decreased loss. Loss no longer changes after 112 iterations, indicating that the model has been trained.

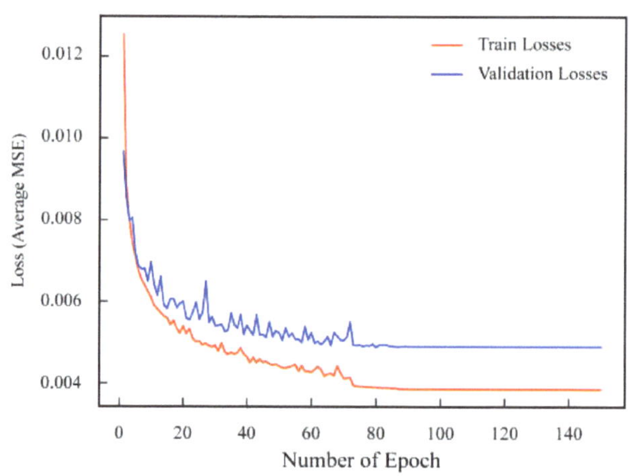

Figure 2. Loss function curves for the training and the validation dataset.

3. Results

3.1. Quality Evaluation of Merging Results

The trained merging model has been tested. The results of comparing GLASS GPP, $CMIP6_{DL}$ GPP, and multi-model ensemble mean with fluxnet data in China are shown in Figure 3. However, the results only represent the accuracy of the ten sites, as there are only ten unevenly distributed flux towers in China. As can be seen from the figure, all three datasets have a wide distribution range in the four indicators. The distribution range in MAE, RMSD and r^2 of the multi-model ensemble mean is wider than that of the other two datasets, indicating that it has a higher uncertainty. Although the distribution range is relatively small on ubRMSD, many outliers exist in the multi-model ensemble mean. The uncertainty of $CMIP6_{DL}$ GPP is lower compared with the multi-model ensemble mean, while the median performs the best in MAE, RMSD, and ubRMSD. The merged data have improved overall (Figure 3a–c), indicating that data are closer to the actual value than the multi-model ensemble mean.

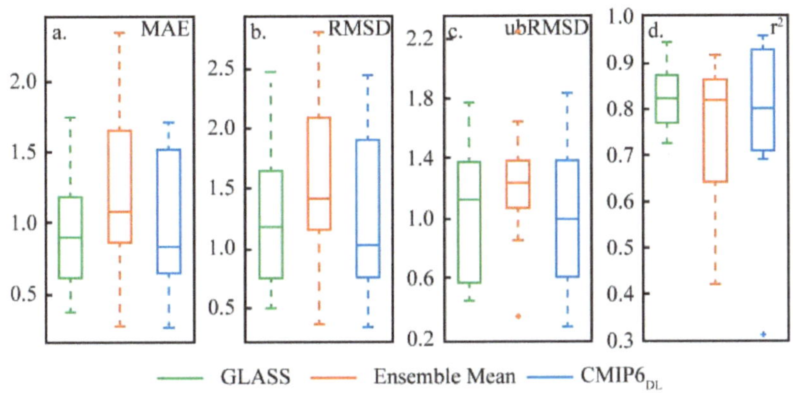

Figure 3. (**a**) MAE (**b**) RMSD (**c**) ubRMSD (**d**) r^2 between GLASS GPP, $CMIP6_{DL}$ GPP, the multi-model ensemble mean, and observations over China. Blue "+" represents outlier of CMIP6DL, orange "+" represents outlier of Ensemble Mean.

Table 2 shows the model's performance on the test dataset taking GLASS GPP as the benchmark, with the results showing that the merged GPP has improved in the aspect of the four indices and has a higher accuracy in China as a whole compared with the multi-model ensemble mean.

Table 2. Mean absolute error (MAE), root mean square deviation (RMSD), unbiased root mean square difference (ubRMSD), and coefficient of determination (r^2) between merged GPP, the multi-model ensemble mean, and GLASS GPP.

	MAE	RMSD	ubRMSD	r^2
$CMIP6_{DL}$ GPP	0.3575	0.7691	0.7668	0.8556
Ensemble mean	1.2378	1.7664	1.5137	0.6632

The normalized Taylor diagram obtained by calculating correlation coefficients, ubRMSD, and standard deviation ratios is used to evaluate further the $CMIP6_{DL}$ GPP, CMIP6 GPP, and muti-model ensemble mean. GLASS GPP has been used as the reference due to its relatively low uncertainties. Figure 4 shows the distribution of correlation coefficient, ubRMSD, and standard deviation ratio between 25 datasets (including $CMIP6_{DL}$ GPP, the multi-model ensemble mean, and 23 CMIP6 datasets) and the reference data in 2013–2014. As shown in Figure 4, the red dot represents GLASS GPP. Since it is used as reference data, the correlation coefficient and standard deviation ratio are equal to 1, and ubRMSD is 0. The black dot represents $CMIP6_{DL}$ GPP data, which is the closest to the red dot in the figure, indicating that the overall performance of $CMIP6_{DL}$ GPP data is better than that of all datasets simulated by other models. The correlation coefficient of all CMIP6 GPP is less than 0.82, and ubRMSD is greater than 0.5. There are significant differences among the 23 model data, of which CAS-ESM2-0 performs the best, even better than the multi-model ensemble mean. The multi-model ensemble mean is still superior to most single CMIP6 data, though it is affected by the high error of model data. In general, $CMIP6_{DL}$ GPP data perform the best in all aspects.

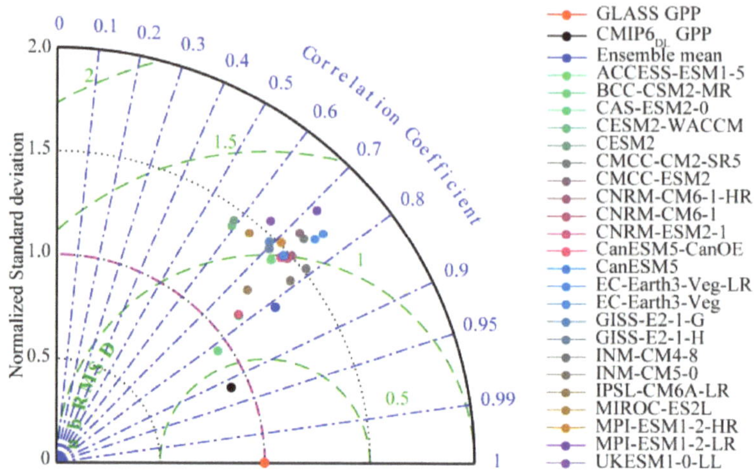

Figure 4. Normalized Taylor plots of $CMIP6_{DL}$ GPP, GLASS GPP, the multi-model ensemble mean, and 23 CMIP6 models for the test dataset.

To evaluate the spatial accuracy of the merged data, the spatial absolute error distributions of $CMIP6_{DL}$ GPP, the multi-model ensemble mean, and GLASS GPP are plotted, respectively. Figure 5a shows the MAE between $CMIP6_{DL}$ GPP and GLASS GPP in the test dataset, with the results showing that the regions with MAE less than 0.5 account for 73.91%. The errors in Northwest, Northeast, and North China are generally minor, less than 0.21 $gCm^{-2}d^{-1}$. However, the effects of $CMIP6_{DL}$ GPP in Central China and the eastern part of Southwest China are not ideal, with the error ranging from 0.5 to 1.5 $gCm^{-2}d^{-1}$, with a total of 7.4% of the regional errors being even more significant than 1 $gCm^{-2}d^{-1}$. Relatively large errors mainly occur in Yunnan Province, with two-thirds of the regional errors being more significant than 0.5 $gCm^{-2}d^{-1}$. Figure 5b shows MAE between the multi-model ensemble mean and GLASS GPP. Except for Northwest China, the error in most regions is beyond 1 $gCm^{-2}d^{-1}$. Specifically, the regions with an error greater than 1 $gCm^{-2}d^{-1}$ account for 45.17%, even more significant than 2 $gCm^{-2}d^{-1}$ in Central, East, and South China. In summary, $CMIP6_{DL}$ GPP generated by merging the model Ynet is highly reliable.

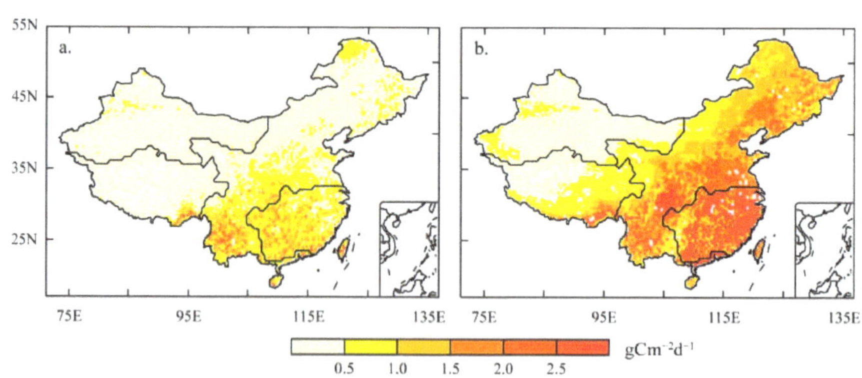

Figure 5. (**a**) MAE between GLASS GPP and $CMIP6_{DL}$ GPP, (**b**) MAE between GLASS GPP and multi-model ensemble mean in the test dataset (2013–2014).

3.2. Assessment of the Variations in the Merged GPP

GLASS GPP is a dataset generated by the light use efficiency model integrating CO_2, radiation, and vapor pressure deficit (VPD), providing a reliable long-term GPP estimation. Figure 6 shows the long-term changes in GLASS GPP, $CMIP6_{DL}$ GPP, and multi-model ensemble mean from 1982 to 2014. As shown in Figure 6a, the variations in GLASS GPP are not apparent and almost remain unchanged in the northwest arid regions. There is an increasing trend in Central China and a significant decrease in the eastern part of Inner Mongolia, the southern part of East China, and South China. The long-term changes in the multi-model ensemble mean are shown in Figure 6c. There is a sharp increasing trend throughout China, except for the western regions. The characteristics of the decrease in the southern coastal areas and Inner Mongolia are not captured, which is inconsistent with the spatial characteristics of the variations in GPP over China. Figure 6b describes the spatial distribution of the long-term variations of the merged GPP data, similar to GLASS GPP. However, the variations are gentler than the reference data since the deep learning merging model is built using a convolution neural network with a "smoothing" effect. That is, it is impossible for the deep learning merging model to learn the extreme value. Compared with the multi-model ensemble mean, the data quality has been significantly improved in eastern Inner Mongolia and South China, dramatically increasing the merged dataset's reliability in the projection period.

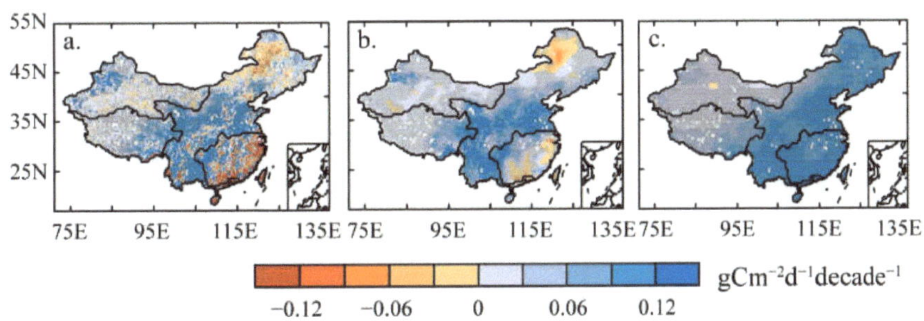

Figure 6. Long-term variation trends of (**a**) GLASS GPP, (**b**) $CMIP6_{DL}$ GPP, and (**c**) multi-model ensemble mean from 1982 to 2014, with the dot indicating passing significance test at 5% level.

To evaluate the differences in the merged results in the time dimension, the grids are randomly selected in four climatic regions over China (arid, semi-arid and semi-humid, humid region, and Qinghai–Tibet Plateau) to test the time series. As can be seen from Figure 7, there are similar seasonal fluctuations in the three datasets. The fluctuation range of the test dataset is highly consistent in the $CMIP6_{DL}$ GPP and the reference data, while it is more significant in the multi-model ensemble mean. GPP has consistently been overestimated throughout the period in the multi-model ensemble mean. As the GPP increases, the error of the multi-model ensemble mean increases. Specifically, the high value of GPP is concentrated in June and July, when the error of the multi-model ensemble mean is the largest. The impacts of temperature, water, solar radiation, and vegetation growth are the main causes of seasonal variation. The temperature, precipitation, vegetation leaf area, and solar radiation of each region reach the highest values during in the year in summer, and then gradually decrease. The comprehensive impact of these factors ultimately leads to an inverted U-shaped change in the average monthly GPP of each vegetation coverage area.

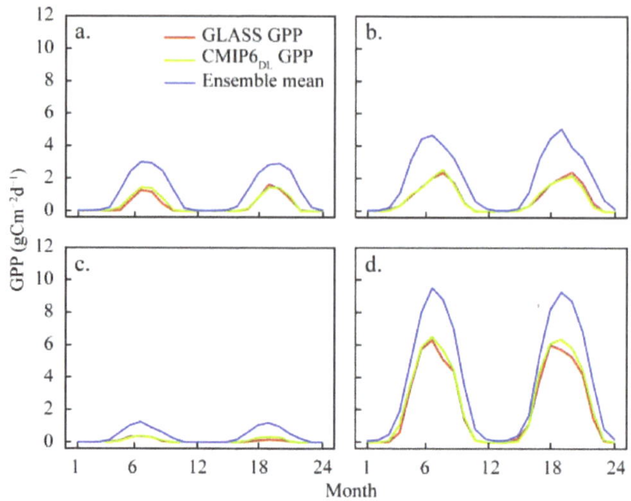

Figure 7. Time series of GPP of test dataset in (**a**) arid region, (**b**) semi-arid and semi-humid region, (**c**) Qinghai–Tibet Plateau, and (**d**) humid region.

3.3. Temporal and Spatial Distribution of Merged GPP in Historical Period

The GPP data over the historical period (1850–2014) and the four climate scenarios (SSP1-2.6, SSP2-4.5, SSP3-7.0, SSP5-8.5) are merged using the trained deep learning merging model based on the data simulated by 23 CMIP6 models. The spatial distribution of GPP in different seasons is shown in Figure 8 to study the seasonal variation characteristics during the historical period (1850–2014). There is a significant spatial distribution gradient of GPP in various seasons in China, mainly due to land use and cover conditions and temperature and water environmental factors. In southern China, the climate is warm and humid, with the best hydrothermal conditions, sufficient sunlight and high light utilization efficiency, high vegetation coverage, and the highest GPP in the growing season. GPP is high in northeast forest regions in summer. However, photosynthesis is inhibited by low temperature in other seasons, resulting in relatively low GPP. In the deserts of the northwest, most areas of the Qinghai–Tibet Plateau, and the grasslands in central and western Inner Mongolia, the plant growing season is relatively short due to water stress and low-temperature constraints, which leads to low productivity and even the lowest GPP. Overall, GPP decreases from southeast to northwest and coastal to inland, consistent with current research results [56–59]. The spatial distribution of GPP is highly compatible with precipitation in summer, indicating that precipitation is an important factor affecting the distribution of GPP in China. In addition, the GPP near the Tianshan Mountains in Xinjiang is relatively high in summer. The prevailing westerly winds from the Atlantic Ocean and the airflow from the Arctic Ocean encounter the uplift of mountain slopes and generate intense precipitation, which promotes a high vegetation coverage in the west and northwest of the Tianshan Mountains [60]. At the same time, the high mountain snowmelt in the Tianshan Mountains also plays a promoting role.

The spatial distribution of GPP during the four seasons is significantly different, mainly due to the comprehensive influences of the monsoon climate and vegetation type. Throughout the whole year, GPP is the highest in summer (June to August), followed by spring (March to May), autumn (September to November), and the lowest in winter, which entirely fits within the ecological definition of the GPP itself. In spring, the high value of GPP is concentrated in the area south of 28° N. As the summer monsoon moves northward, the temperature in most parts increases, with the average GPP in the historical period reaching 13.6 $gCm^{-2}d^{-1}$. The spatial distribution of GPP is highly affected by hydrothermal conditions. Thus, the high value is mainly distributed in Yunnan and

Guizhou. In autumn, the spatial distribution is similar to that in spring due to the decrease in temperature in northern China. GPP is mostly lower than 1 $gCm^{-2}d^{-1}$ in winter, and even 0 in some regions. Due to the minimum precipitation and temperature in winter, most of the vegetation is in the non-growing season, which makes the vegetation coverage the lowest, resulting in the lowest GPP throughout the year. Although crops are sown in the north, they grow slowly due to the low temperature. On the contrary, the GPP in southern Yunnan remains high even in winter (Figure 8d).

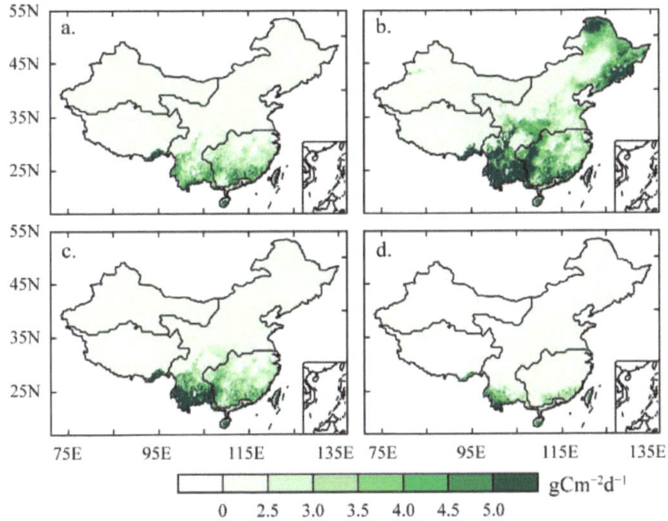

Figure 8. Spatial distribution characteristics of GPP in (**a**) MAM, (**b**) JJA, (**c**) SON, (**d**) DJF over China in the historical period (1850–2014).

The global climate has undergone significant changes in the past few decades, including frequent extreme climate disasters and sustained warming. Ecosystems and terrestrial carbon sinks will be significantly affected by climate change. On the other hand, GPP is greatly influenced by human activity [61]. Although large-scale afforestation benefits vegetation restoration and increases GPP, population growth and rapid economic development have exacerbated urbanization and disrupted the balance of terrestrial ecosystems. These human interventions significantly impact the formation of GPP dynamics [62,63]. Figure 9 shows the spatial distribution of variation trends in GPP in all seasons over China during the historical period (1850–2014). There are increasing trends in most regions; however, changes in these trends exhibit significant heterogeneity in space and time. The trend is the strongest in summer, followed by spring and autumn, and the most gentle in winter. In summer, the variation trend of GPP in more than one-third of the regions is relatively large, concentrated in Northeast China, the south of North China, and the southwest regions. The annual mean value of GPP in these regions is significant, with the variation trends also being substantial. It is mainly related to the superior local water and heat conditions and the suitable climate environment, which results in the overall growth of vegetation. The GPP in the south of Central China and East China shows a weak decreasing trend. It is mainly due to the increase in temperature in the southern region since the 1990s, which increases the frequency of heat waves and drought events [64,65]. There is a negative impact on the terrestrial carbon cycle, resulting in a decline in GPP. Compared with summer, the regions with a substantial increase in autumn tend to reduce; meanwhile, the trend is gentler. In spring, the increasing trends concentrate in Anhui and Guizhou. In winter, there is a trend of widespread decline in Inner Mongolia, Jilin, and Liaoning, accounting for 31.57% of the whole region, while there is an increasing trend in some regions in the south.

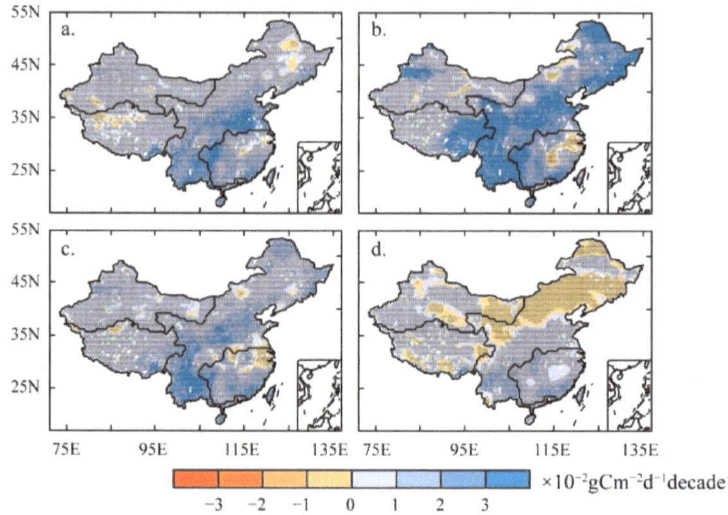

Figure 9. Spatial distribution of GPP variation trends in (**a**) MAM, (**b**) JJA, (**c**) SON, and (**d**) DJF over China during the historical period (1850–2014), with the dot indicating passing significance test at 5% level.

3.4. Long-Term Changes in Merged GPP in Projection Period under Multiple Climate Scenarios

To explore the difference between the changes in GPP in the near future (2021–2040) and the far future (2071–2100), the spatial distribution of the multi-year mean of GPP in each season under different climate scenarios in historical baseline, near, and far future is described in Figure 10 and Figure S2, respectively. Among them, the distributions under different scenarios in the future describe the change in the multi-year mean compared with that during baseline. The positive value represents the increase in GPP, while the negative one represents the decrease. Figure 10a1–a4 shows that the spatial distribution of the mean annual GPP in each season of the baseline GPP increased significantly in the four seasons compared with the seasonal average over the period 1850–2014 (Figure 8). In spring, the average annual GPP greater than 2 $gCm^{-2}d^{-1}$ occupied 22.93% of the region, concentrated in the south of 30° N. In summer, the area with an annual average GPP of more than 2 $gCm^{-2}d^{-1}$ doubled compared with spring, reaching 48.97%, with the maximum yearly peak at 14.19 $gCm^{-2}d^{-1}$. There are similar peaks and spatial distribution in autumn and spring, while it remains almost the same in winter.

There are significant seasonal differences in the variations in GPP based on the historical baseline under multiple scenarios in the near future. In spring, GPP increases in most eastern regions, and rises more with the enhancement of radiation forcing levels. However, GPP decreased significantly in the east of the plain of Henan Province under the high emission scenario. In summer, GPP increased substantially in nearly half of the regions of the country, including the Northeast, North, Southwest, and south of Northwest China. GPP accelerated with the increase in radiation forcing levels, with more than 20% of the regions having values greater than 0.4 $gCm^{-2}d^{-1}$ under SSP5-8.5. In autumn, it increased significantly in Yunnan, while there was a slight weakening trend in the north of Henan. The spatial characteristics distribution maintains consistency under multiple climate scenarios. Winter is the only season when GPP decreases in a large area. Under the four scenarios, the GPP in 21% of regions is lower than the baseline, mainly distributed in the west, Inner Mongolia, and Heilongjiang. The reduction in GPP is the same under different scenarios. Except for the decrease in GPP in Fujian, it generally increases in the south.

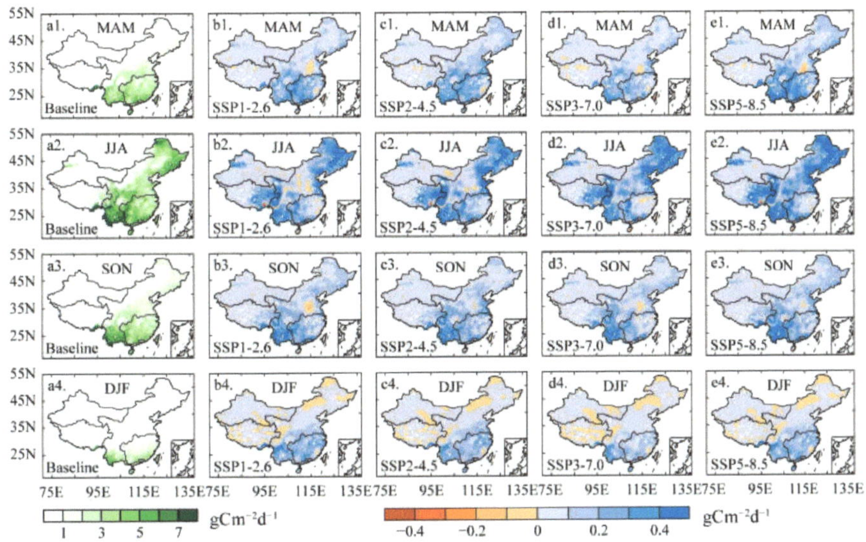

Figure 10. Spatial distribution of the multi-annual mean CMIP6$_{DL}$ GPP in (**a1–a4**) baseline (1995–2014) and near future (2021–2040) under (**b1–b4**) SSP1-2.6, (**c1–c4**) SSP2-4.5, (**d1–d4**) SSP3-7.0, and (**e1–e4**) SSP5-8.5 relative to baseline in (**a1–e1**) MAM, (**a2–e2**) JJA, (**a3–e3**) SON, and (**a4–e4**) DJF.

The change in GPP becomes increasingly intense with the enhancement of radiative forcing levels in the far future, as seen from Figure S2. The variations in GPP are studied with the Heihe–Tengchong line as the boundary. In spring, it can be found that GPP increases more as radiative forcing levels are enhanced in the eastern part of the dividing line. The area with an increase of more than 0.8 $gCm^{-2}d^{-1}$ rises from 0.62% to 23.99%. The change in the west of the boundary is gentle, with an average rise of 0.09 $gCm^{-2}d^{-1}$. The area of GPP increasing by more than 0.8 $gCm^{-2}d^{-1}$ expands from 0.62% to 23.99% of the regions. The change in the western part of the dividing line is relatively gentle, with an average increase of 0.09 $gCm^{-2}d^{-1}$. Summer is the peak photosynthesis season for plants in northern ecosystems compared to other seasons, leading to a more extensive range and broader areas of increase. GPP increases in about 94% of the regions under SSP1-2.6 relative to the baseline. With the enhancement of radiation forcing levels, the rise of GPP in Northeast, North, and the east of Southwest China becomes more and more intense, while that in Henan is turning from decrease to increase. The magnitude of the increase is further accelerated under SSP5-8.5. The reduction in GPP in Henan in autumn only occurs in the low-emission scenario, which differs from spring. In winter, GPP reduces in parts of the west and northeast regions, roughly similar to the scenario in the near future (Figure 10b1–e4). There is a considerable area reduction in GPP in the eastern part of the Qinghai–Tibet Plateau under the SSP1-2.6 scenario. With the increase in radiation forcing levels, the area of GPP reduction gradually decreases. GPP increased south of 30° N, with the maximum increase reaching 2.48 $gCm^{-2}d^{-1}$ in Yunnan under the SSP5-8.5 scenario. Compared with the near future, GPP growth will be more dramatic in the far future.

The multi-year variations in GPP in each pixel can better reflect the characteristics of spatial changes in addition to the regional average. Figure 11 describes the spatial distribution of GPP variation trends in China under four scenarios from 2015 to 2100. GPP generally shows an increasing trend in the projection period, with the trend becoming more and more intense as the radiation forcing levels increase. There is no significant change in GPP in the northwest arid regions under different scenarios, which maintains a low growth. A slight decreasing trend can be observed visually at the junction of Shandong, Henan, Jiangsu, and Anhui provinces under the SSP1-2.6 scenario, which is related to the environmental effects of aerosols [66]. The spatial pattern of GPP variations under the medium and high emission scenario is similar, with GPP increasing in each pixel. Likewise, the variation trends become more intense as radiative forcing levels increase. The most significant increase in GPP occurs in Yunnan and Guizhou under SSP2-4.5, SSP3-7.0, and SSP5-8.5. The regions where GPP is more sensitive to radiation forcing levels are concentrated in the southwest, the east of the northwest, and the northeast. GPP increases rapidly as radiative forcing levels increase. It is found that the increase in the high-emission scenario is even four times that of the low-emission scenario.

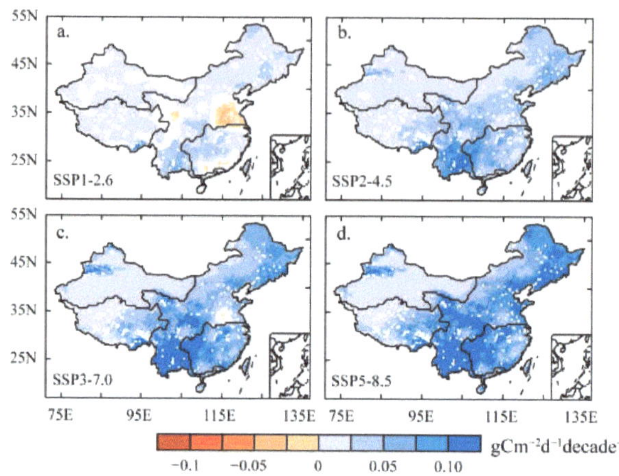

Figure 11. Spatial distribution of the variation trends of annual average $CMIP6_{DL}$ GPP under (**a**) SSP1-2.6, (**b**) SSP2-4.5, (**c**) SSP3-7.0, and (**d**) SSP5-8.5 from 2015 to 2100. Shaded values have passed the significance test at the 5% level.

Considering the characteristics of the future variations in GPP, it is divided into three periods for analysis, including 2021–2040 (near future), 2041–2070 (medium future), and 2071–2100 (far future) due to the long projection period of the merged data. As can be seen from Figure S3, the increasing trend of GPP turns into a decreasing trend as time goes on under the SSP1-2.6 scenario. In the near future, GPP shows a significant increasing trend except for the junction of Shandong, Jiangsu, Henan, and Anhui provinces. In the medium future, the changes in average regional GPP are relatively stable, though there are scattered pixels with increasing and decreasing trends of GPP across the country. In the far future, the regions with reduced trends of GPP nationwide account for about 93% of the country. Under the low emission scenario, the air temperature rises first and remains stable in the far future, while the carbon dioxide emissions begin to decrease in 2020 [67]. This shows that GPP is greatly affected by air temperature. The amount of carbon dioxide directly affects vegetation's carbon sequestration when the temperature is stable, leading to GPP decline. GPP increases in the near future under the SSP2-4.5 scenario; however, the variation trends slow down over time. Under SSP3-7.0 and SSP5-8.5 scenarios, GPP increased over time in all regions, with the trends strengthening gradually.

4. Discussion

It can be seen from Figure 10 and Figure S2 that the variations in GPP in the north and south of China are opposite in winter while increasing in all regions in spring, summer, and autumn. However, this cannot reflect the impact of the reduction in winter on the change in the average GPP across the country. Therefore, Figure 12 shows the inter-annual changes in GPP in China under the historical baseline and different scenarios in the future. The thick grey line represents the line between the baseline and the future. GPP shows an increasing trend from 1.76 $gCm^{-2}d^{-1}$ to 1.84 $gCm^{-2}d^{-1}$ from 1995 to 2014. It may result from a combination of factors such as appropriate climate change, the fertilization effect of CO2, nitrogen deposition, and human activities such as afforestation and agricultural irrigation [68–71]. From 2015 to 2035, GPP increases steadily yearly with similar increasing trends under the four scenarios. From 2036, the difference in the variations in GPP in the four scenarios is noticeable. Among them, the variation speed of GPP is the fastest under SSP5-8.5, with the average reaching 2.49 $gCm^{-2}d^{-1}$ at the end of the 21st century. The second fastest occurs under SSP3-7.0, which is slightly slower than that of the high-emission scenario. The annual average is roughly the same before 2055 as under SSP2-4.5. However, the gap between the two scenarios gradually increases from 2067, when the growth rate under SSP2-4.5 slows down. From 2081 to 2100, it is almost stable and begins declining slowly in 2097. Among the four scenarios, GPP decreases significantly only in the SSP1-2.6 scenario. Although there is a decreasing trend at the end of the century under the SSP2-4.5 scenario (Figure 12), GPP shows an overall increasing trend in each pixel over the period 2015–2100 (Figure 11b). More specifically, it keeps increasing, as do the other three scenarios in the near future, followed by a slight fluctuation around 1.96 $gCm^{-2}d^{-1}$ for a long time in the medium term. It is worth noting that GPP declines significantly in the far future, especially from 2075 onwards. Meteorological factors are the dominant factors in the interannual variation in global GPP. In the context of climate warming, the frequent occurrence of extreme events such as high temperatures and droughts is expected to impact vegetation growth, leading to fluctuations in ecosystem GPP. According to the definition of severe climate, the Intergovernmental Panel on Climate Change [72] pointed out that climate change has led to changes in the frequency, intensity, spatial scope, duration, and time of extreme climate, which may have an unprecedented impact on the terrestrial carbon cycle. In addition, due to global warming, climate change is expected to further increase the frequency, persistence, and intensity of extreme weather in the mid to late 21st century [73–75], making the impact of future climate change on terrestrial ecosystems even more uncertain [76,77].

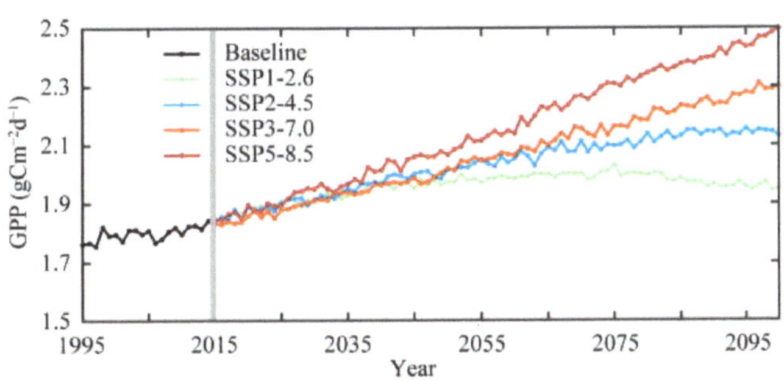

Figure 12. Inter-annual variations in $CMIP6_{DL}$ GPP at baseline (1995–2014) and in the future (2015–2100) under climate scenarios over China.

Large-scale studies using merged GPP can reduce the uncertainty of terrestrial ecosystem carbon cycle research. However, the merging process will also introduce certain uncertainties, mainly affected by input errors, scale effects, and merging algorithms. Input errors are caused by the CMIP6 datasets, GLASS GPP, and flux tower observation GPP. Although model simulation provides an essential research tool for the carbon cycle of large-scale terrestrial ecosystems, there are still significant differences in the simulation results of the models at regional and global scales due to the differences in the structure, parameter settings, input data, and spatial resolution of each CMIP6 model. Schaefer et al. [5] compared GPP simulation differences in North America across 26 models to explore the responses of different model structures and environmental factors to simulations based on data from 39 flux stations. Zhang et al. [78] studied the impact of different parameter settings and meteorological data in the CEVSA2 model on the GPP simulation results in the Changbai Mountain region of China from 2003 to 2005, indicating that the GPP difference caused by parameter setting in the model is 5% to 8%. The reference data GLASS GPP is generated by integrating multiple algorithms using the Bayesian integration method, reducing the uncertainty of a single algorithm. However, there are differences in spatial scales between ground and satellite data, and the algorithm fails to distinguish potential differences in photosynthesis between C3 and C4 crops. In addition, the problem of mixed pixels can also affect the accuracy of GLASS GPP estimation [79]. The GPP observed by the flux tower is considered the benchmark, while the high heterogeneity of the ground can also make the representativeness of the flux station data poor. Therefore, significant uncertainties in the GPP benchmark lead to a lack of consensus on the global GPP distribution [15,80]. Chen et al. [81] showed that the flux-tower-estimated GPP represents 50% to 60% of the actual situation in the case of surface heterogeneity. There is still a problem of insufficient spatial representation compared to the range of flux contribution areas assumed by eddy covariance.

The mismatch between the spatial resolution of the model and the floor area of the flux tower causes the uncertainty imposed by scale effects. Even with the same model and source of input data, the simulation results may still be different due to spatial heterogeneity caused by different spatial resolutions [1]. At the regional scale, improving spatial resolution can reduce the uncertainty of GPP simulation [82]. However, in practical applications, models often need to simulate large-scale and long-time series data. When considering factors such as simulation time, data storage capacity, and analysis difficulty, they often reduce the spatial resolution to improve simulation efficiency. In GPP estimation at different spatial scales, there will be different judgments on the coverage type, leading to incorrect maximum light energy utilization, further resulting in errors. In mid- and low-resolution GPP remote sensing products, most of the pixels under heterogeneous ground surfaces are mixed. For example, there is a significant difference between MODIS surface coverage with 1 km and global 30 m resolution. At different scales, heterogeneity characteristics have an important impact on surface vegetation parameter characteristics [83]. Taking the mixed pixel of forest and grassland as an example, there is a significant difference in the a priori maximum light energy utilization rate between the two types. It may bring about significant errors using a single type of maximum weak energy utilization rate to replace pixel values. Wang et al. [1] discussed the impact of two spatial resolutions of 1 km and 8 km input data on GPP simulation, with the results showing that the difference was mainly due to the difference in LAI caused by mixed pixels within the range of 8km. The difference in simulation results with different spatial resolutions at forest stations was more significant than that at grassland stations. Therefore, it is feasible to appropriately reduce spatial resolution to improve the model's simulation efficiency. However, it is necessary to minimize the error of low spatial resolution in the GPP simulation of forest ecosystems and growing seasons.

In the long sequence data merging, considering time information can effectively enhance model learning capabilities and improve the quality of merged data, which helps to reduce the data uncertainty in the projection period. Since the deep learning merging model focuses more on spatial features and no individual modules can capture temporal characteristics, the temporal information is not fully utilized. In addition, there are other factors contributing to uncertainties. For example, complex vegetation structures, significant dry and wet seasons, and cloud cover can also bring uncertainties [84]. In highly productive regions, methods based on optical remote sensing often underestimate GPP due to the saturation of reflectance measurements in high-density canopy [85]. In addition, optical remote sensing is strongly affected by cloud cover, resulting in data gaps and high uncertainty in areas with frequent cloud cover and high GPP, such as tropical forests [16].

5. Conclusions

In this study, a GPP-merged dataset with high resolution in China has been developed using the deep learning method. It combines the advantages of 23 CMIP6 GPP datasets without relying on any prior knowledge. The lower resolution of the multi-model data has been improved to 0.25° of the merged GPP. From the perspective of evaluation indicators, the performance of merged data is superior to the others in all aspects. In China, the error has been kept at a low level. The merged dataset has significantly improved in temporal and spatial dimensions compared with every single model and the multi-model mean. The merged data are closer to the actual value than the multi-model mean according to the fluxnet data. The merged data greatly reduce the uncertainties and improve the reliability of GPP estimation in the projection period. A further analysis of the spatiotemporal change pattern of GPP in China has resulted in the following findings.

During the baseline, GPP is the highest in summer, followed by spring and autumn, and the lowest in winter due to the combined effects of factors such as the monsoon climate and vegetation types. GPP exhibits significant spatial heterogeneity due to differences in topography, vegetation types, and hydrothermal conditions in different regions. Generally, GPP increases gradually from the northwest inland to the southeast coast. GPP in the southern part of Yunnan remains high in winter. According to the long-term variation trends, GPP is increasing in most regions of China. It is worth noting that the characteristics of the decrease in the southern coastal areas and Inner Mongolia are not captured by the multi-model ensemble mean; however, they are captured by the merged GPP. There are seasonal differences in the variations in GPP relative to the baseline under different scenarios in the future projection period, while the variations are almost the same across different scenarios in the near future. In summer, GPP increased rapidly with the increase in radiation forcing levels. In winter, GPP decreases in a large region mainly distributed in the west, Inner Mongolia, and Heilongjiang. The long-term variation trends of GPP in China under different scenarios are significantly different. The growth rate of GPP has reached the maximum under the SSP5-8.5 scenario since 2036. The increasing trend becomes slightly slower under the SSP3-7.0 scenario than under the high-emission scenario. Furthermore, GPP decreased significantly under the SSP1-2.6 scenario. The annual GPP in each pixel in China shows an increasing trend, which is more intense with the increase in radiation forcing levels. GPP changes from increasing to decreasing over time, since GPP is affected by temperature and carbon dioxide under the SSP1-2.6 scenario. Under the SSP2-4.5 scenario, GPP will increase in the near future; however, the trend will slow down as time goes by. Under the SSP3-7.0 and SSP5-8.5 scenarios, GPP increases with time; the increasing trend rises gradually.

Supplementary Materials: The following supporting information can be downloaded at https://www.mdpi.com/article/10.3390/f14061201/s1, Figure S1: The probability density distribution of GLASS GPP during 1982–2014, Figure S2: Spatial distribution of the multi-annual mean GPP in far future (2071–2100) relative to baseline under (a1–a4) SSP 1-2.6, (b1–b4) SSP2-4.5, (c1–c4) SSP 3-7.0, (d1–d4), and SSP 5-8.5 in (a1–d1) spring, (a2–d2) summer, (a3–d3) autumn, and (a4–d4) winter, Figure S3: Spatial distribution of the variation trends of annual average GPP under four climate scenarios in the near, middle, and far future, with the dot indicating passing significance test at 5% level.

Author Contributions: Conceptualization, J.L. and G.W.; methodology, J.L. and D.F.; software, J.L. and D.F.; validation, D.F. and I.K.N.; formal analysis, J.L.; investigation, J.L.; resources, D.F.; data curation, I.K.N.; writing—original draft preparation, J.L.; writing—review and editing, J.L. and G.W.; visualization, J.L.; supervision, G.W.; project administration, J.L. and G.W.; funding acquisition, J.L. and G.W. All authors have read and agreed to the published version of the manuscript.

Funding: This research was funded by China Meteorological Administration Innovation Development Project (CXFZ2023J071) and the Wuxi University Research Start-up Fund for Introduced Talents (2023r041).

Data Availability Statement: Data are available from https://doi.org/10.6084/m9.figshare.22004504.v1 (accessed on 23 April 2023).

Conflicts of Interest: The authors declare no conflict of interest. The funders had no role in the design of the study; in the collection, analyses, or interpretation of data; in the writing of the manuscript; or in the decision to publish the results.

References

1. Wang, M.; Zhou, L.; Wang, S.; Wang, X.; Sun, L. An analysis of the Gross Primary Productivity simulation difference resulting from the spatial resolution. *Geogr. Res.* **2016**, *35*, 617–626.
2. Zhang, X.; Wang, H.; Yan, H.; Ai, J. Analysis of spatio-temporal changes of gross primary productivity in China from 2001 to 2018 based on Remote Sensing. *Acta Ecol. Sin.* **2021**, *41*, 6351–6362.
3. Ito, A.; Inatomi, M.; Huntzinger, D.N.; Schwalm, C.; Michalak, A.M.; Cook, R.; King, A.W.; Mao, J.; Wei, Y.; Post, W.M.; et al. Decadal trends in the seasonal-cycle amplitude of terrestrial CO_2 exchange resulting from the ensemble of terrestrial biosphere models. *Tellus B Chem. Phys. Meteorol.* **2016**, *68*, 28968. [CrossRef]
4. Chen, M.; Rafique, R.; Asrar, G.R.; Bond-Lamberty, B.; Ciais, P.; Zhao, F.; Reyer, C.P.O.; Ostberg, S.; Chang, J.; Ito, A.; et al. Regional contribution to variability and trends of global gross primary productivity. *Environ. Res. Lett.* **2017**, *12*, 105005. [CrossRef]
5. Schaefer, K.; Schwalm, C.R.; Williams, C.; Arain, M.A.; Barr, A.; Chen, J.M.; Davis, K.J.; Dimitrov, D.; Hilton, T.W.; Hollinger, D.Y.; et al. A model-data comparison of gross primary productivity: Results from the North American Carbon Program site synthesis. *J. Geophys. Res. Biogeosci.* **2012**, *117*, G03010. [CrossRef]
6. Friedlingstein, P.; O'Sullivan, M.; Jones, M.W.; Andrew, R.M.; Hauck, J.; Olsen, A.; Peters, G.P.; Peters, W.; Pongratz, J.; Sitch, S.; et al. Global Carbon Budget 2020. *Earth Syst. Sci. Data* **2020**, *12*, 3269–3340. [CrossRef]
7. Damm, A.; Elbers, J.A.N.; Erler, A.; Gioli, B.; Hamdi, K.; Hutjes, R.; Kosvancova, M.; Meroni, M.; Miglietta, F.; Moersch, A.; et al. Remote sensing of sun-induced fluorescence to improve modeling of diurnal courses of gross primary production (GPP). *Glob. Chang. Biol.* **2010**, *16*, 171–186. [CrossRef]
8. Guo, L.; Gao, J.; Ma, S.; Chang, Q.; Zhang, L.; Wang, S.; Zou, Y.; Wu, S.; Xiao, X. Impact of spring phenology variation on GPP and its lag feedback for winter wheat over the North China Plain. *Sci. Total Environ.* **2020**, *725*, 138342. [CrossRef] [PubMed]
9. Wang, S.; Quan, Q.; Meng, C.; Chen, W.; Luo, Y.; Niu, S.; Yang, Y. Experimental warming shifts coupling of carbon and nitrogen cycles in an alpine meadow. *J. Plant Ecol.* **2021**, *14*, 541–554. [CrossRef]
10. Zhang, Z.; Zhang, Y.; Porcar-Castell, A.; Joiner, J.; Guanter, L.; Yang, X.; Migliavacca, M.; Ju, W.; Sun, Z.; Chen, S.; et al. Reduction of structural impacts and distinction of photosynthetic pathways in a global estimation of GPP from space-borne solar-induced chlorophyll fluorescence. *Remote Sens. Environ.* **2020**, *240*, 111722. [CrossRef]
11. Cheng, Y.-B.; Zhang, Q.; Lyapustin, A.I.; Wang, Y.; Middleton, E.M. Impacts of light use efficiency and fPAR parameterization on gross primary production modeling. *Agric. For. Meteorol.* **2014**, *189–190*, 187–197. [CrossRef]
12. Baldocchi, D.; Ryu, Y.; Keenan, T. Terrestrial Carbon Cycle Variability. *F1000research* **2016**, *5*, 2371. [CrossRef]
13. Nemani, R.R.; Keeling, C.D.; Hashimoto, H.; Jolly, W.M.; Piper, S.C.; Tucker, C.J.; Myneni, R.B.; Running, S.W. Climate-driven increases in global terrestrial net primary production from 1982 to 1999. *Science* **2003**, *300*, 1560–1563. [CrossRef]
14. Speckman, H.N.; Frank, J.M.; Bradford, J.B.; Miles, B.L.; Massman, W.J.; Parton, W.J.; Ryan, M.G. Forest ecosystem respiration estimated from eddy covariance and chamber measurements under high turbulence and substantial tree mortality from bark beetles. *Glob. Chang. Biol.* **2015**, *21*, 708–721. [CrossRef]
15. Anav, A.; Friedlingstein, P.; Beer, C.; Ciais, P.; Harper, A.; Jones, C.; Murray-Tortarolo, G.; Papale, D.; Parazoo, N.C.; Peylin, P.; et al. Spatiotemporal patterns of terrestrial gross primary production: A review. *Rev. Geophys.* **2015**, *53*, 785–818. [CrossRef]
16. Wild, B.; Teubner, I.; Moesinger, L.; Zotta, R.-M.; Forkel, M.; van der Schalie, R.; Sitch, S.; Dorigo, W. VODCA2GPP—A new, global, long-term (1988–2020) gross primary production dataset from microwave remote sensing. *Earth Syst. Sci. Data* **2022**, *14*, 1063–1085. [CrossRef]
17. Chen, Y.; Gu, H.; Wang, M.; Gu, Q.; Ding, Z.; Ma, M.; Liu, R.; Tang, X. Contrasting Performance of the Remotely-Derived GPP Products over Different Climate Zones across China. *Remote Sens.* **2019**, *11*, 1855. [CrossRef]
18. Pei, Y.; Dong, J.; Zhang, Y.; Yang, J.; Zhang, Y.; Jiang, C.; Xiao, X. Performance of four state-of-the-art GPP products (VPM, MOD17, BESS and PML) for grasslands in drought years. *Ecol. Inform.* **2020**, *56*, 101052. [CrossRef]
19. Zheng, Y.; Shen, R.; Wang, Y.; Li, X.; Liu, S.; Liang, S.; Chen, J.M.; Ju, W.; Zhang, L.; Yuan, W. Improved estimate of global gross primary production for reproducing its long-term variation, 1982–2017. *Earth Syst. Sci. Data* **2020**, *12*, 2725–2746. [CrossRef]
20. Gu, Y.; Wylie, B.K.; Howard, D.M.; Phuyal, K.P.; Ji, L. NDVI saturation adjustment: A new approach for improving cropland performance estimates in the Greater Platte River Basin, USA. *Ecol. Indic.* **2013**, *30*, 1–6. [CrossRef]
21. Lawrence, D.M.; Fisher, R.A.; Koven, C.D.; Oleson, K.W.; Swenson, S.C.; Bonan, G.; Collier, N.; Ghimire, B.; van Kampenhout, L.; Kennedy, D.; et al. The Community Land Model Version 5: Description of New Features, Benchmarking, and Impact of Forcing Uncertainty. *J. Adv. Model. Earth Syst.* **2019**, *11*, 4245–4287. [CrossRef]
22. Jung, M.; Schwalm, C.; Migliavacca, M.; Walther, S.; Camps-Valls, G.; Koirala, S.; Anthoni, P.; Besnard, S.; Bodesheim, P.; Carvalhais, N.; et al. Scaling carbon fluxes from eddy covariance sites to globe: Synthesis and evaluation of the FLUXCOM approach. *Biogeosciences* **2020**, *17*, 1343–1365. [CrossRef]
23. Jung, M.; Reichstein, M.; Bondeau, A. Towards global empirical upscaling of FLUXNET eddy covariance observations: Validation of a model tree ensemble approach using a biosphere model. *Biogeosciences* **2009**, *6*, 2001–2013. [CrossRef]
24. Jung, M.; Reichstein, M.; Margolis, H.A.; Cescatti, A.; Richardson, A.D.; Arain, M.A.; Arneth, A.; Bernhofer, C.; Bonal, D.; Chen, J.; et al. Global patterns of land-atmosphere fluxes of carbon dioxide, latent heat, and sensible heat derived from eddy covariance, satellite, and meteorological observations. *J. Geophys. Res.* **2011**, *116*, 1–16. [CrossRef]
25. Stocker, B.D.; Zscheischler, J.; Keenan, T.F.; Prentice, I.C.; Penuelas, J.; Seneviratne, S.I. Quantifying soil moisture impacts on light use efficiency across biomes. *New Phytol.* **2018**, *218*, 1430–1449. [CrossRef] [PubMed]

26. Tramontana, G.; Ichii, K.; Camps-Valls, G.; Tomelleri, E.; Papale, D. Uncertainty analysis of gross primary production upscaling using Random Forests, remote sensing and eddy covariance data. *Remote Sens. Environ.* **2015**, *168*, 360–373. [CrossRef]
27. Tramontana, G.; Jung, M.; Schwalm, C.R.; Ichii, K.; Camps-Valls, G.; Ráduly, B.; Reichstein, M.; Arain, M.A.; Cescatti, A.; Kiely, G.; et al. Predicting carbon dioxide and energy fluxes across global FLUXNET sites with regression algorithms. *Biogeosciences* **2016**, *13*, 4291–4313. [CrossRef]
28. Wei, S.; Yi, C.; Fang, W.; Hendrey, G. A global study of GPP focusing on light-use efficiency in a random forest regression model. *Ecosphere* **2017**, *8*, e01724. [CrossRef]
29. O'Sullivan, M.; Smith, W.K.; Sitch, S.; Friedlingstein, P.; Arora, V.K.; Haverd, V.; Jain, A.K.; Kato, E.; Kautz, M.; Lombardozzi, D.; et al. Climate-Driven Variability and Trends in Plant Productivity Over Recent Decades Based on Three Global Products. *Glob. Biogeochem. Cycles* **2020**, *34*, e2020GB006613. [CrossRef]
30. Gilabert, M.; Sánchez-Ruiz, S.; Moreno, Á. Annual Gross Primary Production from Vegetation Indices: A Theoretically Sound Approach. *Remote Sens.* **2017**, *9*, 193. [CrossRef]
31. Alemohammad, S.H.; Fang, B.; Konings, A.G.; Aires, F.; Green, J.K.; Kolassa, J.; Miralles, D.; Prigent, C.; Gentine, P. Water, Energy, and Carbon with Artificial Neural Networks (WECANN): A statistically-based estimate of global surface turbulent fluxes and gross primary productivity using solar-induced fluorescence. *Biogeosciences* **2017**, *14*, 4101–4124. [CrossRef] [PubMed]
32. Chen, J.M.; Mo, G.; Pisek, J.; Liu, J.; Deng, F.; Ishizawa, M.; Chan, D. Effects of foliage clumping on the estimation of global terrestrial gross primary productivity. *Glob. Biogeochem. Cycles* **2012**, *26*, 1–18. [CrossRef]
33. Ryu, Y.; Baldocchi, D.D.; Kobayashi, H.; van Ingen, C.; Li, J.; Black, T.A.; Beringer, J.; van Gorsel, E.; Knohl, A.; Law, B.E.; et al. Integration of MODIS land and atmosphere products with a coupled-process model to estimate gross primary productivity and evapotranspiration from 1 km to global scales. *Glob. Biogeochem. Cycles* **2011**, *25*, GB4017. [CrossRef]
34. Yuan, W.; Liu, S.; Liang, S.; Tan, Z.; Liu, H.; Young, C. Estimations of Evapotranspiration and Water Balance with Uncertainty over the Yukon River Basin. *Water Resour. Manag.* **2012**, *26*, 2147–2157. [CrossRef]
35. Yuan, W.; Liu, S.; Dong, W.; Liang, S.; Zhao, S.; Chen, J.; Xu, W.; Li, X.; Barr, A.; Andrew Black, T.; et al. Differentiating moss from higher plants is critical in studying the carbon cycle of the boreal biome. *Nat. Commun.* **2014**, *5*, 4270. [CrossRef]
36. Dunne, J.P.; John, J.G.; Adcroft, A.J.; Griffies, S.M.; Hallberg, R.W.; Shevliakova, E.; Stouffer, R.J.; Cooke, W.; Dunne, K.A.; Harrison, M.J.; et al. GFDL's ESM2 Global Coupled Climate–Carbon Earth System Models. Part I: Physical Formulation and Baseline Simulation Characteristics. *J. Clim.* **2012**, *25*, 6646–6665. [CrossRef]
37. Gent, P.R.; Danabasoglu, G.; Donner, L.J.; Holland, M.M.; Hunke, E.C.; Jayne, S.R.; Lawrence, D.M.; Neale, R.B.; Rasch, P.J.; Vertenstein, M.; et al. The Community Climate System Model Version 4. *J. Clim.* **2011**, *24*, 4973–4991. [CrossRef]
38. Sun, Z.; Wang, X.; Zhang, X.; Tani, H.; Guo, E.; Yin, S.; Zhang, T. Evaluating and comparing remote sensing terrestrial GPP models for their response to climate variability and CO_2 trends. *Sci. Total Environ.* **2019**, *668*, 696–713. [CrossRef]
39. Bentsen, M.; Bethke, I.; Debernard, J.B.; Iversen, T.; Kirkevåg, A.; Seland, Ø.; Drange, H.; Roelandt, C.; Seierstad, I.A.; Hoose, C.; et al. The Norwegian Earth System Model, NorESM1-M—Part 1: Description and basic evaluation of the physical climate. *Geosci. Model Dev.* **2013**, *6*, 687–720. [CrossRef]
40. Williams, M.; Richardson, A.D.; Reichstein, M.; Stoy, P.C.; Peylin, P.; Verbeeck, H.; Carvalhais, N.; Jung, M.; Hollinger, D.Y.; Kattge, J.; et al. Improving land surface models with FLUXNET data. *Biogeosciences* **2009**, *6*, 1341–1359. [CrossRef]
41. Beer, C.; Reichstein, M.; Tomelleri, E.; Ciais, P.; Jung, M.; Carvalhais, N.; Rodenbeck, C.; Arain, M.A.; Baldocchi, D.; Bonan, G.B.; et al. Terrestrial gross carbon dioxide uptake: Global distribution and covariation with climate. *Science* **2010**, *329*, 834–838. [CrossRef]
42. Bo, Y.; Li, X.; Liu, K.; Wang, S.; Zhang, H.; Gao, X.; Zhang, X. Three Decades of Gross Primary Production (GPP) in China: Variations, Trends, Attributions, and Prediction Inferred from Multiple Datasets and Time Series Modeling. *Remote Sens.* **2022**, *14*, 2564. [CrossRef]
43. Zhang, Y.; Ye, A. Would the obtainable gross primary productivity (GPP) products stand up? A critical assessment of 45 global GPP products. *Sci. Total Environ.* **2021**, *783*, 146965. [PubMed]
44. Song, L.; Li, Y.; Ren, Y.; Wu, X.; Guo, B.; Tang, X.; Shi, W.; Ma, M.; Han, X.; Zhao, L. Divergent vegetation responses to extreme spring and summer droughts in Southwestern China. *Agric. For. Meteorol.* **2019**, *279*, 107703. [CrossRef]
45. Yuan, W.; Liu, S.; Yu, G.; Bonnefond, J.-M.; Chen, J.; Davis, K.; Desai, A.R.; Goldstein, A.H.; Gianelle, D.; Rossi, F.; et al. Global estimates of evapotranspiration and gross primary production based on MODIS and global meteorology data. *Remote Sens. Environ.* **2010**, *114*, 1416–1431. [CrossRef]
46. Alexandrov, G.A. CMIP6 simulations of GPP growth satisfy the constraint imposed by increasing CO_2 seasonal-cycle amplitude. *IOP Conf. Ser. Earth Environ. Sci.* **2020**, *606*, 012003. [CrossRef]
47. Eyring, V.; Cox, P.M.; Flato, G.M.; Gleckler, P.J.; Abramowitz, G.; Caldwell, P.; Collins, W.D.; Gier, B.K.; Hall, A.D.; Hoffman, F.M.; et al. Taking climate model evaluation to the next level. *Nat. Clim. Chang.* **2019**, *9*, 102–110. [CrossRef]
48. Ichii, K.; Suzuki, T.; Kato, T.; Ito, A.; Hajima, T.; Ueyama, M.; Sasai, T.; Hirata, R.; Saigusa, N.; Ohtani, Y.; et al. Multi-model analysis of terrestrial carbon cycles in Japan: Limitations and implications of model calibration using eddy flux observations. *Biogeosciences* **2010**, *7*, 2061–2080. [CrossRef]
49. Knutti, R.; Furrer, R.; Tebaldi, C.; Cermak, J.; Meehl, G.A. Challenges in Combining Projections from Multiple Climate Models. *J. Clim.* **2010**, *23*, 2739–2758. [CrossRef]
50. Ershadi, A.; McCabe, M.F.; Evans, J.P.; Chaney, N.W.; Wood, E.F. Multi-site evaluation of terrestrial evaporation models using FLUXNET data. *Agric. For. Meteorol.* **2014**, *187*, 46–61. [CrossRef]

51. Dong, C.; Loy, C.C.; Tang, X. *Accelerating the Super-Resolution Convolutional Neural Network*; Springer International Publishing: Cham, Germany, 2016.
52. Liu, Y.; Ganguly, A.R.; Dy, J. Climate Downscaling Using YNet: A Deep Convolutional Network with Skip Connections and Fusion. In Proceedings of the 26th ACM SIGKDD International Conference on Knowledge Discovery & Data Mining, Virtual Event, CA, USA, 6–10 July 2020; Association for Computing Machinery: Virtual Event, CA, USA, 2020; pp. 3145–3153.
53. Feng, D.; Wang, G.; Wei, X.; Amankwah, S.O.Y.; Hu, Y.; Luo, Z.; Hagan, D.F.T.; Ullah, W. Merging and Downscaling Soil Moisture Data From CMIP6 Projections Using Deep Learning Method. *Front. Environ. Sci.* **2022**, *10*, 425. [CrossRef]
54. Cao, Y.J.; Song, Z.H.; Wu, Z.T.; Du, Z.Q. Spatio-temporal dynamics of gross primary productivity in China from 1982 to 2017 based on different datasets. *Chin. J. Appl. Ecol.* **2022**, *33*, 2644–2652.
55. Odena, A.; Dumoulin, V.; Olah, C. Deconvolution and Checkerboard Artifacts. *Distill* **2016**, *1*, e3. [CrossRef]
56. Zan, M.; Zhou, Y.; Ju, W.; Zhang, Y.; Zhang, L.; Liu, Y. Performance of a two-leaf light use efficiency model for mapping gross primary productivity against remotely sensed sun-induced chlorophyll fluorescence data. *Sci. Total Environ.* **2018**, *613–614*, 977–989. [CrossRef] [PubMed]
57. Jia, B.; Luo, X.; Cai, X.; Jain, A.; Huntzinger, D.N.; Xie, Z.; Zeng, N.; Mao, J.; Shi, X.; Ito, A.; et al. Impacts of land use change and elevated CO_2 on the interannual variations and seasonal cycles of gross primary productivity in China. *Earth Syst. Dyn.* **2020**, *11*, 235–249. [CrossRef]
58. Li, X.; Liang, S.; Yu, G.; Yuan, W.; Cheng, X.; Xia, J.; Zhao, T.; Feng, J.; Ma, Z.; Ma, M.; et al. Estimation of gross primary production over the terrestrial ecosystems in China. *Ecol. Modell.* **2013**, *261–262*, 80–92. [CrossRef]
59. Yan, H.; Wang, S.Q.; Wang, J.B.; Cao, Y.; Xu, L.L.; Wu, M.X.; Cheng, L.; Mao, L.X.; Zhao, F.H.; Zhang, X.Z.; et al. Multi-model analysis of climate impacts on plant photosynthesis in China during 2000–2015. *Int. J. Climatol.* **2019**, *39*, 5539–5555. [CrossRef]
60. Chen, X.; Fu, B.; Shi, P.; Guo, Q. Vegetation dynamics in response to climate change in Tianshan, Central Asia from 2000 to 2016. *Arid Land Geogr.* **2019**, *42*, 162–171.
61. Zhang, Y.; Zhang, C.; Wang, Z.; Chen, Y.; Gang, C.; An, R.; Li, J. Vegetation dynamics and its driving forces from climate change and human activities in the Three-River Source Region, China from 1982 to 2012. *Sci. Total Environ.* **2016**, *563–564*, 210–220. [CrossRef]
62. Luo, L.; Ma, W.; Zhuang, Y.; Zhang, Y.; Yi, S.; Xu, J.; Long, Y.; Ma, D.; Zhang, Z. The impacts of climate change and human activities on alpine vegetation and permafrost in the Qinghai-Tibet Engineering Corridor. *Ecol. Indic.* **2018**, *93*, 24–35. [CrossRef]
63. Jiang, Y.; Guo, J.; Peng, Q.; Guan, Y.; Zhang, Y.; Zhang, R. The effects of climate factors and human activities on net primary productivity in Xinjiang. *Int. J. Biometeorol.* **2020**, *64*, 765–777. [CrossRef]
64. Jia, J.; Hu, Z. Spatial and Temporal Features and Trend of Different Level Heat Waves Over China. *Adv. Earth Sci.* **2017**, *32*, 546–559.
65. Ye, D.; Yin, J.; Chen, Z.; Zheng, Y.; Wu, R. Spatiotemporal Change Characteristics of Summer Heatwaves in China in 1961–2010. *Clim. Chang. Res.* **2013**, *9*, 15–20.
66. Gao, M. Environmental Effect Condition (Air Temperature) of Aerosols on Gross Primary Productivity of Vegetation. *Int. J. Ecol.* **2020**, *09*, 210–222. [CrossRef]
67. Zhou, B.; Qian, J. Changes of weather and climate extremes in the IPCC AR6. *Clim. Chang. Res.* **2021**, *17*, 713–718.
68. Piao, S.; Sitch, S.; Ciais, P.; Friedlingstein, P.; Peylin, P.; Wang, X.; Ahlstrom, A.; Anav, A.; Canadell, J.G.; Cong, N.; et al. Evaluation of terrestrial carbon cycle models for their response to climate variability and to CO_2 trends. *Glob. Chang. Biol.* **2013**, *19*, 2117–2132. [CrossRef] [PubMed]
69. Lu, C.; Tian, H.; Liu, M.; Ren, W.; Xu, X.; Chen, G.; Zhang, C. Effect of nitrogen deposition on China's terrestrial carbon uptake in the context of multifactor environmental changes. *Ecol. Appl.* **2012**, *22*, 53–75. [CrossRef] [PubMed]
70. Ouyang, Z.; Zheng, H.; Xiao, Y.; Polasky, S.; Liu, J.; Xu, W.; Wang, Q.; Zhang, L.; Xiao, Y.; Rao, E.; et al. Improvements in ecosystem services from investments in natural capital. *Science* **2016**, *352*, 1455–1459. [CrossRef] [PubMed]
71. Yuan, W.; Li, X.; Liang, S.; Cui, X.; Dong, W.; Liu, S.; Xia, J.; Chen, Y.; Liu, D.; Zhu, W. Characterization of locations and extents of afforestation from the Grain for Green Project in China. *Remote Sens. Lett.* **2014**, *5*, 221–229. [CrossRef]
72. IPCC. *Managing the Risks of Extreme Events and Disasters to Advance Climate Change Adaption*; IPCC: Cambridge, UK, 2012; p. 582.
73. IPCC. *Climate Change 2013: The Physical Science Basis. Contribution of Working Group I to the Fifth Assessment Report of the Intergovernmental Panel on Climate Change*; IPCC: Cambridge, UK; New York, NY, USA, 2013; p. 1535.
74. Niu, X.; Wang, S.; Tang, J.; Lee, D.-K.; Gutowski, W.; Dairaku, K.; McGregor, J.; Katzfey, J.; Gao, X.; Wu, J.; et al. Ensemble evaluation and projection of climate extremes in China using RMIP models. *Int. J. Climatol.* **2018**, *38*, 2039–2055. [CrossRef]
75. Sui, Y.; Lang, X.; Jiang, D. Projected signals in climate extremes over China associated with a 2 °C global warming under two RCP scenarios. *Int. J. Climatol.* **2018**, *38*, e678–e697. [CrossRef]
76. Samaniego, L.; Thober, S.; Kumar, R.; Wanders, N.; Rakovec, O.; Pan, M.; Zink, M.; Sheffield, J.; Wood, E.F.; Marx, A. Anthropogenic warming exacerbates European soil moisture droughts. *Nat. Clim. Chang.* **2018**, *8*, 421–426. [CrossRef]
77. Yao, R.; Wang, L.; Huang, X.; Chen, X.; Liu, Z. Increased spatial heterogeneity in vegetation greenness due to vegetation greening in mainland China. *Ecol. Indic.* **2019**, *99*, 240–250. [CrossRef]
78. Zhang, L.; Yu, G.; Gu, F.; He, H.; Zhang, L.; Han, S. Uncertainty analysis of modeled carbon fluxes for a broad-leaved Korean pine mixed forest using a process-based ecosystem model. *J. For. Res.* **2012**, *17*, 268–282. [CrossRef]
79. Yu, T.; Sun, R.; Xiao, Z.; Zhang, Q.; Liu, G.; Cui, T.; Wang, J. Estimation of Global Vegetation Productivity from Global LAnd Surface Satellite Data. *Remote Sens.* **2018**, *10*, 327. [CrossRef]

80. Parazoo, N.C.; Bowman, K.; Fisher, J.B.; Frankenberg, C.; Jones, D.B.; Cescatti, A.; Perez-Priego, O.; Wohlfahrt, G.; Montagnani, L. Terrestrial gross primary production inferred from satellite fluorescence and vegetation models. *Glob. Chang. Biol.* **2014**, *20*, 3103–3121. [CrossRef] [PubMed]
81. Chen, B.; Coops, N.C.; Fu, D.; Margolis, H.A.; Amiro, B.D.; Black, T.A.; Arain, M.A.; Barr, A.G.; Bourque, C.P.A.; Flanagan, L.B.; et al. Characterizing spatial representativeness of flux tower eddy-covariance measurements across the Canadian Carbon Program Network using remote sensing and footprint analysis. *Remote Sens. Environ.* **2012**, *124*, 742–755. [CrossRef]
82. Cai, W.; Yuan, W.; Liang, S.; Zhang, X.; Dong, W.; Xia, J.; Fu, Y.; Chen, Y.; Liu, D.; Zhang, Q. Improved estimations of gross primary production using satellite-derived photosynthetically active radiation. *J. Geophys. Res. Biogeosci.* **2014**, *119*, 110–123. [CrossRef]
83. Yu, W.; Li, J.; Liu, Q.; Zeng, Y.; Yin, G.; Zhao, J.; Xu, B. Extraction and Analysis of Land Cover Heterogeneity over China. *Adv. Earth Sci.* **2016**, *31*, 1067–1077.
84. Lin, S.; Li, J.; Liu, Q. Overview on estimation accuracy of gross primary productivity with remote sensing methods. *J. Remote. Sens.* **2018**, *22*, 234–254.
85. Turner, D.P.; Ritts, W.D.; Cohen, W.B.; Gower, S.T.; Running, S.W.; Zhao, M.; Costa, M.H.; Kirschbaum, A.A.; Ham, J.M.; Saleska, S.R.; et al. Evaluation of MODIS NPP and GPP products across multiple biomes. *Remote Sens. Environ.* **2006**, *102*, 282–292. [CrossRef]

forests

Article

Semi-Supervised Tree Species Classification for Multi-Source Remote Sensing Images Based on a Graph Convolutional Neural Network

Xueliang Wang [1,*], Jian Wang [2], Zuozheng Lian [3] and Nan Yang [4]

1 Network Information Center, Qiqihar University, Qiqihar 161006, China
2 National Asset Management Office, Qiqihar University, Qiqihar 161006, China
3 School of Computer and Control Engineering, Qiqihar University, Qiqihar 161006, China
4 School of Architecture and Civil Engineering, Qiqihar University, Qiqihar 161006, China
* Correspondence: wangxueliang@qqhru.edu.cn

Abstract: As a current research hotspot, graph convolution networks (GCNs) have provided new opportunities for tree species classification in multi-source remote sensing images. To solve the challenge of limited label information, a new tree species classification model was proposed by using the semi-supervised graph convolution fusion method for hyperspectral images (HSIs) and multispectral images (MSIs). In the model, the graph-based attribute features and pixel-based features are fused to deepen the correlation of multi-source images to improve accuracy. Firstly, the model employs the canonical correlation analysis (CCA) method to maximize the correlation of multi-source images, which explores the relationship between information from various sources further and offers more valuable insights. Secondly, convolution calculations were made to extract features and then map graph node fusion, which not only reduces redundancy features but also enhances compelling features. Finally, the relationship between representative descriptors is captured through the use of hyperedge convolution in the training process, and the dominant features on the graph are fully mined. The tree species are classified through two fusion feature operations, leading to improved classification performance compared to state-of-the-art methods. The fusion strategy can produce a complete classification map of the study areas.

Keywords: hypergraph convolution; data fusion; classification of tree species

Citation: Wang, X.; Wang, J.; Lian, Z.; Yang, N. Semi-Supervised Tree Species Classification for Multi-Source Remote Sensing Images Based on a Graph Convolutional Neural Network. *Forests* **2023**, *14*, 1211. https://doi.org/10.3390/f14061211

Academic Editor: Feng Zhao

Received: 19 April 2023
Revised: 25 May 2023
Accepted: 9 June 2023
Published: 11 June 2023

1. Introduction

Recently, achieving accurate and reliable tree species classification from a large number of trees has gained more attention. Multi-source products typically provide more trustworthy information than a single product of ground surface covering [1,2]. Hyperspectral images (HSIs) are an essential part of multi-source data learning and can reflect the spectral characteristics of forest mapping, which is crucial for understanding forest cover [3]. Multispectral data images (MSIs) contain high-resolution spatial information, which is also helpful when analyzing forest tree species. By integrating multi-source data, data fusion can overcome the limitations of a single data source [4]. Current fusion methods for HSIs and MSIs rely on feature extraction and feature fusion, respectively, to leverage the correlation between the two data sources [5]. To leverage diverse information from multiple sources of data, it is necessary to implement strategies that enable the effective extraction, integration, and analysis of data [6–8].

Deep learning has been applied to feature fusion to improve its performance in terms of feature fusion and has achieved satisfactory results [9–11]. Li et al. proposed an effective CNN (PPF-CNN) based on pixel features [12] in combination with a small number of existing samples, which enabled data enhancement to optimize the classification results. A multi-region CNN (MRCNN) [13] algorithm is proposed to mine spectral–spatial information, which improves the mining performance. However, with the further development of the network, limited labeled samples may lead to overfitting or performance degradation. Fortunately, a semi-supervised approach can be used to mitigate this shortcoming. The insufficiency of labels in remote sensing image classification can now be addressed with the recent emergence of graph convolution networks. The recent advancement of GCNs has provided a promising solution to the insufficient label problem in hyperspectral or multispectral image (HSI/MSI) classification. Unlike traditional methods, GCNs operate on a graph and require only a small amount of labeled data to establish the relationships between multi-source nodes. By effectively aggregating and transforming features from a node's neighborhood, GCNs provide an efficient pathway for multi-source image classification [14]. GCNs are particularly suitable for handling non-Euclidean data, which refers to datasets that do not adhere to the principles and assumptions of Euclidean geometry, and by learning node features through hidden layers, they better capture the local features, resolving the issue of missing class boundary information. The Chebyshev polynomial [15] parameterized differentiable graph convolution algorithm is used by GCNs to transport the node information after using the feature construction of all samples to create the topological structure (G-Conv). The whole learning process of this method does not need manual intervention. By excavating the structural information of many unlabeled samples in the feature space, the deviation of learning trained with labeled samples is corrected. The potential value of unlabeled samples is fully utilized, and the 'small sample' problem in terms of classification is effectively solved. Not only is this approach applicable to non-Euclidean statistics, but it also has broad applicability to standard domains [16]. In reality, there is much research on the use of GCNs in relation to remote sensing images. For example, Qin developed a spectral–spatial GCN (S2GCN) by employing current pixel spatial information [17], which has made significant improvement to the original GCN. However, at the end of the above network, the SoftMax function is usually used to analyze the extracted features, which generates a probability vector that reflects the category of the pixel. This method lacks intraclass compactness, which reduces classification performance [18]. Spectral and spatial information were extracted to construct adjacent matrices, and an innovative prototype layer was designed. This prototype layer contains distance-based cross-entropy loss function and novel temporal entropy-based regularization, which can not only generate more low-level features, such as separable between species and compact within species, but also represent the prototypes belonging to each species [19].

Most methods extract features and then combine them using various techniques. Additionally, low-rank model methods are used to convert multiple sources of features into a common space through low-rank sparse representation. Feature fusion strategies are used to convert multi-source features into a unified fused feature, but the process of feature extraction and fusion are separate, which may result in changes to the original information contained in the features. In graph representation learning, taking into account both the global and local structure of the data can make the graph representation model more robust against the effects of noisy and sparse data. However, there have been only a limited number of GCN models that have prioritized preserving both the local and global structures of the data concurrently.

Existing graph/hypergraph-based neural networks suffer from a significant limitation in that they only make use of the initial graph/hypergraph structures and do not account for dynamic modifications that may occur in the feature embedding process. This limitation hinders the network's ability to adapt to changing input data.

To address this issue, it is crucial to develop approaches that can account for the modifications of graph/hypergraph structures and ensure that the original information in the features remains intact throughout the fusion process. A semi-supervised graph model is proposed based on an extraction fusion network for HSIs and MSIs, to fully use the correlation of multi-source data. The feature extraction method is directed by the model via feature fusion. The model directly outputs unified fusion features from multi-source data as input. For feature fusion, a multimodal graph is built, and feature extraction is constrained using the graph-based loss function. The innovation points of the paper are as follows:

(1) To extract the discriminative features, a common subspace is explored and found by CCA operations on HSI and MSI, and the correlation is maximized between HSI and MSI inputs.
(2) For the information fusion between HSI and MSI, both the node features and hypergraph features are integrated to improve the ability of global information extraction, and the ability to express the relationship between all vertices becomes more robust. During the initialization of hypergraph convolution, feature fusion is performed on the nodes, and the hyperedge features are fused in the process of hypergraph convolution learning.
(3) Compared with other state-of-the-art converged networks, it is more efficient and achieves better classification results.

2. Materials and Methods

2.1. Study Area

The areas were studied in the Tahe Forestry Bureau (Figure 1), which is located in the Daxing'an Mountains, northwest of Heilongjiang Province, China (123° to 125° E and 52° to 53° N). The studied areas have a borderline of 173 km and a total area of 14,420 km^2. The climate of the studied areas is a cold–temperate continental climate and experiences severe climatic changes, with short hot, humid summers and long, dry, cold winters. The annual average temperature of the area is −2.4 °C, and the average yearly precipitation is 463.2 mm, occurring mainly in July and August. The forest, with a storage capacity of 53.4 million m^3, covers 81% of the total area. Dominant tree species include Birch, Larch, Spruce, Mongolica Pine, Willow, and Poplar [20].

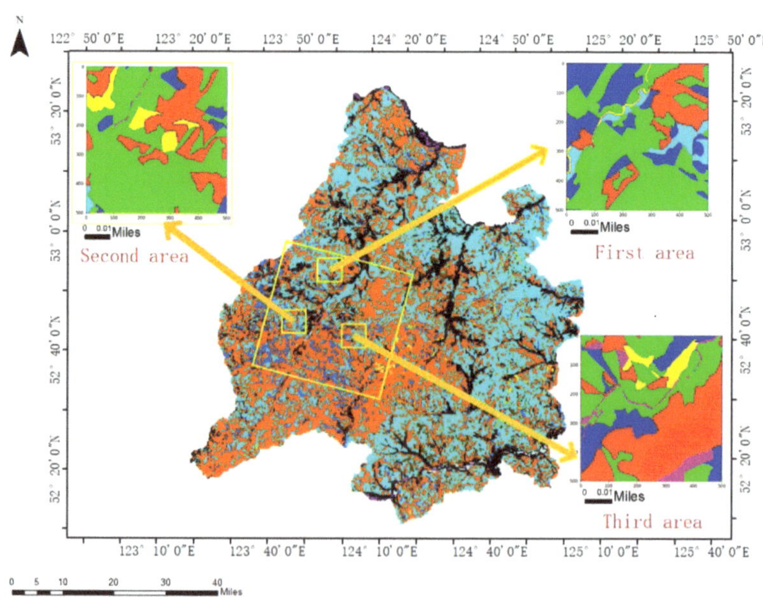

Figure 1. Map of the study area.

2.2. Data

To classify the tree species, we used data taken from HJ-1A and Sentinel-2. Figure 1 displays the HSI data for HJ-1A and the MSI data for Sentinel-2A, collected from the China Center for Resources Satellite Data and Application and the USGS, respectively. The

HJ-1A satellite has a high-speed imaging system with 115 bands and a spatial resolution of 100 m [13], while Sentinel-2A offers 13 spectral bands with a spatial resolution of 10 m, providing rich data for coastal and land remote sensing [20]. We used ENVI 5.1 software to enhance the resolution of the HJ-1A/HSI images (collected on 20 August 2016) to match the MSI spatial resolution and fill the gap concerning the relatively low HSI resolution. The interpolation method was used to resample the experimental HSI data. The Tahe Forestry Bureau conducted a survey in 2018 and used the results to classify major forest species in the research region. The study areas were 500 × 500 × 115 pixels and 500 × 500 × 13 pixels for HSI and MSI data, respectively. We selected the area with the most species as the research object, which included Birch, Larch, Spruce, Mongolia, Willow, and Poplar. Table 1 lists the three study areas used in this work, where the training samples comprise approximately one-third of the total samples.

Table 1. List of 6 tree species samples of the three study areas.

	Birch	Larch	Mongolia	Poplar	Spruce	Willow
First area	130,124	39,216	57,620	3019	15,330	3492
Second area	150,771	58,829	11,412	2175	17,048	1067
Third area	99,082	82,746	38,114	1013	13,460	1,515,486

2.3. Classification Method

2.3.1. Hypergraph

The Hypergraph Neural Network, which is commonly referred to as HGNN [21], has been visually depicted in Figure 2. Each dataset in the multimodal dataset contains numerous nodes with features. Then, using the complex correlation of multimodal data sets, several parts of hyperedge features are constructed. The hypergraph adjacency matrix and node features are input into HGNN to output the pixel features classification map [22]. Hyperedge convolution is computed as follows:

$$X^{k+1} = \sigma\left(D_v^{-\frac{1}{2}} HWD_e^{-1} H^T D_v^{-\frac{1}{2}} X^k \Theta^k\right) \quad (1)$$

where $X^k \in \mathbb{R}^{N\times C}$ is the feature of the l th layer. $X^0 = X$, and σ is the nonlinear activation function. The initial node features X^k are learned through filtering matrix Θ^k to extract the dimensional feature C_2. Then, according to the node features of the hyperedge $\mathbb{R}^{E\times C_2}$, the hyperedge feature is realized via $H^T \in \mathbb{R}^{E\times N}$. The output node feature is then produced by multiplying the hyperedge features that are associated with it, and the hyperedge feature is produced from the matrix H. Dv and De in the Hyperedge convolution play the role of normalization [21]. Therefore, through hyperedge convolution, the HGNN layer can successfully extract the high-order correlation of the hypergraph.

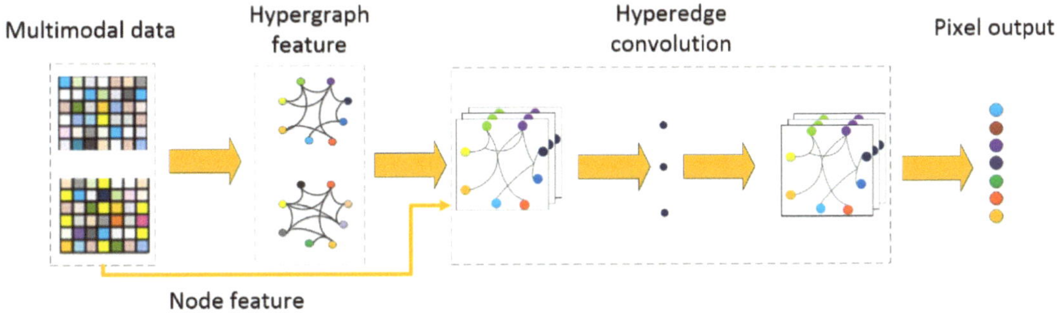

Figure 2. Hypergraph neural network (HGNN).

2.3.2. Overall Architecture

The graph neural network employs an undirected graph to model the data and utilizes graph convolution for feature extraction by calculating various data relationships. Building on this method, we propose a tree species classification model that leverages the distinct framework of hyperspectral and multispectral data modules for feature fusion. The model takes in a multi-source remote sensing image as input and produces unified fusion features as output, as illustrated in Figure 3. The framework encompasses association feature extraction, hypergraph convolution learning, and classifier classification.

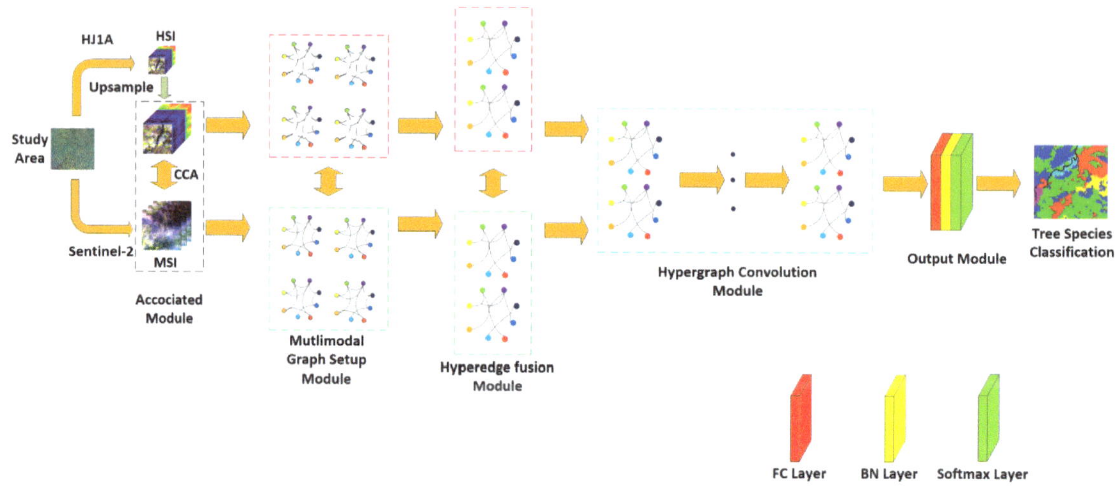

Figure 3. Flowchart of multi-source fusion hypergraph convolution network.

The fusion module is designed to extract and merge features from both HSI and MSI data. The weight matrices of HSI and MSI are merged to generate the incidence matrix of the multimodal graph, which accounts for complementary information and correlations between the two data sources. The feature extraction and fusion network is trained using a loss function that incorporates graph embedding, enabling the network to effectively capture the features of interest. Finally, the SoftMax classifier is used to categorize the tree species map at the pixel level.

To lower the dimensionality of the HSI data from 115 to 12, we first employ the KPCA approach. This generates a vector that represents each pixel in the data collection. The complete image's vector is then fed as input to the network. In this setup, X^H and X^L correspond to the HSI and MSI data, respectively,

$$X^H = \left\{X_1^H, X_2^H, \ldots, X_n^H\right\},\ X_i^H \in \mathbb{R}^h \tag{2}$$

$$X^L = \left\{X_1^L, X_2^L, \ldots, X_n^L\right\},\ X_i^L \in \mathbb{R}^m \tag{3}$$

where h and m are the numbers of spectral channels for the HSI and MSI, respectively, and X is the vector representing the i-th pixel. Therefore, the input of the network is as follows:

$$X = \{X_1, X_2, \ldots, X_n\},\ X_i \in \mathbb{R}^{h+m} \tag{4}$$

where $X_i = \mathrm{CAT}\left(X_i^H,\ X_i^L\right)$, and CAT() represents concatenate operation. Next, we feed X into the network for feature extraction and fusion. Although the network structure is not the primary focus of our research, the multimodal graph and graph loss are crucial for feature extraction and fusion. We employ the Smish method [23] as the activation function in this study. In practice, the foundation of the feature extraction network can be substituted with other networks, such as a convolution layer, because the network's input consists of multimodal images. The network outputs unified fused features using the loss function based on the multimodal graph, which are then provided to the classifier for pixel recognition. We adopt the FC layer and Softmax layer as the output layer of the proposed network to demonstrate the potential of multimodal and loss-based graphs function.

2.3.3. Associated Feature Module

The primary objective of multi-source learning is to establish the connection between various data, which is crucial for comprehending the relationship depicted in multi-source remote sensing images. By exploiting the relationship between different viewpoints, we can improve the final interpretation performance [24,25]. This research area has received increasing attention in the field of data mining over the past decade [26,27]. In this part, we focus on multi-perspective learning from the perspective of feature fusion and classification methods. We use the common subspace approach, which maximizes the correlation between two inputs, as explored via the CCA method. This standard two-view subspace learning approach is employed to achieve our research objectives.

For a multi-source learning problem, hyperspectral and multispectral images are represented as $\alpha \in R^{L \times W \times H}$, $\beta \in R^{L \times W \times M}$, respectively, where L represents the length, W represents the width, and H and M represent the number of bands in the two data sources, respectively. Then, α and β are transformed into $V^{v \times H}$ and $V^{v \times M}$, respectively, $v = L \times W$. We assume that the linear representation of α and β are represented as follows,

$$U_H = r_1(\alpha) \tag{5}$$

$$U_M = r_2(\beta) \tag{6}$$

r_1, r_2 represent the projection directions of HSI and MSI, respectively. CCA is obtained by maximizing the correlation between α and β. The first projection direction can be obtained by optimizing the following equation, and r_1, r_2 represent the HSI and MSI projection axes, respectively. By maximizing the correlation between α and β with the vector generated by CCA, the following equation can be optimized to yield the initial projection direction,

$$\begin{aligned} &\max\rho(r_1, r_2) = r_1 s_{HM} r_2 \\ &s.t. r_1 s_{HH} r_1 = 1, r_2 s_{MM} r_2 = 1 \end{aligned} \tag{7}$$

S_{HM} is the covariance matrix of the HSI and MSI among them. The Lagrangian multiplier operator can be used to maximize the objective function and find the optimal solution sum of r_1^* and r_2^* for the problem.

$$U_H^* = r_1^*(\alpha) \tag{8}$$

$$U_M^* = r_2^*(\beta) \tag{9}$$

Multi-source image categorization involves assigning the same space to data from various sources, as based on Equations (8) and (9). The use of a sum representation enhances the relevance of the data and features, which is highly beneficial for the multi-source classification of tree species. This approach not only processes the initial input but also reduces its redundancy and complexity. However, the rate of convergence for deep learning is slow [28]. By providing HSI- and MSI-related features, this approach enables the development of the depth model, which can lead to further improvements in classification performance.

2.3.4. Multi-Source Hypergraph Fusion

To efficiently integrate information across multimodal images, input pixels are represented using a graph structure. Compared to CNNs, the graph structure offers a higher capacity to capture the relationship between all of the vertices, as the size of the convolution kernel in a CNN limits the extraction of global information.

Both the association features of HSIs and MSIs processed by the CCA algorithm are $Q_h^{L \times W \times H} = U_H^*$, $Q_m^{L \times W \times M} = U_M^*$, respectively. Each pixel was represented as a vertex of the hypergraph, and their dimensions are transformed into $X^{n\times H}$ and $Y^{n\times M}$, where $n = L \times W = |V|$ is the number of hypergraph vertices, and H and M represent the spectral dimensions of the HSIs and MSIs, respectively. Their features are extracted as $X_i^{n\times H_j}$ and $Y_i^{n\times H_j}$. For each vertex $v \in V$ and the hyperedge $e \in E$, the incidence matrix generated from the selected k nearest neighbors is $H^{|V|\times|E|}$, where, $|V| = |E| = n$.

$$h(i,j) = \begin{cases} e^{-\frac{n\sigma||x_i - x_j||^2}{\sum_{j=1}^{n} d(x_i, x_j)}}, x_i \in N_k(x_j) \\ 0 \end{cases} \tag{10}$$

where σ is an adjustable hyper-parameter, $d(X_i, X_j)$ is the Euclidean distance between the two vertices X_i and X_j. The mean value is used to regulate the multimodal distance and simplify the process of adjusting the hyperparameters.

It was assumed that $[f_1, f_2, \ldots, f_n]$ is a multimodal feature vector. According to Equation (10), the incidence matrix $[H_1^h, H_2^h, \ldots, H_n^h]$ and $[H_1^m, H_2^m, \ldots, H_n^m]$ of HSIs and MSIs are calculated, respectively. Then, the fused features are obtained as $H_f^h = CAT\left(H_1^h, H_2^h, \ldots, H_n^h\right)$, $H_f^m = CAT(H_1^m, H_2^m, \ldots, H_n^m)$, where $CAT()$ represents the multi vector connection operation. Then, the obtained hyperedge features are further studied.

2.3.5. Hyperedge Learning

To obtain fused hyperedges from multimodal features, we connect their incidence matrices. This process enables the hypergraph convolution in Equation (1) to be applied, which becomes

$$X^{l+1} = \sigma\left(D_v^{-\frac{1}{2}} H_f W_f D_e^{-1} H_f^T D_v^{-\frac{1}{2}} X^l \Theta^l\right) \tag{11}$$

In the case of without considering regularization [29], the equation is simplified as follows:

$$X^{l+1} = \sigma\left(H_f W_f H_f^T X^l \Theta^l\right) \tag{12}$$

Since H and W are diagonal matrices, the equation becomes

$$X^{l+1} = \sigma\left(\left(H_1 W_1 H_1^T + \ldots H_n W_n H_n^T\right) X^l \Theta^l\right) \tag{13}$$

For multi-source remote sensing images, each node has many characteristics [30]. The hyperedges of its hypergraph are first learned, respectively, then they are integrated. The objective function in backpropagation is calculated via cross-entropy loss function and the final feature map outputs with pixel-level SoftMax function. The Algorithm 1 is presented as follows:

Algorithm 1 Pseudo code of hypergraph feature fusion for HSI and MSI

Input: HSI associated feature X_H, MSI associated feature X_M, neighbor node number k, iteration number of layer n, number of graph convolution layer g.
1: Generate X'_H and X'_M by flatting X_H and X_M, respectively
2: Generate X by connecting X'_H and X'_M horizontally
3: Generate the fusion incidence matrix of HSI and MSI as H, according to Equations (8) and (9)
4: Calculate the degree diagonal matrix De of the hyperedge and the degree diagonal matrix Dv of the vertex
5: Initialization parameters W and Θ
6: for i = 1 to n
7:   for j = 1 to g
8:     Calculate characteristic X according to Equation (10)
9:     X_{pre} = SoftMax(BN(FC(Hconv(X))))
10:   Calculate losses L, update W and Θ
11:   Gradient back propagation
12: end for
13: end for
14: Output tree species classification map based pixel node

2.3.6. Evaluation Indicators

To test the tree species classification accuracy of the proposed method, the OA, average accuracy (AA), and Kappa coefficient (kappa), were determined using Equations (14)–(16), respectively.

$$\mathrm{OA} = \frac{\sum_{i=1}^{k} \mathrm{C(i,i)}}{\mathrm{M}}, \tag{14}$$

$$\mathrm{AA} = \frac{\sum_{i=1}^{k} \mathrm{OA}}{\mathrm{K}}, \tag{15}$$

$$kappa = \frac{\mathrm{M}\sum_{i=1}^{k} \mathrm{C(i,i)} - \sum_{i=1}^{k}(\mathrm{C(i,+)C(+,i)})}{\mathrm{M}^2 - \sum_{i=1}^{k}(\mathrm{C(i,+)C(+,i)})}, \tag{16}$$

where i and k represent i-th tree species and the size, respectively. OA represents the proportion of correctly classified samples in the whole test sample, AA denotes the average accuracy of every tree species, and kappa is a statistical measure that reflects the consistency between the ground truth and classified ground maps.

3. Results

3.1. Experimental Setup

The experiment uses HJ-1A and Sentinel2A images as datasets, which were introduced in Section 2.2. Several compared models are as follows:

SpectralNET [31]: A deep learning method for spectral clustering by embedding input data points into the eigenspace of their associated graph Laplacian matrix and subsequently clusters them.

FuNet [32]: A new minibatch GCN was proposed by training large-scale GCNs in a mini-batch mode. The method has the ability to predict data that is not part of the training set without the need to retrain the networks.

MFDF [33]: A classification model based on decision fusion between multiple features and super-pixel segmentation, which integrated 2D and 3D Gabor features of multi-source datasets.

DMULN [6]: end-to-end pattern model which integrates the multi-view features, and the view union pool was proposed by associating with the feature extractor, and the fused features are input into the classifier.

The proposed model and other compared methods were evaluated using 10%, 20%, and 30% of the samples as randomly chosen training sets. For the other samples, we allocated 30% of the samples as validation sets and the rest as test sets randomly. Parameter settings have a great impact on performance. Although the resolution of the datasets is different, the resolution of the geomap is fixed. The experiment is implemented in Python 3. The parameters of graph convolution are set in Table 2, where 'Hconv' refers to the hypergraph convolution layer.

Table 2. Detailed layers and shape in multi-source fusion hypergraph convolution model.

Layer	Shape	Layer	Shape
Input	(500 × 500 × 115)	Input	(500 × 500 × 12)
CCA	(500 × 500 × 17)	(500 × 500 × 12)	
Calculate W_h		Calculate W_m	
Normalization		Normalization	
Hconv	128	Hconv	128
Smish		Smish	
Fusion hypergraph Hconv FC Layer BN Layer Softmax			

After hypergraph fusion, the proposed model consisted of two FC-BN layers and two active layers. The patch size was set to 7, and we set both the learning rate and weight decay to 0.005. We used the KNN method (k = 10) to construct the initial graph for the datasets, with k values set to [5,10,15,20] and the number of convolution layers set to 15. We initialized the weights of all methods using the Glorot method. Adam was utilized as an optimizer, with a maximum of 1000 epochs. To ensure the optimal performance of other comparative models, we consulted the relevant literature. The method was repeated 100 times, with the average outcome for 10 iterations and the corresponding standard deviation used as the result. The training procedure was terminated if the loss did not decrease for 100 consecutive epochs.

3.2. Classification Performance Comparison

The average accuracy (AA) of tree species classification in terms of the three multi-source datasets is shown in Figure 4. The proposed method achieves the highest performance, followed by MFDF, FuNet, DMULN, and SpectralNet, as shown. The proposed model outperforms MFDF, FuNet, DMULN, and SpectralNet by 0.67, 0.46, 0.3, and 0.7, respectively, in terms of OA, as shown in Figure 5. Figure 6 shows the KAPPA values of the proposed model are 0.38, 0.23, 0.16, and 0.69 higher than those of MFDF, FuNet, DMULN, and SpectralNET, respectively. These results indicate that the proposed method is superior to the other methods. The performance of the proposed model is further demonstrated in the three figures.

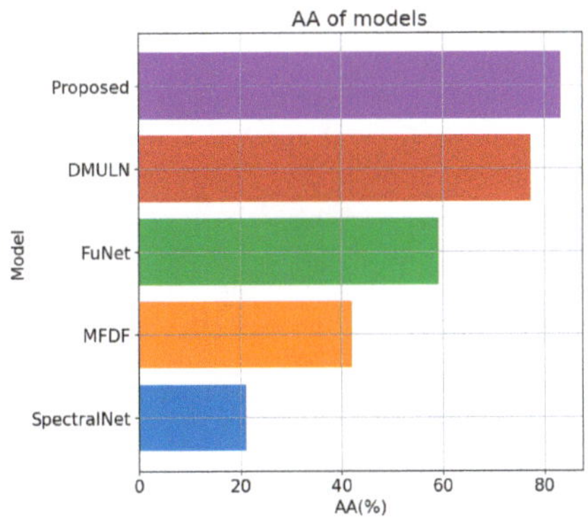

Figure 4. AA of all methods across the tree species dataset.

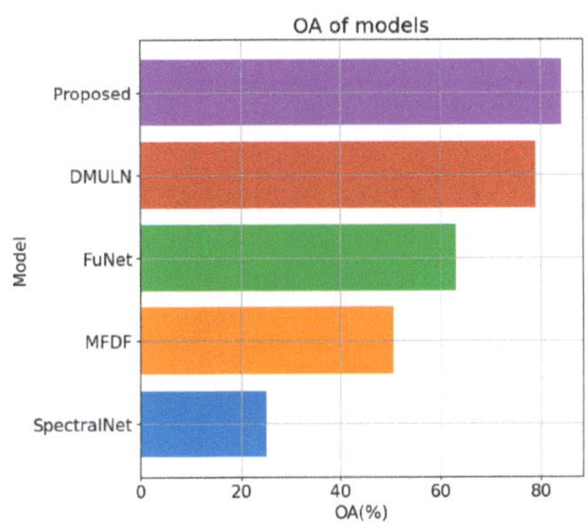

Figure 5. OA of all methods across three datasets.

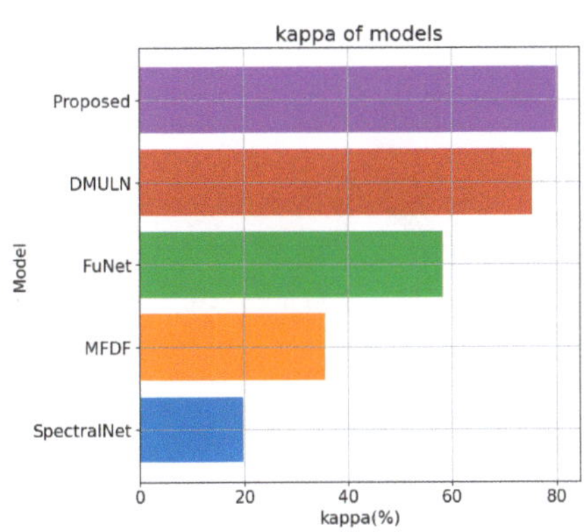

Figure 6. Kappa of all methods across three datasets.

Tables 3–6 illustrate the confusion matrix of five models used for tree species classification, and MFDF, FuNet, DMULN, and SpectralNET are unsatisfactory compared to the proposed model. Spruce is particularly challenging to classify, but the proposed model has a higher recognition rate for Spruce than that in the other models, leading to an overall increase in OA. The classification effect of the SpectralNET model is inadequate, as it identifies almost no other tree species except for Larch and Birch. Other models also lack the advanced ability to identify multiple tree species. The DMULN model mistakenly classified almost 30% of Spruce trees as Larch, while the recognition rate of Poplar was 0. The OA of MFDF for Poplar is 0.61, which is better than DMULN. The classification performance of FuNet for other tree species is slightly better than DMULN, except for Spruce. As shown in Figure 5 and Table 7, the recognition rate of the proposed model for Spruce, Mongolian, and Willow is better than that of MFDF, but the recognition rate for Spruce is not significantly improved.

Table 3. Confusion matrix of tree species classification using the SpectralNet method. Tree species code, column (ground truth code), and row (prediction code).

Tree Species	Code	0	1	2	3	4	5
Birch	0	462	513	101	99	2	0
Larch	1	428	2535	701	311	0	0
Spruce	2	433	1411	525	103	0	0
Mongolica	3	81	153	52	49	0	0
Willow	4	4	100	0	0	0	0
Poplar	5	2	88	0	0	0	0
	Precision	32.76	52.81	38.07	8.71	0	0

Table 4. Confusion matrix of tree species classification when using the DMULN method.

Tree Species	Tree Species Code	0	1	2	3	4	5
Birch	0	805	342	4	25	0	0
Larch	1	297	3600	16	51	1	0
Spruce	2	186	11.21	507	44	0	0
Mongolica	3	70	103	14	146	0	0
Willow	4	1	68	0	1	32	0
Poplar	5	15	67	1	5	0	0
	Precision	68.30	90.01	27.29	43.30	31.69	0

Table 5. Confusion matrix of tree species classification when using the FuNet method.

Tree Species	Tree Species Code	0	1	2	3	4	5
Birch	0	859	240	42	34	0	1
Larch	1	314	35.66	31	47	0	3
Spruce	2	63	850	930	14	0	1
Mongolica	3	56	126	0	151	0	0
Willow	4	0	70	1	2	30	0
Poplar	5	1	33	0	0	0	55
	Precision	73.03	89.98	50.0	45.19	29.29	61.14

Table 6. Confusion matrix of tree species classification when using the MFDF method.

Tree Species	Tree Species Code	0	1	2	3	4	5
Birch	0	10.81	71	1	21	0	1
Larch	1	390	35.25	7	35	2	3
Spruce	2	39	496	13.00	24	0	0
Mongolica	3	39	32	1	261	0	0
Willow	4	0	13	1	7	82	0
Poplar	5	3	5	0	0	0	81
	Precision	91.90	88.95	69.85	78.13	78.72	90.28

Table 7. Confusion matrix of tree species classification when using the proposed method.

Tree Species	Tree Species Code	0	1	2	3	4	5
Birch	0	10.82	71	1	21	0	1
Larch	1	286	3601	74	2	1	0
Spruce	2	48	79	1714	6	5	5
Mongolica	3	12	13	14	290	2	2
Willow	4	7	2	5	0	90	0
Poplar	5	1	3	4	1	0	80
	Precision	92.07	90.82	92.14	86.78	85.87	91.05

The results of the various methods used to generate tree species classification maps in three regions are presented in Figure 7. The proposed method employs the fusion map convolution method using HJ-1A and Sentinel-2 data, achieving an OA of 0.88, an AA of 0.85, and a Kappa of 0.82 in the consistent areas. The proposed model outperforms other methods in identifying Spruce and Larch, which have commercial value due to their rarity. The SpectralNet method performs poorly, followed by DMULN, FuNet, and MFDF. Other methods have blurred edges, low recognition rates, and a high rate of misclassification and fragmentation. Overall, the proposed method accurately identifies all of the tree species and yields favorable results, surpassing the compared methods. The superior performance is attributed to the proposed strategy based on depth hypergraph convolution fusion and hyperedge convolution fusion.

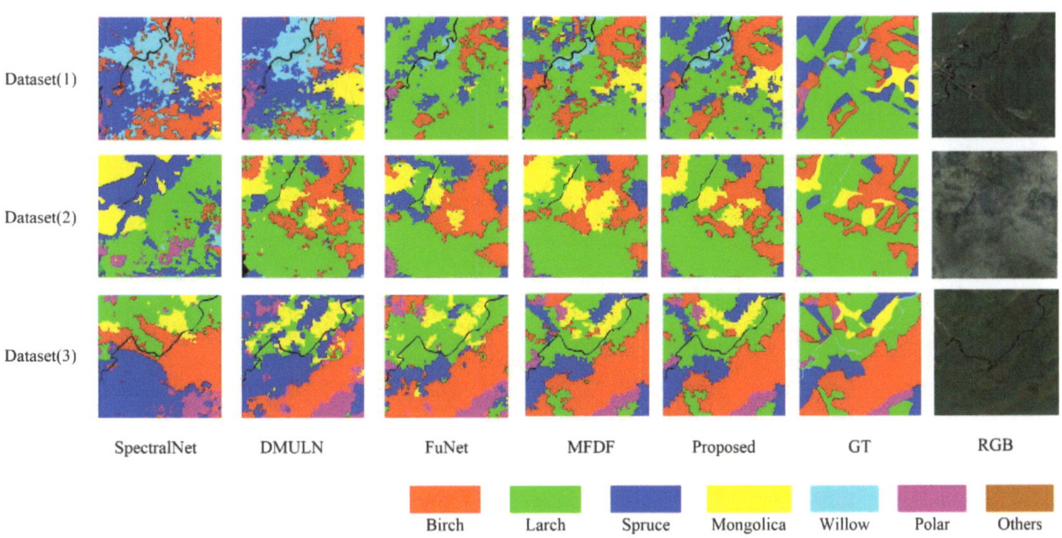

Figure 7. Tree species classification map of multi-source datasets in three datasets. (**1**) SpectralNet, (**2**) DMULN, (**3**) FuNet.

3.3. Parameter Analysis

In this section, we utilize three tree species datasets from Section 2.2 to analyze the key parameters that affect classification performance. These parameters include the labeling ratio (partial labeling of the total datasets), K value, and depth. We conduct tests and analyses to examine the impact of these parameters on classification performance. Figure 8 displays the tree species classification accuracy of the five models with varying label rates. The classification accuracy of all five models increases as the label rates increase. However, the proposed model achieves desirable accuracy and outperforms the other methods significantly.

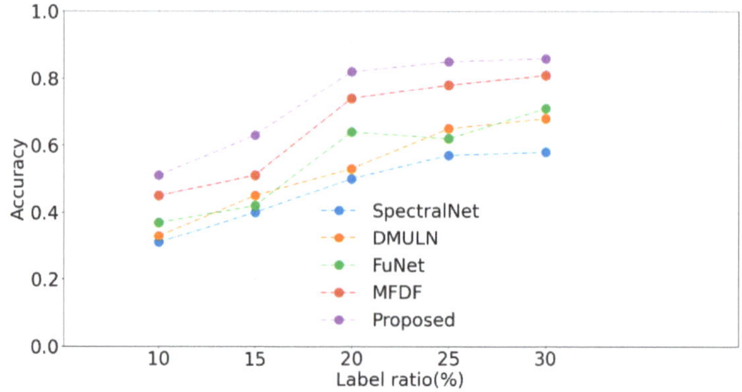

Figure 8. The accuracy of tree species classification with different label rates.

To verify the robustness of the method, it is advisable to strive for consistency in the selection of k across different modal features while minimizing any potential impact on accuracy. This approach allows for a thorough assessment of the method's resilience, particularly in terms of its ability to handle variations across modalities. By maintaining a consistent value of K, the performance of the method can be effectively evaluated, and the robustness can be determined in terms of achieving accurate results while considering the unique characteristics of each modality. Figure 9 illustrates the classification accuracy of three models for different K values ($K \in \{5, 10, 15, 20, 25, 30\}$). As only three of the compared models have a K value parameter, the results show that the accuracy of the three methods varies with K. The accuracy tends to increase as the K value is set between 5 and 15. However, when K is set between 15 and 30, the accuracy starts to decrease.

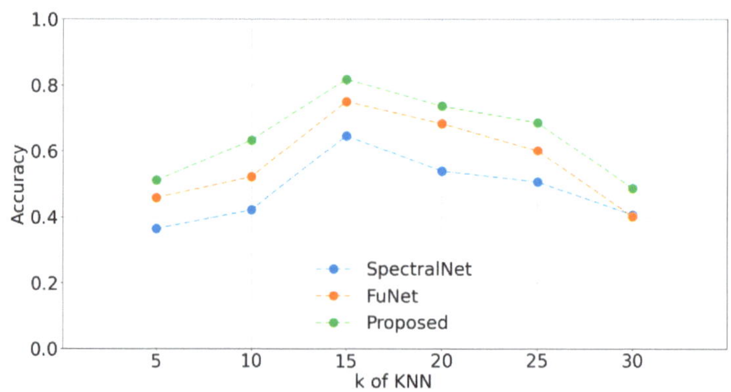

Figure 9. The accuracy of tree species classification with k Values (k value in KNN).

The experimental results demonstrate that the proposed method achieves the best performance when K is set to 15. These findings lead to two main conclusions: (1) A small K value may fail to capture the neighborhood of the data, while an increasing K value could result in incorrect neighborhood samples that render the relationship between samples less discriminative. (2) The proposed fusion learning method is sensitive to the choice of the K value.

To investigate the influence of the depth of the proposed model, we set the range of the DHCN layers to {5, 10, 15, 20, 25}. Figure 10 demonstrates that the accuracy of the classification result with DHCN is highest when the K value is set to 15, and the method is not extremely sensitive to the number of layers. However, as the number of layers increases beyond 15, performance slightly degrades with excessive smooth curve.

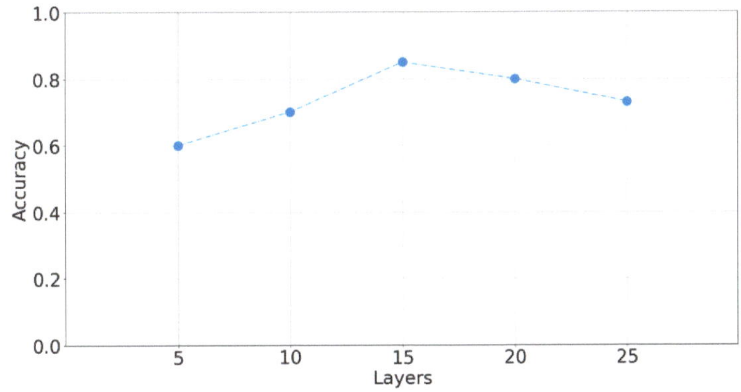

Figure 10. The accuracy of tree species classification with different depths of the proposed model.

As shown in Figures 9 and 10, the main factors that affect the computation time are the complexity of the datasets, the number of categories, the number of spectral channels, and the image size. The computation time is influenced by various factors, including image size, data complexity, and the model parameters. Figures 11 and 12 illustrate the RAM usage and running time of different models in the classification experiments. Larger images and more complex datasets necessitate increased memory and computation time. Comparatively, the GCN-based methods require more memory and time compared to the CNN-based method, primarily due to the time-consuming computation of the adjacency matrix. However, the proposed model, with its fusion graph structure, outperforms other GCN methods in terms of speed. This is achieved by eliminating the utilization of ineffective features, resulting in improved overall system efficiency. By removing irrelevant or redundant features from the data, the model can concentrate on the most informative aspects of the input, leading to enhanced performance and faster computation times.

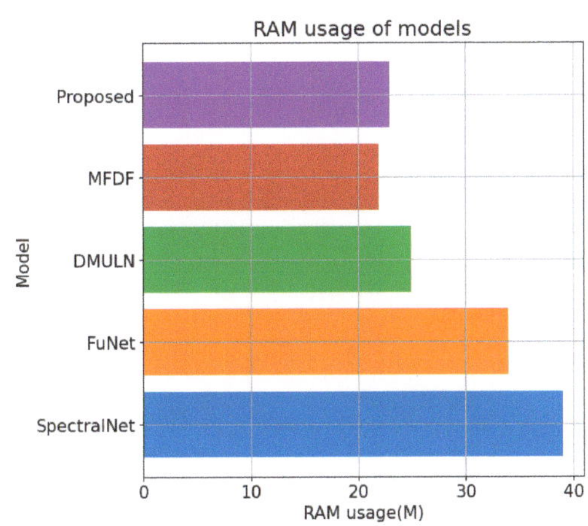

Figure 11. RAM usage of different models.

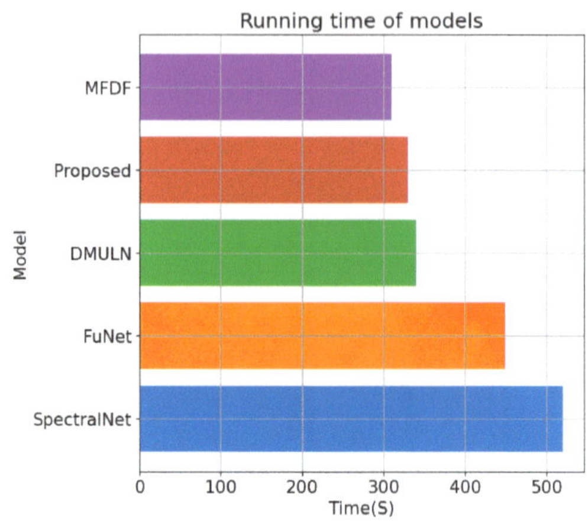

Figure 12. Running time of different models.

4. Discussion

The proposed model first performs typical association analysis on the two data sources used as the input, maximizes the correlation of multi-source data, performs convolution calculation on the generated vector to extract features, and then fuses the nodes of its graph structure. This process not only reduces redundant information but also strengthens the effective features. Finally, hyperedge convolution is introduced into the graph convolution training process to adaptively mine the relationship between the representative descriptors and fully integrate the node and attribute features.

The SpectralNet model exhibits significant statistical advantages when using the spectral clustering method as it overcomes the scalability and generalization of the spectral embedding. However, in our experiment, the SpectralNet method displayed severe shortcomings in terms of coniferous forest species classification, with almost no Willow identified and other tree species misrecognized as Larch. The DMULN method [32] utilizes an encoder–decoder network to input the features related to the two data sources separately. Its recognition ability in terms of Mongolian, Poplar, and Willow is better than the other three methods, owing to the benefits of deep multi-view learning and view pooling. The DMULN method proved to be superior to SpectralNet in terms of tree species classification performance as it can learn both spectral and spatial modal features simultaneously during the experiment. However, it is inferior to FuNet and MFDF. FuNet utilizes mini batches of non-European features in graph convolution processing as well as European features CNN processing, which are then fused together. This approach has demonstrated impressive performance in a single hyperspectral data source. However, in the present experiment, FuNet did not perform as well when using multi-source tree species datasets. MFDF, which is based on Gabor wavelet feature representation, utilizes a two-dimensional Gabor filter [31], making it more suitable for feature extraction when using multi-source datasets. As we use Sentinel-2 data as the multispectral data, which is more effective when extracting spatial features, the Gabor extraction of spatial features is slightly worse, resulting in a lower OA in terms of tree species classification compared to the proposed method [34]. However, its recognition rate for Spruce is significantly lower than the proposed method, which uses a graph structure to represent the higher-order features. Hyperedge learning also integrates the features of the graph structure from the two data sources, thus improving the recognition of Spruce and the overall recognition rate more accurately [33]. The proposed hypergraph fusion structure can transfer the complex high-order correlation between HSIs and MSIs, and better represent the underlying data interrelation between them than the basic graph structure. Additionally, the proposed method has the advantage of fusing multimodal information into the same data structure with flexible hyperedges, owing to the existence of multimodal features. Through hypergraph fusion and hyperedge convolution fusion, the multi-source graph convolution model proposed significantly reduces computation time while improving learning efficiency.

Our model outperforms the compared models in the classification of six tree species, yielding a higher AA. The proposed model has several advantages:

(1) The model utilizes multiple graph learning and multi-source fusion, where each graph provides complementary information that is unique from the other graphs. By removing the noise hyperedges present in tiny graphs, the model improves tree species classification performance.
(2) Multi-graph learning is proven to be feasible in tree species classification, and our model considers both the global and local features of multi-source data simultaneously with regularization.
(3) Compared to other models, our proposed method is more effective in classifying tree species by using the fusion of multi-source data. The utilization of multimodal graph learning enhances the effectiveness of the classification process.

5. Conclusions

In this paper, we proposed a novel model for tree species classification by designing a multi-source fusion graph neural network. The proposed model first calculates the pixel-based correlation between HSIs and MSIs, generating two types of hypergraph structures. Both the HSI graph structure and MSI graph structure are saved in each initial graph and fused with each other in the hyperedge learning process. The proposed model fuses the two data sources twice, capturing the global graph from the low-dimensional space of the original high-dimensional data. We propose a new fusion method that combines complementary and common information to correctly capture the graph structure inherent in the data. We evaluated our method using a tree species dataset and compared it with state-of-the-art approaches. The experimental results show that the proposed method is effective in improving the accuracy of tree species classification.

In the future, our research aims to investigate multi-source feature fusion algorithms based on self-supervised learning methods. Additionally, we intend to explore tree species classification in higher-resolution remote sensing images. These endeavors will further enhance the accuracy and capabilities of our classification models, enabling us to tackle more complex and detailed datasets in the field of tree species classification.

Author Contributions: X.W.: conceptualization, methodology, software, data curation, funding acquisition, writing—original draft—writing—review and editing; J.W.: writing—review and editing—supervision; Z.L.: supervision; N.Y.: writing—review and editing—supervision. All authors have read and agreed to the published version of the manuscript.

Funding: This research was supported by the Basic Scientific Research Project of Heilongjiang Provincial Universities (Grant number 145109219).

Institutional Review Board Statement: Not applicable.

Informed Consent Statement: Not applicable.

Data Availability Statement: Not applicable.

Conflicts of Interest: The authors declare no conflict of interest.

References

1. Hao, S.; Zhou, Y.; Guo, Y. A brief survey on semantic segmentation with deep learning. *Neurocomputing* **2020**, *406*, 302–321. [CrossRef]
2. Ran, Y.H.; Li, X.; Lu, L.; Li, Z.Y. Large-scale land cover mapping with the integration of multi-source information based on the Dempster–Shafer theory. *Int. J. Geogr. Inf. Sci.* **2012**, *26*, 169–191. [CrossRef]
3. Hua, T.; Zhao, W.; Liu, Y.; Wang, S.; Yang, S. Spatial consistency assessments for global land-cover datasets: A comparison among GLC2000, CCI LC, MCD12, GLOBCOVER and GLCNMO. *Remote Sens.* **2018**, *10*, 1846. [CrossRef]

4. Hou, W.; Hou, X. Data fusion and accuracy analysis of multi-source land use/land cover datasets along coastal areas of the Maritime Silk Road. *ISPRS Int. J. Geo-Inf.* **2019**, *8*, 557. [CrossRef]
5. Liu, K.; Xu, E. Fusion and Correction of Multi-Source Land Cover Products Based on Spatial Detection and Uncertainty Reasoning Methods in Central Asia. *Remote Sens.* **2021**, *13*, 244. [CrossRef]
6. Liu, X.; Jiao, L.; Li, L.; Cheng, L.; Liu, F.; Yang, S.; Hou, B. Deep multiview union learning network for multisource image classification. *IEEE Trans. Cybern.* **2020**, *52*, 4534–4546. [CrossRef]
7. Ma, Z.; Jiang, Z.; Zhang, H. Hyperspectral Image Classification Using Feature Fusion Hypergraph Convolution Neural Network. *IEEE Trans. Geosci. Remote Sens.* **2021**, *60*, 5517314. [CrossRef]
8. Mu, C.; Dong, Z.; Liu, Y. A Two-Branch Convolutional Neural Network Based on Multi-Spectral Entropy Rate Superpixel Segmentation for Hyperspectral Image Classification. *Remote Sens.* **2022**, *14*, 1569. [CrossRef]
9. See, L.M.; Fritz, S. A method to compare and improve land cover datasets: Application to the GLC-2000 and MODIS land cover products. *IEEE Trans. Geosci. Remote Sens.* **2006**, *44*, 1740–1746. [CrossRef]
10. Yu, Y.; Liang, S.; Samali, B.; Nguyen, T.N.; Zhai, C.; Li, J.; Xie, X. Torsional capacity evaluation of RC beams using an improved bird swarm algorithm optimised 2D convolutional neural network. *Eng. Struct.* **2022**, *273*, 115066. [CrossRef]
11. Yu, Y.; Li, J.; Li, J.; Xia, Y.; Ding, Z.; Samali, B. Automated damage diagnosis of concrete jack arch beam using optimized deep stacked autoencoders and multi-sensor fusion. *Dev. Built Environ.* **2023**, *14*, 100128. [CrossRef]
12. Chao, G.; Sun, S. Semi-supervised multi-view maximum entropy discrimination with expectation laplacian regularization. *Inf. Fusion* **2019**, *45*, 296–306. [CrossRef]
13. Zhu, P.; Hu, Q.; Hu, Q.; Zhang, C.; Feng, Z. Multi-view label embedding. *Pattern Recognit.* **2018**, *84*, 126–135. [CrossRef]
14. Qin, A.; Shang, Z.; Tian, J.; Wang, Y.; Zhang, T.; Tang, Y.Y. Spectral–spatial graph convolutional networks for semisupervised hyperspectral image classification. *IEEE Geosci. Remote Sens. Lett.* **2018**, *16*, 241–245. [CrossRef]
15. Defferrard, M.; Bresson, X.; Vandergheynst, P. Convolutional Neural Networks on Graphs with Gast Localized Spectral Filtering. In Proceedings of the Conference on Neural Information Processing Systems, Barcelona, Spain, 5–10 December 2016; pp. 3844–3852.
16. Tang, C.; Chen, J.; Liu, X.; Li, M.; Wang, P.; Wang, M.; Lu, P. Consensus learning guided multi-view unsupervised feature selection. *Knowl.-Based Syst.* **2018**, *160*, 49–60. [CrossRef]
17. Ding, S.; Cong, L.; Hu, Q.; Jia, H.; Shi, Z. A multiway p-spectral clustering algorithm. *Knowl.-Based Syst.* **2019**, *164*, 371–377. [CrossRef]
18. Li, S.; Liu, H.; Tao, Z.; Fu, Y. Multi-view graph learning with adaptive label propagation, in: Big Data (Big Data). In Proceedings of the 2017 IEEE International Conference on Big Data (Big Data), Boston, MA, USA, 11–14 December 2017; pp. 110–115.
19. Xi, B.; Li, J.; Li, Y.; Song, R.; Xiao, Y.; Du, Q.; Chanussot, J. Semisupervised cross-scale graph prototypical network for hyperspectral image classification. *IEEE Trans. Neural Netw. Learn. Syst.* **2022**. Available online: https://ieeexplore.ieee.org/abstract/document/9740412 (accessed on 17 April 2023). [CrossRef] [PubMed]
20. Wang, L.; Fan, W.Y. Identification of forest dominant tree species group based on hyperspectral remote sensing data. *J. Northeast. For. Univ.* **2015**, *43*, 134–137.
21. Feng, Y.; You, H.; Zhang, Z.; Ji, R.; Gao, Y. Hypergraph neural networks. In Proceedings of the AAAI conference on artificial intelligence, Hilton, HI, USA, 27 January–1 February 2019; Volume 33, pp. 3558–3565.
22. Drusch, M.; Del Bello, U.; Carlier, S.; Colin, O.; Fernandez, V.; Gascon, F.; Hoersch, B.; Isola, C.; Laberinti, P.; Martimort, P.; et al. Sentinel-2: ESA's optical high-resolution mission for GMES operational services. *Remote Sens. Environ.* **2012**, *120*, 25–36. [CrossRef]
23. Wang, X.; Ren, H.; Wang, A. Smish: A novel activation function for deep learning methods. *Electronics* **2022**, *11*, 540. [CrossRef]
24. Zhou, D.; Huang, J.; Schölkopf, B. Learning with Hypergraphs: Clustering, Classification, and Embedding. In Proceedings of the Conference on Neural Information Processing Systems, Vancouver, BC, Canada, 3–6 December 2007; pp. 1601–1608.
25. Wang, X.; Ren, H. DBMF: A Novel Method for Tree Species Fusion Classification Based on Multi-Source Images. *Forests* **2021**, *13*, 33. [CrossRef]
26. Xue, X.; Nie, F.; Li, Z.; Wang, S.; Li, X.; Yao, M. A multiview learning framework with a linear computational cost. *IEEE Trans. Cybern.* **2017**, *48*, 2416–2425.
27. Xie, G.S.; Zhang, X.Y.; Yan, S.; Liu, C.L. Hybrid CNN and dictionary-based models for scene recognition and domain adaptation. *IEEE Trans. Circuits Syst. Video Technol.* **2015**, *27*, 1263–1274. [CrossRef]
28. Xie, G.S.; Zhang, X.Y.; Shu, X.; Yan, S.; Liu, C.L. Task-driven feature pooling for image classification. In Proceedings of the IEEE International Conference on Computer Vision, Santiago, Chile, 7–13 December 2015; pp. 1179–1187.
29. Pei, J.; Huang, Y.; Sun, Z.; Zhang, Y.; Yang, J.; Yeo, T.S. Multiview synthetic aperture radar automatic target recognition optimization: Modeling and implementation. *IEEE Trans. Geosci. Remote Sens.* **2018**, *56*, 6425–6439. [CrossRef]
30. Hardoon, D.R.; Szedmak, S.; Shawe-Taylor, J. Canonical correlation analysis: An overview with application to learning methods. *Neural Comput.* **2004**, *16*, 2639–2664. [CrossRef] [PubMed]
31. Shaham, U.; Stanton, K.; Li, H.; Nadler, B.; Basri, R.; Kluger, Y. Spectralnet: Spectral clustering using deep neural networks. *arXiv* **2018**, arXiv:1801.01587.
32. Hong, D.; Gao, L.; Yao, J.; Zhang, B.; Plaza, A.; Chanussot, J. Graph convolutional networks for hyperspectral image classification. *IEEE Trans. Geosci. Remote Sens.* **2020**, *59*, 5966–5978. [CrossRef]

33. Jia, S.; Zhan, Z.; Zhang, M.; Xu, M.; Huang, Q.; Zhou, J.; Jia, X. Multiple feature-based superpixel-level decision fusion for hyperspectral and LiDAR data classification. *IEEE Trans. Geosci. Remote Sens.* **2020**, *59*, 1437–1452. [CrossRef]
34. Wang, X.; Yang, N.; Liu, E.; Gu, W.; Zhang, J.; Zhao, S.; Sun, G.; Wang, J. Tree Species Classification Based on Self-Supervised Learning with Multisource Remote Sensing Images. *Appl. Sci.* **2023**, *13*, 1928. [CrossRef]

forests MDPI

Article

Improvement and Assessment of Convolutional Neural Network for Tree Species Identification Based on Bark Characteristics

Zhelin Cui, Xinran Li, Tao Li and Mingyang Li *

Co-Innovation Center for Sustainable Forestry in Southern China, Nanjing Forestry University, Nanjing 210037, China; 3210100096@njfu.edu.cn (Z.C.); xinran0212@njfu.edu.cn (X.L.); litao3014@njfu.edu.cn (T.L.)
* Correspondence: lmy196727@njfu.edu.cn; Tel.: +86-025-8542-7327

Abstract: Efficient tree species identification is of great importance in forest inventory and management. As the textural properties of tree barks vary less notably as a result of seasonal change than other tree organs, they are more suitable for the identification of tree species using deep learning models. In this study, we adopted the ConvNeXt convolutional neural network to identify 33 tree species using the BarkNetV2 dataset, compared the classification accuracy values of different tree species, and performed visual analysis of the network's visual features. The results show the following trends: (1) the pre-trained network weights exhibit up to 97.61% classification accuracy for the test set, indicating that the network has high accuracy; (2) the classification accuracy values of more than half of the tree species can reach 98%, while the confidence level of correct identification (probability ratio of true labels) of tree species images is relatively high; and (3) there is a strong correlation between the network's visual attractiveness and the tree bark's biological characteristics, which share similarities with humans' organization of tree species. The method suggested in this study has the potential to increase the efficiency of tree species identification in forest resources surveys and is of considerable value in forest management.

Keywords: tree species identification; convolutional neural network; bark image; visual attractiveness

Citation: Cui, Z.; Li, X.; Li, T.; Li, M. Improvement and Assessment of Convolutional Neural Network for Tree Species Identification Based on Bark Characteristics. *Forests* **2023**, *14*, 1292. https://doi.org/10.3390/f14071292

Academic Editor: Peter F. Newton

Received: 23 May 2023
Revised: 16 June 2023
Accepted: 21 June 2023
Published: 23 June 2023

1. Introduction

Forest resource inventory is an essential component of forestry, reflecting the quantity, quality, and dynamic changes of forest resources. Identifying tree species in the survey area is a primary goal of forest resource inventory. To identify tree species in survey areas, surveyors typically rely on visual methods based on external characteristics, such as roots, stems, leaves, flowers, and fruits. For trees that cannot be visually identified, surveyors usually need to collect specimens and consult reference materials. This manual identification process requires solid expertise in dendrology and is often time consuming, costly, and inefficient. Machine learning can be applied to identify tree species and improve the efficiency of forest resource inventory. In machine learning and cognitive science, neural networks are models that simulate the structures and functions of biological neural networks [1,2]. After training, deep neural networks can automatically learn to extract features from large-scale, diverse, high-dimensional complex data and perform efficient classification, prediction, and pattern recognition [3].

Early studies of tree species identification usually used artificial neural networks or support vector machines to analyze the hyper-spectral features of trees, and there were also multiple methods that used texture feature extraction and descriptors to assist in tree classification [4–13]. Although these methods reduced some manual costs of classification, traditional machine learning methods usually required human design or selection of texture features and descriptors, which were subjective and often could not fully express the complexity and diversity of textures, and lost some information during post-processing operations, such as dimensionality reduction [14–16]. Therefore, the efficiency of traditional machine learning methods needs to be improved.

Deep learning is a machine learning method that uses deep neural networks to extract features and automatically learn data representations. Compared to traditional machine learning methods, deep learning can better discover and extract the feature information in datasets, thus having a higher generalization ability. The development of deep learning algorithms led to the increase in network layers, which makes convolutional neural networks or multi-feature recognition networks ever-more applicable in species classification and identification based on forest images [17–20]. For tree species identification, scholars who use deep learning models usually focus on leaves, flowers, fruits, or tree shapes. The classifier of the network usually recognizes the entire image as a whole [21], and the mixed background value can easily confuse the original representation of the network. Tree organs, such as leaves and flowers, are usually difficult to distinguish from the background (noise) during image acquisition, making it challenging to obtain precise image information for leaves and flowers found on tall trees. In addition, flowers and fruits are only present at certain times of the year, while leaf information is unavailable during defoliation. Consequently, there are many challenges involved in species identification using images of tree organs in forest resource inventory. The morphological characteristics of the bark, which is the outermost layer of the stems and roots of woody plants, are important features involved in distinguishing tree species. Employing tree bark as an identifier has several advantages compared to leaves. Most tree bark shapes are stable, unless subjected to irreversible disasters (e.g., forest fires), and tree bark textures do not change or change little with seasonal change [11].

There are few existing studies of tree species identification based on bark texture features that used deep learning algorithms. Most of these studies used ResNet [22] network as the basis for analysis and adopted deeper convolutional layers to achieve higher accuracy when performing bark texture recognition. Carpentier, M. et al. publicized a bark dataset called BarkNet 1.0, which is also the largest publicly available dataset, and a high tree species classification accuracy of 93.88% using ResNet18 and ResNet34 [23] was achieved. Misra et al. implemented an alternative classification method using patch-based convolutional neural networks that fine-tuned the network's patch predictions and determined the image category via majority voting with an ensemble-based classifier [24]. Robert et al. developed DeepBark, which is a model capable of detecting bark surfaces under high background brightness [25]. Faizal achieved promising results on BarkVN-50 using a deeper network called ResNet101 [26]. Kim et al. trained VGG-16 and EfficientNet, obtained an identification accuracy value above 90%, and applied class activation mapping (CAM) aggregation to identify the critical classification features for each tree species [27]. Therefore, the image recognition technology of the convolutional neural network has great practical value for bark identification, which can quickly and accurately identify tree species, making forest resource surveys more intelligent and efficient. Using this technique, forestry workers can collect tree bark images on-site and upload them to the server for identification. By optimizing data collection and image processing, this identification technique can meet the needs of forest survey fieldwork teams in terms of efficiency and accuracy.

Most existing studies on tree species identification use pre-trained weights of networks that were trained on ImageNet, rather than using bark images as the pre-training data [22–27]. This method could lead to some misclassification and performance degradation. Many researchers use relatively backward networks to work on image classification of tree species [22–26] (like ResNet). Compared to some algorithms developed in recent years, the performance of traditional convolutional neural networks is inferior. Furthermore, due to the difficulty in demonstrating the identification processes of deep convolutional neural networks on images, only a few studies combine network vision with the biological features of tree species, thus failing to reveal the effectiveness of network classification results.

In this paper, we pre-train three ConvNeXt networks using different depths on the bark dataset. The three research objectives are as follows: (1) to compare the performances of these ConvNeXt networks; (2) to analyze the biological features that led to the discrepancies in classification accuracy values between different tree species; and (3) to explore the relationships between the visual attractiveness and biological features of bark images of different tree species.

2. Materials and Methods

2.1. Datasets

Most existing bark datasets for tree species identification based on deep learning experience categorization problems, such as insufficient data, blurred images, or limited species diversity. Collecting and producing high-quality bark datasets for deep learning is time consuming and labor intensive, which impedes academic research in this field. Therefore, we use a dataset combining previous bark image datasets with new bark images that we collected in Nanjing, China (Table 1).

Table 1. Partially available public bark image datasets.

Collaborators	Dataset Name	Species	Number of Images	Dataset Size	Creation Year
Truong Hoang	BarkVN 50	50	5578	185 MB	2020
Rémi Ratajczak	Bark 101	101	2592	317 MB	2019
Matic Švab	TRUNK 12	12	360	1.1 GB	2014
Tae Kyung	BARK-KR	54	6918	9.77 GB	2021
Carpentier	BarkNet 1.0	23	23616	30.1 GB	2017
Cui	BarkNJ	10	7671	21.4 GB	2023

We selected the BarkNet 1.0 dataset (hereafter referred to as BarkNet) as our research data for the following reasons: (1) BarkNet contains the largest data in terms of the number of tree species and images, with a total of 23 tree species and 23,616 images. Other datasets, such as BarkVN50, although covering a relatively larger number of tree species, had a relatively smaller number of images per tree species. (2) Regarding image quality, most of the images in the BarkNet dataset were manually cropped to remove noise and irrelevant background values and only contained clear bark images. Most image backgrounds in other datasets were not cropped, and some bark images had noisy data, such as shadows, lighting changes, and halos. Thus, the BarkNet dataset was more suitable for pre-training the network.

To further expand the dataset used to pre-train the network, we collected some common tree species in Nanjing using mobile devices, such as mobile phones. The image data we collected contained 7671 images of 10 tree species found at Nanjing Forestry University, and we named it BarkNJ. We used a pre-processing method similar to BarkNet to optimize the images in the dataset, retaining only the clear bark images after noise removal, and named the combined dataset BarkNetV2. Table 2 shows some details of the dataset used in this paper. We sorted the Latin names of each tree species and then selected a clear bark image of each tree species as an example image of that species, as shown in Figure 1.

Table 2. Basic information of BarkNetV2 used in study. For convenience of training, we named each tree species in form of a species index, which is index list of tree species.

ID	Species	Common Name	Species Index	Number of Trees	Number of Images
1	*Abies balsamea*	Balsam fir	SAB	41	922
2	*Acer platanoides*	Norway maple	ERB	1	70
3	*Acer rubrum*	Red maple	ERR	64	1676
4	*Acer saccharum*	Sugar maple	ERS	81	1911
5	*Betula alleghaniensis*	Yellow birch	BOJ	43	1255
6	*Betula papyrifera*	White birch	BOP	32	1285
7	*Camptotheca acuminata*	Campo tree	CAAA	28	620
8	*Cedrus deodara*	Deodar cedar	CSDA	43	874
9	*Cinnamomum camphora*	Camphor wood	CMCA	49	947
10	*Cupressus funebris*	Cypress wood	CSFS	38	710
11	*Fagus grandifolia*	American beech	HEG	41	840
12	*Fraxinus americana*	White ash	FRA	61	1472
13	*Juniperus chinensis*	Round cypress	JSCS	52	927
14	*Koelreuteria paniculata*	Golden rain tree	KAPA	35	627
15	*Larix laricina*	Tamarack	MEL	77	1874
16	*Liriodendron chinense*	Liriodendron	LNCE	33	562
17	*Metasequoia glyptostroboides*	Redwood	MAGS	50	743
18	*Ostrya virginiana*	American hophornbeam	OSV	29	612
19	*Picea abies*	Norway spruce	EPO	72	1324
20	*Picea glauca*	White spruce	PIR	44	596
21	*Picea mariana*	Black spruce	EPN	44	885
22	*Picea rubens*	Red spruce	EPR	27	740
23	*Pinus rigida*	Pitch pine	PID	4	123
24	*Pinus resinosa*	Red pine	EPB	29	596
25	*Pinus strobus*	Eastern white pine	PIB	39	1023
26	*Platanus acerifolia*	Plane tree	PSAA	47	705
27	*Populus canadensis*	Canadian poplar	PSCS	69	1044
28	*Populus grandidentata*	Big-tooth aspen	PEG	3	64
29	*Populus tremuloides*	Quaking aspen	PET	58	1037
30	*Quercus rubra*	Northern red oak	CHR	109	2724
31	*Thuja occidentalis*	Northern white cedar	THO	38	746
32	*Tsuga canadensis*	Eastern hemlock	PRU	45	986
33	*Ulmus americana*	American elm	ORA	24	767
Total	NA	NA	NA	1398	31,287

Note: For convenience of distinction, images named with three capital letters are images located in BarkNet dataset. Images named with four capital letters are located in BarkNJ dataset that we collected. Tree species mentioned in BarkNet are typical of eastern coastal forests of Canada (sub-boreal coniferous forest climate and humid continental climate), whereas species mentioned in BarkNJ are common in eastern China (subtropical monsoon climate).

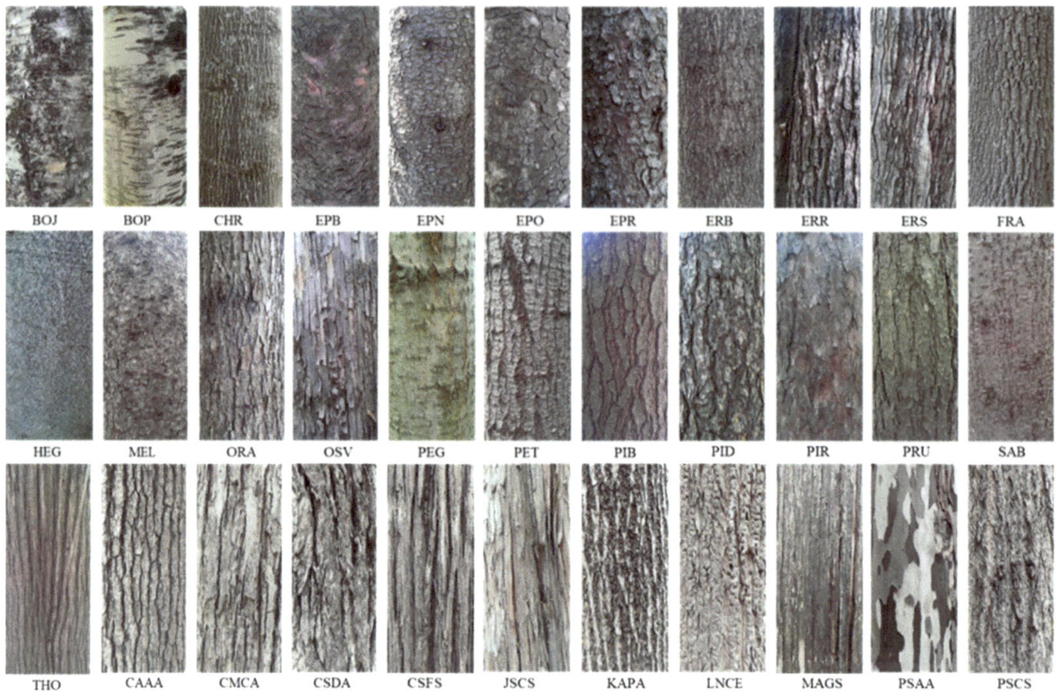

Figure 1. Bark sample images and index of different tree species in BarkNet dataset.

2.2. Methodology

2.2.1. Selection of the Networks

Facebook AI Research and UC Berkeley jointly proposed a pure convolutional neural network called ConvNeXt [28], which achieved the highest classification accuracy on ImageNet with convolutional structure alone, surpassing the accuracy of the Swin Transformer tool proposed during the same period [29]. To optimize the network performance, the architecture of ConvNeXt uses a series of Vit strategies, which is similar to the Swin Transformer architecture. The ConvNeXt architecture is shown in Figure 2.

The modifications in the ConvNeXt architecture can be specified as follows: (1) the ratio of stacked block layers is adjusted to 1:1:9:1, with block 3 containing the highest percentage of stacked blocks. This modification is made to balance the computational requirements with performance. (2) The ResNet stem is modified to adopt the Patchify backbone used in Transformer to improve the network's performance. (3) Grouped convolution in ResNet is replaced with depth-wise convolution, with the number of groups being the same as the number of channels. (4) The Bottleneck module of ResNet is substituted for a modified version of the MobileNetV2 Inverted Bottleneck module to optimize network performance. (5) The depth-wise convolutional layers are moved up to match the placement of the MHA (Multi-Headed Attention) module before the MLP (Multiple Layer Perceptron) layers in the Swin Transformer architecture. Additionally, the size of the convolutional kernel is changed from 3×3 to 7×7 to align with the Swin Transformer. (6) The activation function of ConvNeXt is replaced with Gelu, which is the same as the Swin Transformer. Additionally, the number of normalization layers and activation functions in a block is reduced, with only normalization layers retained after a depth-wise convolution operation. (7) The down-sampling layer is created by combining a layer normalization and a convolutional layer with a kernel size of 2 and a stride size of 2.

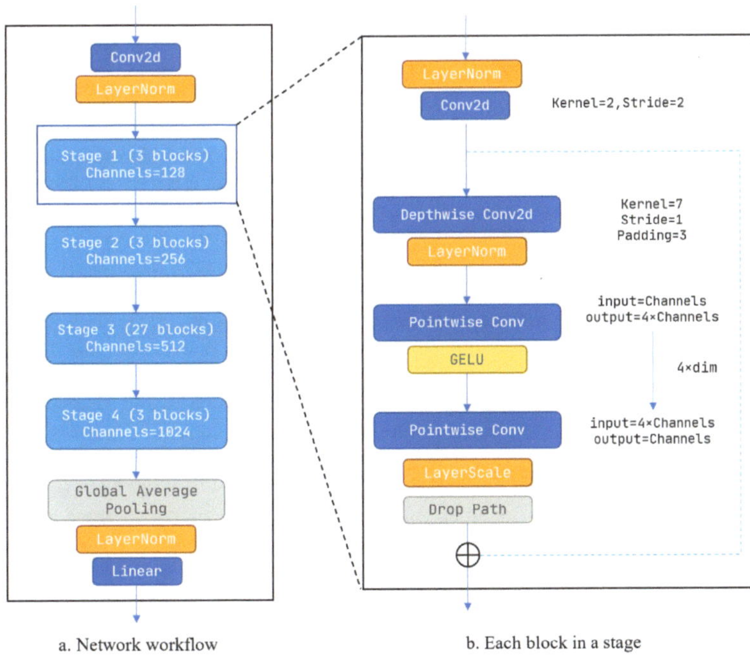

Figure 2. ConvNeXt network architecture (ConvNeXt-B as an example). (**a**) shows network workflow, while (**b**) depicts architecture of each block in a stage of network.

By adjusting the feature dimensions and the number of individual blocks, ConvNeXt can be divided into four networks of different depths, which correspond to the computational complexity of different Swin Transformer versions. The network parameters for each type are given in Table 3. It should be noted that the smaller networks of ConvNeXt-T and ConvNeXt-S have the same feature dimensions, though the ratio for the number of each block in the stage is different. Liu et al. believe that retaining the same number of stacked layers as in the Swin Transformer can achieve a better effect; thus, the results of ConvNeXt-T, in our experiments, are only used for reference [28,29]. Using a network with high computational complexity on a small dataset may lead to the network becoming overfitted. The number of images used in our dataset is considerably smaller than that of ImageNet, and the computational complexity of ConvNeXt-L is excessive for this amount of data. Furthermore, as Table 3 shows, despite having more than double the parameters, ConvNeXt-L provides only a 0.4% accuracy gain over ConvNeXt-B. Therefore, we omitted ConvNeXt-L from our experiments.

Table 3. Main parameters of four ConvNeXt networks.

Network	Parameters (M)	Channels	Stage	Flops (G)	Accuracy (ImageNet-1k)
ConvNeXt-T(tiny)	28.59	(96, 192, 384, 768)	(3, 3, 9, 3)	4.46	82.1%
ConvNeXt-S(small)	50.22	(96, 192, 384, 768)	(3, 3, 27, 3)	8.69	83.1%
ConvNeXt-B(base)	88.59	(128, 256, 512, 1024)	(3, 3, 27, 3)	15.36	85.1%
ConvNeXt-L(large)	197.77	(192, 384, 768, 1536)	(3, 3, 27, 3)	34.37	85.5%

Note: Parameters denotes number of network parameters, Channels denote number of channel dimensions, Stage denotes proportion of blocks per stage in network structure, and Flops denotes floating-point operations per second.

2.2.2. Setting of the Network Parameters

All training of ConvNeXt was performed via a PyTorch 1.11.0+cu113 with Python version 3.9.9. In contrast to Carpentier et al.'s study, we did not download pre-trained weights from ImageNet for transfer learning [23]. Instead, we selected the ConvNeXt network, which was a superior-performing convolutional neural network used for pre-training of the bark dataset.

Due to the significant differences between the ImageNet and BarkNetV2 networks, we recalculated the channel means and variances of all images and randomly split the training, validation, and test sets by 70%, 20%, and 10%, respectively, using the random seed setting. As the resolution of bark images in our datasets is high, training directly with bark images may degrade the network's performance. Thus, the images of the training and validation sets were randomly cropped and centered, respectively, and the processed images were horizontally randomly rotated. The size of the images sent to the network for training was uniformly set to 224 × 224 pixels to maximize the capability of the graphics card.

Paszke and Sebastian used theoretical deduction and experimental verification to prove that choosing a multiple of two or eight as a batch size made no significant difference in practice [30,31]. Therefore, to maximize the efficiency of the graphics card while keeping the variables consistent, we set the batch size to 32, and the total number of training cycles was set to 50 epochs. Weight decay is a form of regularization that can effectively reduce the overfitting of deep networks; thus, we chose AdamW to be the optimizer and set the weight decay rate to 0.05. As the network needs to use the bark dataset for pre-training, we chose an appropriate learning rate of 0.001, which meant that the weights changed less with each iteration. The network took more time to reach its optimal value while finding the optimal value of the loss function. We set the first 10 epochs of the total training period as warm-up learning, during which stage the learning rate gradually increased with the pre-set value to stabilize the network. After the network was relatively stable, training was conducted according to the pre-set learning rate, which could make the network more quickly converge and have a better training effect. The main parameters for pre-training three ConvNeXt networks with different depths are listed in Table 4.

Table 4. Pre-training parameters for three ConvNeXt networks.

Network	Batch Size	Image Size	Learning Rate	Learning Rate Schedule	Training Epochs	Warm-Up Epochs
ConvNeXt-T	32	224 × 224	1×10^{-3}	Cosine decay	50	10
ConvNeXt-S	32	224 × 224	1×10^{-3}	Cosine decay	50	10
ConvNeXt-B	32	224 × 224	1×10^{-3}	Cosine decay	50	10

2.2.3. Visualization of the Network Workflow

Despite the excellent performance of convolutional neural networks in various types of image identification tasks, the process of network recognition is often considered a black box, preventing the observation of the internal working mechanism. Therefore, we tried to use some visualization methods to reveal the correlation between the biological characteristics of different tree species and the visual mechanisms of deep convolutional neural networks. The following three methods were used in this study to visualize the network workflow: (1) Integrated Gradients, which attributed the projection of a deep network to its input, rather than a particular layer, by invoking the gradient operator [32]. There are several methods for visualizing the internal components of deep neural networks. Integrated Gradients is a widely applicable approach with a robust theoretical foundation. (2) Smooth Grad CAM++ combined with Grad-CAM and Smooth Grad, which is a technique that can model and visualize a subset of feature maps or neurons at each neural network level [33–35], was also used. The output visualization presented a hierarchical feature that effectively incorporated the elements of visual appeal, localization, and object-like capture. (3) Deep Feature Decomposition, which provides insight into clustering patterns in feature space and presents results as heat maps, was also used. The maps used different colors to differentiate between concepts and adjusted the intensity to highlight semantically similar image regions [36].

3. Results and Analysis

3.1. Comparison of Identification Results

Figure 3 demonstrates the identification accuracy and loss value of three ConvNeXt networks with varying depths during training. Figure 3a shows that the identification accuracy of ConvNeXt-S on the 22nd epoch validation set exceeded 90%, indicating that the network achieved a high accuracy rate. In contrast to the rapid improvement in identification accuracy during initial training, the identification accuracy of the network on the validation set increased from 90 to 97% between the 22nd and 40th epochs. After about 20 epochs of training, the accuracy only increased by about 7%. The network gradually stabilized after the 40th epoch, with minimal change in accuracy. At the 50th epoch, the network's Top-1 accuracy on the validation set reached 97.71%. It is worth noting that the lightweight network of ConvNeXt-S increased the Top-1 accuracy on the test set to 97.61%, surpassing that of ConvNeXt-B (97.58%). In Figure 3b, the training process of ConvNeXt-B and ConvNeXt-S was similar. However, due to the more significant number of parameters, the training curve of ConvNeXt-B was relatively more stable than ConvNeXt-S. The identification accuracy of ConvNeXt-B first exceeded 90% at the 24th epoch. The network gradually stabilized after training reached 40 epochs, reaching the maximum Top-1 accuracy on the validation and test sets at the 50th epoch, with 97.79% and 97.58%, respectively. In Figure 3c, the shallower depth of ConvNeXt-T resulted in a more extended period in which the identification accuracy of the validation set first reached 90% (26th epoch) and more fluctuations occurred compared to other networks. After 50 epochs of training, the Top-1 accuracy of ConvNeXt-T on the validation and test sets reached 97.49% and 97.29%, respectively.

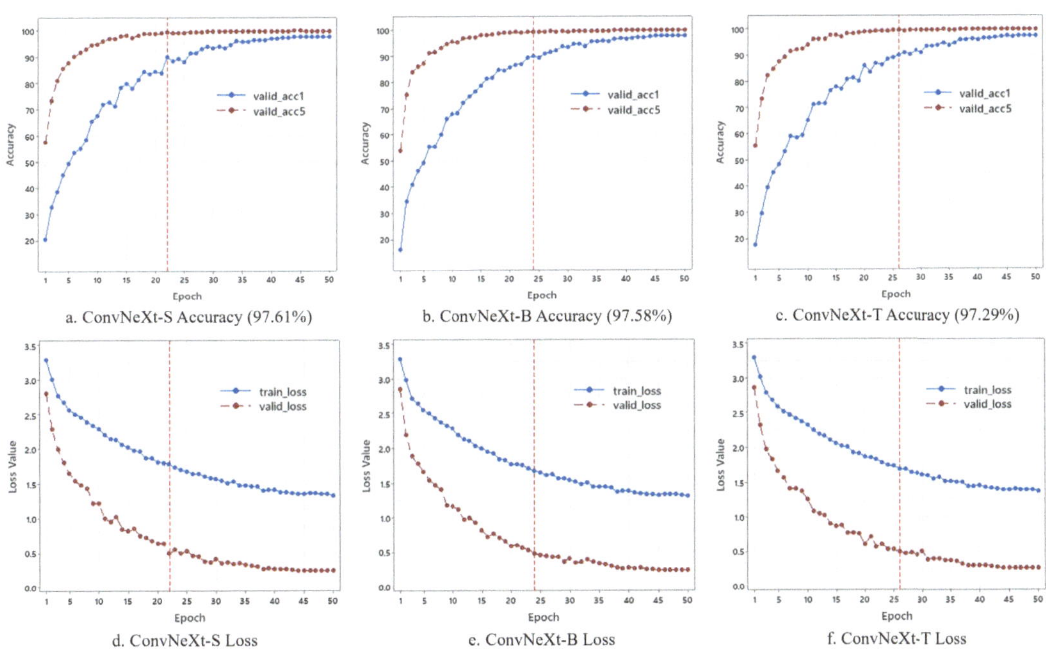

a. ConvNeXt-S Accuracy (97.61%) b. ConvNeXt-B Accuracy (97.58%) c. ConvNeXt-T Accuracy (97.29%)

d. ConvNeXt-S Loss e. ConvNeXt-B Loss f. ConvNeXt-T Loss

Figure 3. Training process of ConvNeXt network. In (**a–c**), vaild_acc1 and vaild_acc5 represent Top-1 and Top-5 accuracies on validation set, respectively, and value in parentheses in subheadings indicates highest Top-1 accuracy of network on test set. In (**d–f**), train_loss and test_loss indicate loss function values of network on training and validation sets, respectively. A reference line is added to indicate moment when Top-1 accuracy of pre-training network initially reaches 90%.

Figure 3d shows that the ConvNeXt-S loss function values changed during training. Due to the small initial learning rate of the warm-up phase, the curve decreased rapidly during the first 10 epochs. When the network first achieved 90% identification accuracy on the validation set, the corresponding loss value was 0.494. It should be noted that the curve did not fluctuate much over a long period after reaching 90% identification accuracy, and the curve eventually stabilized at the 40th epoch. Subsequently, the fluctuation of the loss value curve was almost negligible. Figure 3e,f show that the training processes for the ConvNeXt-B and ConvNeXt-T were, in principle, very similar to those of the ConvNeXt-S. The loss values of the three networks of different depths decreased at basically the same downward trend. However, the deeper network of ConvNeXt-B could fit earlier, and its training process was more stable than those of the other two networks. When the training reached about 40 epochs, the loss function curves of the three networks stabilized and experienced no significant fluctuations.

In general, deeper networks are typically associated with superior non-linear representation capabilities, enabling the learning of more complex transformations and accommodating more complicated feature inputs. However, our experiments demonstrate the opposite results: deeper networks exhibit a slightly lower degree of classification accuracy compared to shallower networks. This result probably occurs because the bark images, which mainly consist of texture and color, are much simpler than images in multi-classification datasets, such as CIFAR and ImageNet, and shallow learners can effectively distinguish these bark features. Therefore, using deep networks can result in gradient instability and network degradation, which inevitably reduces the learning capacity of specific deep layers of uncomplicated datasets. Consequently, this issue leads to a reduction in identification performance on a new dataset.

3.2. Identification Precision by Species

A confusion matrix is a table that shows the extent to which a model predicted the correct class for each test datum, as well as the extent to which it made mistakes. The confusion matrix in Figure 4 shows the average identification accuracy of different tree species on the test set using the pre-trained weights obtained by pre-training the ConvNeXt-S network (without finetune) on the BarkNetV2 dataset. The results demonstrate that ConvNeXt-S achieved high identification accuracy (above 98%) for about half of the tree species. However, the identification accuracy of some tree species was relatively low, with an average accuracy rate of approximately 95%. The tree species with relatively low identification accuracy for ConvNeXt-S were EPR (*Acer rubrum*), ORA (*Ulmus americana*), OSV (*Ostrya virginiana*), and SAB (*Abies balsamea*). It should be noted that due to the limited number of bark images available for ERB (*Acer platanoides*), PEG (*Populus grandidentata*), and PID (*Pinus rigida*), the pre-trained weights obtained may not apply to the practical classification tasks for these tree species, despite there being no misidentifications of these species in the experiment.

When performing an identification task, a network typically uses a classical voting method to select the label that corresponds to the highest value of category probability as the predicted class output, as with ConvNeXt. The confusion matrix can indicate the identification accuracy of the network for each tree species, but it cannot effectively express the prediction correctness (i.e., the proportion of the tree species with the highest probability value). To address this issue, we loaded the pre-trained optimal weights for multi-label prediction onto the BarkNetV2 test set. Next, we obtained the predicted labels and corresponding category probabilities of 33 tree species found on the network (Table 5). In Table 5, the category probabilities indicate the confidence level of the network for accurate identification, while the mean and standard deviation of category probabilities indicate the average categorical confidence and the dispersion of the category probabilities, respectively.

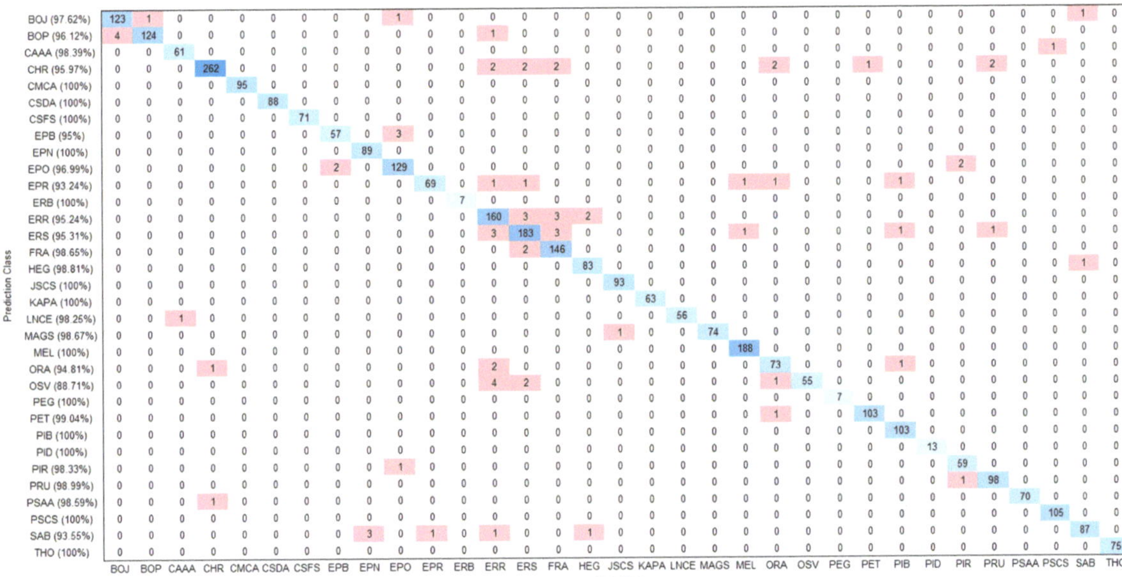

Figure 4. ConvNeXt-S confusion matrix on test set. True Class and Prediction Class represent tree species index and the network prediction, respectively. Average classification accuracy of each tree species is shown in parentheses next to Prediction Class.

Table 5. Identification confidence of ConvNeXt-S.

Species Index	Scientific Name	Test Image	Mean	Std	Accuracy
BOJ	*Betula alleghaniensis*	126	0.7952	0.1451	97.62%
BOP	*Betula papyrifera*	129	0.7899	0.1701	96.12%
CAAA	*Camptotheca acuminata*	62	0.8049	0.1104	98.39%
CHR	*Quercus rubra*	273	0.7528	0.1713	95.97%
CMCA	*Cinnamomum camphora*	95	0.8344	0.0354	100.00%
CSDA	*Cedrus deodara*	88	0.8262	0.0328	100.00%
CSFS	*Cupressus funebris*	71	0.7988	0.0697	100.00%
EPB	*Pinus resinosa*	60	0.7465	0.2051	95.00%
EPN	*Picea mariana*	89	0.8049	0.0794	100.00%
EPO	*Picea abies*	133	0.7760	0.1517	96.99%
EPR	*Picea rubens*	74	0.7925	0.1594	93.24%
ERB	*Acer platanoides*	7	0.7662	0.1027	100.00%
ERR	*Acer rubrum*	168	0.7654	0.1646	95.24%
ERS	*Acer saccharum*	192	0.7215	0.2241	95.31%
FRA	*Fraxinus americana*	148	0.8041	0.0846	98.65%
HEG	*Fagus grandifolia*	84	0.7993	0.1348	98.81%
JSCS	*Juniperus chinensis*	93	0.8190	0.0936	100.00%
KAPA	*Koelreuteria paniculata*	63	0.8299	0.0307	100.00%
LNCE	*Liriodendron chinense*	57	0.7917	0.0907	98.25%
MAGS	*Metasequoia glyptostroboides*	75	0.7926	0.1030	98.67%
MEL	*Larix laricina*	188	0.8451	0.0323	100.00%
ORA	*Ulmus americana*	77	0.7246	0.1991	94.81%
OSV	*Ostrya virginiana*	62	0.7519	0.1867	88.71%
PEG	*Populus grandidentata*	7	0.9040	0.0805	100.00%

Table 5. *Cont.*

Species Index	Scientific Name	Test Image	Mean	Std	Accuracy
PET	*Populus tremuloides*	104	0.8061	0.0889	99.04%
PIB	*Pinus strobus*	103	0.7855	0.1454	100.00%
PID	*Pinus rigida*	13	0.7917	0.1131	100.00%
PIR	*Picea glauca*	60	0.7904	0.1546	98.33%
PRU	*Tsuga canadensis*	99	0.8027	0.1131	98.99%
PSAA	*Platanus acerifolia*	71	0.8106	0.0573	98.59%
PSCS	*Populus canadensis*	105	0.8280	0.0295	100.00%
SAB	*Abies balsamea*	93	0.7572	0.1913	93.55%
THO	*Thuja occidentalis*	75	0.8108	0.0501	100.00%

Note: Test image indicates number of images in test set for each category. Mean and Std indicate mean and standard deviation of category probability, respectively. When top-1 category is true label, we record its probability. Otherwise, its probability is recorded as zero. Means and variances in category probabilities are obtained via statistics on classification results of all images in test set. Accuracy indicates identification accuracy of each tree species.

A higher degree of variability in the category probabilities implies a lower level of confidence in the network's ability to generate valid predictions. Moreover, the identification accuracy of a tree species on the network would be adversely affected by the high degree of fluctuation in its category probabilities. The pre-trained weights of ConvNeXt-S achieved excellent identification results for PEG (*Populus grandidentata*), MEL (*Larix laricina*), and CMCA (*Cinnamomum camphora*), with high confidence probabilities noted for the categories, as well as an overall identification accuracy of almost 100%. In contrast, for ERS (*Acer saccharum*), SAB (*Abies balsamea*), ORA (*Ulmus americana*), and others, the pre-trained weights were significantly less confident than the average figure, and the overall recognition accuracy for these species was generally in the range of 90% to 95%. It is worth mentioning that ConvNeXt-S achieved higher identification accuracy on the species in our BarkNJ dataset than in BarkNet, and the confidence probabilities of the categories are also higher than in BarkNet, thus showing the advantages of our dataset.

3.3. *Visualization of Network Identification Process*

3.3.1. Selection of Sample Images

All three networks showed some misidentified images, although their overall identification accuracies were slightly different. Tree species from the same family or genus are often difficult to distinguish in terms of morphological features; even deep learning networks may confuse these species. Many closely taxonomically related tree species mentioned in BarkNetV2 were used in this study, such as *Betula alleghaniensis* (yellow birch) and *Betula papyrifera* (white birch) in the genus *Birch* of the family *Betulaceae*; *Pinus resinosa* (red pine) and *Picea rubens* (red spruce), which are both in the family *Pinaceae*; and *Pinus strobus* (eastern white pine) and *Tsuga canadensis* (eastern hemlock) in the family *Pinaceae*. Furthermore, sun illumination, bark occlusion, or camera shaking during image acquisition may cause noises in the images. These noises may cause the network to misidentify features of the tree species during the identification process.

The workflow of neural networks is widely considered to be a black box, making it difficult to observe the identification process in detail. To further investigate how the ConvNeXt network identifies bark images, eight photos were selected as sample images for the visualization of the network workflow, and the details of the sample images are shown in Table 6. To ensure the effectiveness of the visualization work, we followed the below procedures to select sample images: four images with high classification accuracy (high overall accuracy for the category and high confidence for a single prediction) and four images with poor classification accuracy (i.e., each of the three networks misidentified the image as another tree species) were selected. The visualization section utilizes the optimal weights of the ConvNeXt-S network. The four selected high-identification precision images were sourced from the following tree species: MEL (*Larix laricina*), EPO (*Picea abies*), LCNE

(*Liriodendron chinense*), and MAGS (*Metasequoia glyptostroboides*). In contrast, the four selected low-identification precision images were sourced from the following tree species: SAB (*Abies balsamea*), ERS (*Acer saccharum*), OSV (*Ostrya virginiana*), and CAAA (*Camptotheca acuminata*). The details of the selected eight images are shown in Table 6, which shows the Top-4 prediction categories with the highest probabilities and the species identified by the network (the classes corresponding to Top-1 Accuracy).

Table 6. Details of sample image.

True Class	Prediction Top-K Accuracy				
	1	2	3	4	Sum of Top-n
MEL (*Larix laricina*)	0.865 (MEL)	0.007 (ERR)	0.007 (ERS)	0.007 (ERR)	0.133
EPO (*Picea abies*)	0.830 (EPO)	0.014 (MEL)	0.012 (EPR)	0.009 (ERR)	0.117
LNCE (*Liriodendron chinense*)	0.834 (LNCE)	0.012 (ERR)	0.011 (CHR)	0.008 (PET)	0.136
MAGS (*Metasequoia glyptostroboides*)	0.856 (MAGS)	0.007 (MEL)	0.007 (FRA)	0.006 (SAB)	0.123
SAB (*Abies balsamea*)	0.569 (EPN)	0.272 (SAB)	0.020 (EPO)	0.011 (PIR)	0.128
ERS (*Acer saccharum*)	0.644 (FRA)	0.131 (CHR)	0.080 (ERS)	0.018 (ERR)	0.127
OSV (*Ostrya virginiana*)	0.853 (ERR)	0.012 (PIB)	0.011 (FRA)	0.010 (LNCE)	0.114
CAAA (*Camptotheca acuminata*)	0.770 (PSCS)	0.028 (FRA)	0.020 (ERR)	0.013 (THO)	0.169

Note: Prediction Top-K Accuracy denotes prediction probability of top-k items of network weights on image, and Sum of Top-n denotes sum of probabilities of categories other than top-4 items.

3.3.2. Integrated Gradient Visualization

The results of integrated gradient visualization in the format of dotted raster images can be seen in Figure 5, in which the original image and the bark outline predicted via network vision can roughly be seen. By applying this method to a neural network, it is possible to generate saliency maps that highlight the regions of an image that are most relevant for identification.

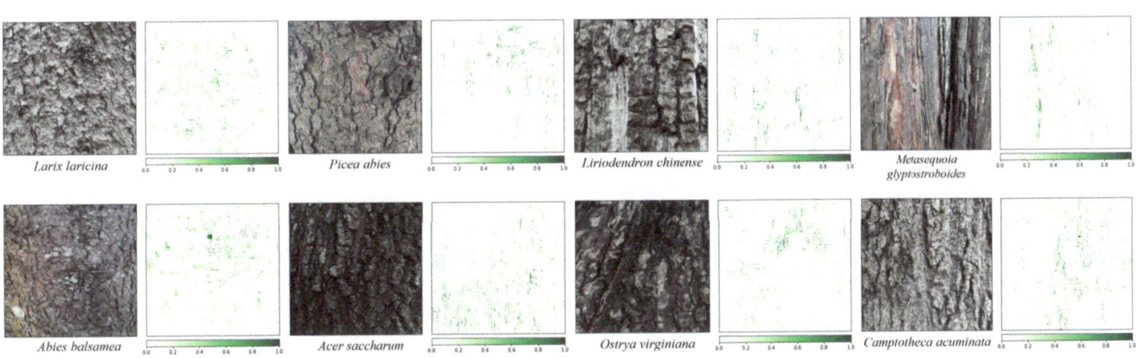

Figure 5. Integrated gradient visualization for network prediction. Left image presents original image, right image is visualization of deep network visual attractiveness captured after adding Gaussian noise smoothing, green color is location of high attractiveness, and deeper color indicates region's importance within network vision.

The bark image of *Larix laricina* has a network visual distribution that is mainly on brick-red bark under the thin scales, while the bark image of *Picea abies* is mainly composed of rough paper-like scales, which are deep red–brown in color and accompanied by cracks and trim pieces that partially stem from the lenticel. The visual distribution of the network is more concentrated at the joints between thick scales and the location where the lenticel grow. The bark image of *Liriodendron chinense* is dominated by vertical texture, which mainly highlights vertical stripes and slab exfoliation. Similarly, the visual distribution of the network of this bark image is more concentrated at the joints between thick scales. The bark image of *Metasequoia glyptostroboides* is dark brown, cracked, and flaked, and the texture of the bark image shows tapered branching and ascending branches, with evident visual attraction at cracks and spalls. In the four bark images with lower identification accuracy rates, the network visualizations are roughly concentrated on the scale cracks and the visually prominent exposed bark. However, the location of the concentrated visual attraction of the biological characteristics of identified tree species slightly deviates from the correct tree species. Therefore, the visualization shows that pre-trained weights derived from our experiments enable a human-like approach to bark perception by focusing attention on regions exhibiting distinct biological features.

3.3.3. Class Activation Mapping Hot Spots

Smooth Grad-CAM++ was utilized to produce heat maps that corresponded to three neural network stages (excluding stage one due to its limited representational capacity). This approach allowed us to visualize the spatial distribution and importance of visual features in each network layer. The heat map was obtained by weighting the feature maps with the corresponding convolutional weights and computing the average across entire feature maps. The resulting heat map visualizes the regions on the image that most strongly influence the final identification decision. The overlay of the original image with the heat map allows us to observe the change in the network's visual recognition of the actual image (Figure 6).

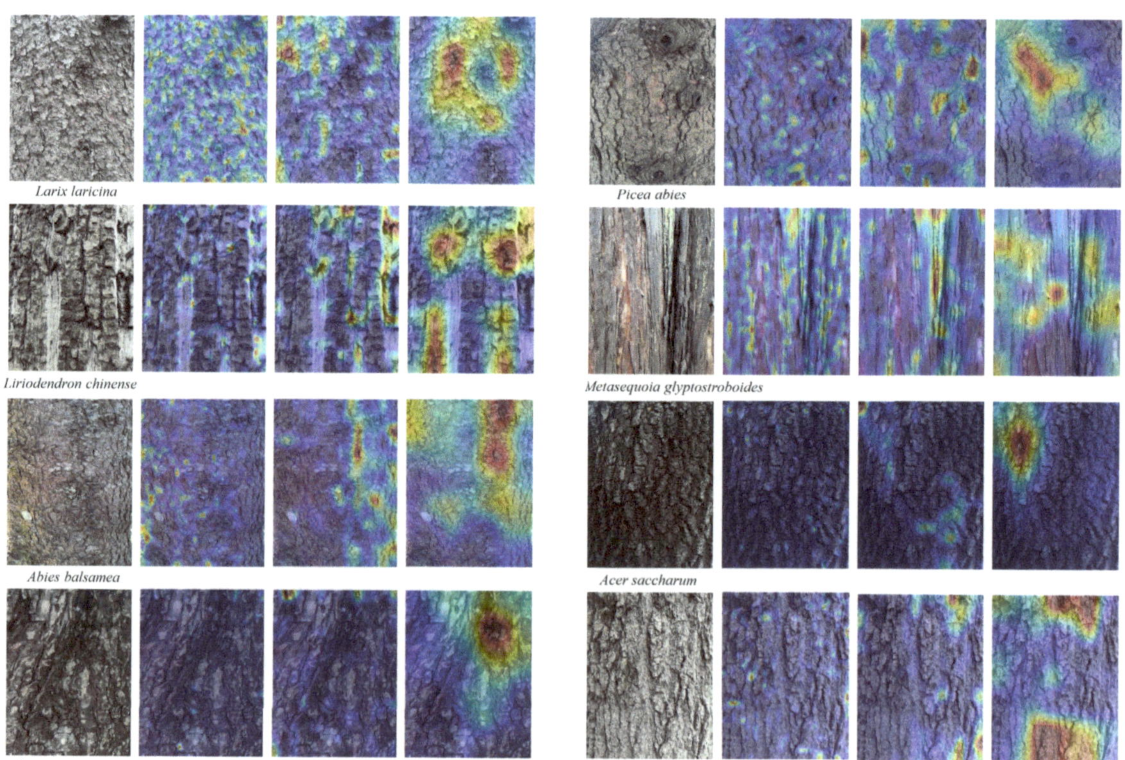

Figure 6. Class activation mapping generated via Smooth Grad CAM++. Visual appeal is positioned to simulate three stages of ConvNeXt-S network, representing high and low visual attractiveness on a heat map. Distribution map of cold spots on bark image is made by superimposing original image to heat map.

For the four images with high recognition accuracy, we found that the network visual attractiveness heat map produced via Class activation mapping was basically located at some positions with distinct tree species features. For example, the exposed brick-red bark of *Larix laricina* and *Picea abies*, as well as the vertical spalls and cracks of *Liriodendron chinense*, were similar to the results produced via Integrated Gradient visualization. However, the heat maps exhibited imprecise positioning in the four images with lower classification accuracy rates. For instance, the extraction algorithm failed to effectively capture the texture features within the seams of *Acer saccharum* and *Ostrya virginiana* bark images. Additionally, the network generated for *Abies balsamea* and *Camptotheca acuminata* displayed incorrect visual locations, resulting in inaccurate coverage.

3.3.4. Image Depth Feature Decomposition

Using deep feature decomposition to perform visualization of category probabilities on four misidentified images, we obtained visualization results similar to those obtained via semantic segmentation (Figure 7). Based on this result, we know which bark image that blocks the network identifies as which tree species. On the bark image of *Larix laricina*, most areas are identified with the correct label, though there are still a few areas with potential identification errors, i.e., misidentified as CHR (*Quercus rubra*). The identification precision of *Picea abies* is relatively higher, with only a tiny portion (one of the lenticel) being misidentified as FRA (*Fraxinus americana*). Both bark images of *Liriodendron chinense* and *Metasequoia glyptostroboides* have relatively noticeable vertical peeling and cracks. Therefore, the decomposition results of these two bark images show a vertical distribution, with only a few misidentification errors appearing on the edge of the bark images, which are misidentified as ERS (*Acer saccharum*) and CHR (*Quercus rubra*), respectively. On the bark image of *Abies balsamea*, only the area in the lower part of the image was correctly identified as SAB, while other areas were identified as EPN (*Picea mariana*) and CHR (*Quercus rubra*). The identification results of *Acer saccharum* are even worse, and the feature decomposition identifies all areas of the image as FRA (*Fraxinus americana*), indicating that the network's feature extractor failed to obtain adequate classification information. On the bark image of *Ostrya virginiana*, there is a large misidentified area in which the tree species is mislabeled as FRA (*Fraxinus americana*) and ERR (*Acer rubrum*), in different places. In contrast to the above cases, the mislabeled places on the bark image of *Camptotheca acuminata* exhibit an extensive vertical distribution, in which the tree species is primarily mislabeled as PSCS (*Populus canadensis*).

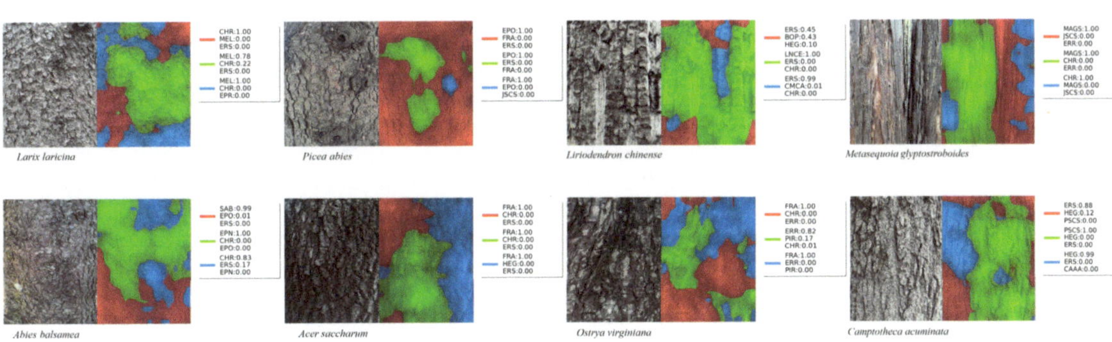

Figure 7. Visualization of network depth feature decomposition. Original image is shown on left. Right image is semantically segmented via deep feature decomposition into regions reflecting different category features, using different colors to mark feature regions. Title contains tree species IDs and corresponding predicted category probabilities.

4. Discussion

Our experiments show that the ConvNeXt network can accurately identify 33 tree species, with an average accuracy rate of 97.61% on the test set, using bark images in the BarkNetV2 dataset. Compared to the identification results obtained using other models on the BarkNet dataset, the ConvNeXt network used in our experiments outperformed the identification results of previous studies.

Transfer learning is a technique where a model trained to perform one task is adapted for another related task. The advantage of transfer learning is that it can save time and computational resources by leveraging the knowledge contained within a large dataset (such as ImageNet) for use in a smaller dataset (such as CIFAR-10). Despite the fact that the network weights obtained via ImageNet are generally applicable, there may be cases where the network performance was not as expected for specific identification tasks. Taxonomically similar trees often show a high degree of similarity in appearance; thus, it is difficult to identify species with very similar taxonomic attributes based on bark images, as demonstrated by our experimental results. The network frequently fails to accurately distinguish tree species with similar identification features or low identification confidence. It is worth mentioning that the same tree species often undergo several phases of changes during the growth process. Although the bark changes much less than other tree organs, some morphological changes may also occur, such as plate shedding, groove deepening, and plate scale thickening. Therefore, if the network pays too much attention to local features on bark images, instead of capturing the common biological patterns at the species level, the network may experience overfitting, and the overall identification performance of the network will be reduced. For these identification tasks, network pre-training can incorporate similar features from different categories more effectively than transfer learning. Compared to the results obtained in the study by Carpentier and Kim, who utilized ImageNet pre-trained weights, the network pre-trained weights obtained through our training on BarkNetV2 have better identification capabilities [23,27].

During the training process, the accuracy and loss function curves of the three types of networks were generally similar. Although ConvNeXt-B has more parameters, ConvNeXt-S performed better on the test set, possibly because having too many parameters may decrease the network's generalization ability, impairing its predictive performance. The reason for this issue lies in the fact that increasing the number of parameters in the network may cause it to experience overfitting of the noise in the training data, instead of capturing the potential signal, leading to a model that is highly accurate on the training set but performs poorly on a new dataset. Although the overall identification accuracy is considerably high, the confusion matrix results show that it is often difficult to distinguish some taxonomically similar tree species. Pre-trained weights of the network often produce misidentification among tree species in the same family or genus. For example, BOJ (*Betula alleghaniensis*) is often confused with BOP (*Betula papyrifera*), while EPR (*Acer platanoides*), ERS (*Acer saccharum*), and ERR (*Acer rubrum*) are misidentified many times. Moreover, EPO (*Picea abies*) and EPR (*Picea rubens*) are often difficult to distinguish. The identification of taxonomically similar trees is a challenging task for convolutional neural networks (CNNs). The high degree of similarity between such tree species often leads to misidentifications, which is a critical issue in image recognition and identification tasks. When the CNNs are pre-trained for one category and tested via another closely related category, such as different species of plants or animals, misidentifications are likely to occur due to the inherent variations in features and patterns. It is also worth mentioning that the result of the confusion matrix in our experiment shows that the network pre-trained weights rarely show misidentifications for the tree species in BarkNJ, and the overall identification accuracy is greater than that of the tree species in BarkNet with higher identification confidence. This result occurs because the bark images collected in this paper exclude the effects of lighting and shadows, and the subsequent processing also removes the images containing noise, which makes our BarkNJ dataset superior to BarkNet in terms of quality.

Tree bark provides essential information about tree species and their environmental conditions. Different kinds of trees possess unique bark properties, such as smooth or rough textures, that can be used to differentiate them. In the case of tree identification, the network would be fed with data about different tree species, including information about their taxonomic attributes and bark characteristics. These data would be used to train the network to recognize patterns unique to each species, as well as to use those patterns to identify new samples of trees. Feature extraction is an essential technique in CNNs that allows the network to focus on the most critical aspects of the image and ignore irrelevant or redundant features. For example, the network architecture can be designed to extract bark-related features, such as texture, pattern, or thickness, to distinguish different tree species based on their bark characteristics. However, the neural network feature extraction process is often called a black box. The network learns to identify and extract features from the input data most relevant to the task, and these features can be highly abstract and difficult to interpret or visualize. During our visualization, we found that the pre-trained neural network weights selectively focused on regions of the image that exhibited distinct features, including grooves, cracks, and lenticels, which strongly resemble the ways in which humans recognize tree species based on bark image. Our findings suggest that the network's identification mechanism is closely linked to tree species' taxonomic attributes and their bark's unique biological features, as indicated based on the patterns revealed during our visualization experiment.

Integrated Gradients, Grad CAM, and Deep Feature Decomposition are three visualization methods that aim to explain the predictions of deep convolutional neural networks by highlighting the regions or features that contribute to the output. However, these methods also have some limitations [37–39]. For example, Integrated Gradients may not be able to capture certain types of relationships between the input and output of a model. When applied to some models, Grad CAM may produce noisy or blurry heat maps. Deep Feature Decomposition requires a pre-trained autoencoder to reconstruct the input image from the decomposed features, which may introduce reconstruction errors or artifacts. Therefore, in our future research, deep network visualization techniques must be further improved to better reveal the principles and characteristics of the network workflow.

The image recognition technology that uses convolutional neural network has significant practical value in identifying tree species based on bark images, which can improve the intelligence and efficiency of forest resource surveys through fast and precise identification of tree species. The traditional methods of tree species identification in the field are less efficient, especially for tall trees, whose leaves are visually difficult to differentiate. For deciduous tree species, it requires significant effort to collect samples of leaves or flowers for identification during the fall or winter. The application of handheld mobile devices makes tree species identification in the field based on bark images more convenient and simplifies the sampling process. Combined with lightweight convolutional neural networks, the efficiency of field identification of tree species can be greatly improved. The technology proposed in our study enables forest workers to collect bark images in the field and upload them to a back-end server for identification. Through the optimization of data collection and image processing, the tree species identification technology proposed in our study can meet the needs of field workers in terms of identification efficiency and accuracy. In addition, through both the accumulation of a large number of bark images and network optimization, the accuracy and generalization performance of the neural network can be continuously improved, further improving the predictability of the model in different environmental settings.

In traditional deep learning processes, it is often necessary to use large amounts of data to train a model to achieve good performance, which usually entails high costs and complicated data acquisition processes. In contrast, few-shot learning techniques are able to use the prior experiences as input data, thus significantly reducing data requirements and providing a more cost-effective solution. In addition, few-shot learning techniques can also improve the robustness of models, allowing machine learning algorithms to maintain a high level of accuracy in the face of more complex scenarios. In future research, we will explore the application of few-shot learning to tree species identification to obtain a model with better generalization ability, while reducing the amount of data required for training. In addition, we will continue to collect bark images to build a large-scale bark dataset for deep learning.

5. Conclusions

Based on the BarkNetV2 bark dataset, we used three ConvNeXt networks with different depths for tree species identification through bark images. In our experiments, the networks could identify 33 different tree species with an identification accuracy of 97.62%, exceeding the performances reported in previous works. Confusion matrices and category probability tables showed that more than half of the tree species could be accurately distinguished with high identification confidence. We described the workflow of network feature extraction by integrating gradients and other visualization methods, and we analyzed the correlation between the visual attractions of the network and the biological features of the tree bark. The results show that the location and importance of the visual attractions of the network are closely related to the biological characteristics of the tree bark. Additionally, we created a new dataset called BarkNJ, which consisted of images of a higher quality than those located in BarkNet. During the experiments, the tree species in the BarkNJ dataset achieved almost complete correct identification. Based on these results, transfer learning and fine-tuning of neural networks can further expand their application scenarios and create considerable potential application value in forest resource surveys both in China and abroad.

Author Contributions: Conceptualization, Z.C. and M.L.; formal analysis, Z.C.; funding acquisition, M.L.; methodology, Z.C. and M.L.; resources, T.L. and X.L.; writing—original draft preparation, Z.C. and X.L.; writing—review and editing, Z.C., M.L., T.L. and X.L. All authors have read and agreed to the published version of this manuscript.

Funding: This research was funded by the National Natural Science Foundation [30972298] and the Top-Notch Academic Programs Project of Jiangsu Higher Education Institutions, China [PPZY2015A062].

Data Availability Statement: The data underlying this article will be shared on reasonable request to the corresponding authors.

Acknowledgments: We would like to express our sincere gratitude to Nanjing Forestry University and the Forest Management Laboratory for providing us with the facilities and equipment required to conduct this study. We are also grateful to all colleagues and technicians who assisted us during the data preparation and analysis stages. Their expertise and support were invaluable for this project's success.

Conflicts of Interest: The authors declare no conflict of interest.

References

1. Puri, M.; Pathak, Y.; Sutariya, V.K.; Tipparaju, S.; Moreno, W. *Artificial Neural Network for Drug Design, Delivery and Disposition*; Academic Press: Cambridge, UK, 2015.
2. Brahme, A. *Comprehensive Biomedical Physics*; Brahme, A., Ed.; Elsevier: Amsterdam, The Netherlands, 2014; Volume 10.
3. Xu, Z.Q.J.; Zhang, Y.Y.; Xiao, Y.Y. Training behavior of deep neural network in frequency domain. In Proceedings of the 26th International Conference on Neural Information Processing (ICONIP), Sydney, NSW, Australia, 12–15 December 2019; pp. 264–274. [CrossRef]
4. Fiel, S.; Sablatnig, R. Automated identification of tree species from images of the bark, leaves or needles. In Proceedings of the 16th Computer Vision Winter Workshop, Graz, Austria, 3 February 2010; pp. 67–74.
5. Wang, Y.; Zhang, W.; Gao, R.; Jin, Z.; Wang, X. Recent advances in the application of deep learning methods to forestry. *Wood Sci. Technol.* **2021**, *55*, 1171–1202. [CrossRef]
6. Chi, Z.R.; Li, H.Q.; Wang, C. Plant species recognition based on bark patterns using novel Gabor filter banks. In Proceedings of the International Conference on Neural Networks and Signal Processing, Nanjing, China, 14–17 December 2003; Volume 2, pp. 1035–1038. [CrossRef]
7. Kim, S.J.; Kim, B.W.; Kim, D.P. Tree recognition for land scape using by combination of features of its leaf, flower and bark. In Proceedings of the SICE Annual Conference, Tokyo, Japan, 13–18 September 2011; pp. 1147–1151.

8. Blaanco, L.J.; Travieso, C.M.; Quinteiro, J.M.; Hernandez, P.V.; Dutta, M.K.; Singh, A. A bark recognition algorithm for plant classification using a least square support vector machine. In Proceedings of the Ninth International Conference on Contemporary Comput ing (IC3), Noida, India, 11–13 August 2016; pp. 1–5. [CrossRef]
9. Boudra, S.; Yahiaoui, I.; Behloul, A. Bark identification using improved statistical radial binary patterns. In Proceedings of the International Conference on Content-Based Multimedia Indexing (CBMI), La Rochelle, France, 4–6 September 2018; pp. 1–6. [CrossRef]
10. Reměs, V.; Haindl, M. Bark recognition using novel rotation ally invariant multispectral textural features. *Pattern Recognit. Lett.* **2019**, *125*, 612–617. [CrossRef]
11. Bertrand, S.; Ameur, R.B.; Cerutti, G.; Coquin, D.; Valet, L.; Tougne, L. Bark and leaf fusion systems to improve auto matic tree species recognition. *Ecol. Inform.* **2018**, *46*, 27–73. [CrossRef]
12. Ratajczak, R. Efficient bark recognition in the wild. In Proceedings of the International Conference on Computer Vision Theory and Applications—Volume 4 (VISAPP), Prague, Czech Republic, 25–27 February 2019; pp. 240–248. [CrossRef]
13. Fekri-Ershad, S. Bark texture classification using improved local ternary patterns and multilayer neural network. *Expert Syst. Appl.* **2020**, *158*, 113509. [CrossRef]
14. Alzubaidi, L.; Zhang, J.; Humaidi, A.J.; Al-dujaili, A.; Duan, Y.; Al-Shamma, O.; Santamaría, J.; Fadhel, M.A.; Al-Amidie, M.; Farhan, L. Review of deep learning: Concepts, CNN architectures, challenges, applications, future directions. *J. Big Data* **2021**, *8*, 53. [CrossRef] [PubMed]
15. Chui, K.T.; Gupta, B.B.; Chi, H.R.; Zhao, M. Convolutional Neural Network and Deep One-Class Support Vector Machine with Imbalanced Dataset for Anomaly Network Traffic Detection. In Proceedings of the International Conference on Cyber Security, Privacy and Networking (ICSPN 2022), Fairfield, OH, USA, 3–11 July 2022; Volume 599, pp. 248–256. [CrossRef]
16. Xin, M.Y.; Wang, Y. Research on image classification model based on deep convolution neural network. *J. Image Video Process.* **2019**, *40*, 1–11. [CrossRef]
17. Choi, S. Plant Identification with Deep Convolutional Neural Network: SNUMedinfo at LifeCLEF Plant Identification Task 2015. In Proceedings of the Conference and Labs of the Evaluation Forum, Toulouse, France, 8–11 September 2015.
18. Lee, S.H.; Chan, C.; Wilkin, P.; Remagnino, P. Deep-plant: Plant identification with convolutional neural networks. In Proceedings of the 2015 IEEE International Conference on Image Processing (ICIP), Quebec City, QC, Canada, 27–30 September 2015; pp. 452–456. [CrossRef]
19. Grinblat, G.L.; Uzal, L.C.; Larese, M.G.; Granitto, P.M. Deep learning for plant identification using vein morphological patterns. *Comput. Electron. Agric.* **2016**, *127*, 418–424. [CrossRef]
20. Sun, Y.; Liu, Y.; Wang, G.; Zhang, H. Deep Learning for Plant Identification in Natural Environment. *Comput. Intell. Neurosci.* **2017**, *6*, 1–6. [CrossRef] [PubMed]
21. Rawat, W.; Wang, Z.H. Deep Convolutional Neural Networks for Image Classification: A Comprehensive Review. *Neural Comput.* **2017**, *29*, 2352–2449. [CrossRef] [PubMed]
22. He, K.M.; Zhang, X.; Ren, S.; Sun, J. Deep Residual Learning for Image Recognition. In Proceedings of the 2016 IEEE Conference on Computer Vision and Pattern Recognition (CVPR), Las Vegas, NV, USA, 26 June–1 July 2016; pp. 770–778. [CrossRef]
23. Carpentier, M.; Giguère, P.; Gaudreault, J. Tree Species Identification from Bark Images Using Convolutional Neural Networks. In Proceedings of the 2018 IEEE/RSJ International Conference on Intelligent Robots and Systems (IROS), Madrid, Spain, 1–5 October 2018; pp. 1075–1081. [CrossRef]
24. Misra, D.; Crispim, C.F.; Tougne, L. Patch-Based CNN Evaluation for Bark Classification. *Eur. Conf. Comput. Vis.* **2020**, *12540*, 197–212. [CrossRef]
25. Robert, M.; Dallaire, P.; Giguère, P. Tree bark reidentification using a deep-learning feature descriptor. In Proceedings of the 17th Conference on Computer and Robot Vision (CRV), Ottawa, ON, Canada, 13–15 May 2020; pp. 25–32. [CrossRef]
26. Faizal, S. Automated Identification of Tree Species by Bark Texture Classification Using Convolutional Neural Networks. *Int. J. Res. Appl. Sci. Eng. Technol.* **2022**, *10*, 1384–1392. [CrossRef]
27. Kim, T.K.; Hong, J.; Ryu, D.; Kim, S.; Byeon, S.; Huh, W.; Kim, K.; Baek, G.H.; Kim, H.S. Identifying and extracting bark key features of 42 tree species using convolutional neural networks and class activation mapping. *Sci. Rep.* **2022**, *12*, 4772. [CrossRef] [PubMed]
28. Liu, Z.; Mao, H.; Wu, C.; Feichtenhofer, C.; Darrell, T.; Xie, S. A ConvNet for the 2020s. In Proceedings of the 2022 IEEE/CVF Conference on Computer Vision and Pattern Recognition (CVPR), New Orleans, LA, USA, 18–24 June 2022; pp. 11966–11976. [CrossRef]
29. Liu, Z.; Lin, Y.; Cao, Y.; Hu, H.; Wei, Y.; Zhang, Z.; Lin, S.; Guo, B. Swin Transformer: Hierarchical Vision Transformer using Shifted Windows. In Proceedings of the 2021 IEEE/CVF International Conference on Computer Vision (ICCV), Montreal, QC, Canada, 11–17 October 2021; pp. 9992–10002. [CrossRef]
30. Paszke, A.; Gross, S.; Massa, F.; Lerer, A.; Bradbury, J.; Chanan, G.; Killeen, T.; Lin, Z.; Gimelshein, N.; Antiga, L.; et al. PyTorch: An Imperative Style, High-Performance Deep Learning Library. *Neural Inf. Process. Syst.* **2019**. *submitted*. [CrossRef]
31. Sebastian, R. *Machine Learning with PyTorch and Scikit-Learn: Develop Machine Learning and Deep Learning Models with Python*; Liu, Y.X., Vahid, M., Dmytro, D., Eds.; Packt Publishing: Birmingham, UK, 2022.
32. Sundararajan, M.; Taly, A.; Yan, Q. Axiomatic Attribution for Deep Networks. In Proceedings of the 34th International Conference on Machine Learning (ICML 17), Sydney, NSW, Australia, 6–11 August 2017; Volume 70. [CrossRef]

33. Omeiza, D.; Speakman, S.; Cintas, C.; Weldemariam, K. Smooth Grad-CAM++: An Enhanced Inference Level Visualization Technique for Deep Convolutional Neural Network Models. *Intell. Syst. Conf.* **2019**. *submitted*. [CrossRef]
34. Selvaraju, R.R.; Das, A.; Vedantam, R.; Cogswell, M.; Parikh, D.; Batra, D. Grad-CAM: Why did you say that? Visual Explana tions from Deep Networks via Gradient-based Localization. *arXiv* **2019**, arXiv:1610.02391. [CrossRef]
35. Smilkov, D.; Thorat, N.; Kim, B.; Viégas, F.B.; Wattenberg, M. SmoothGrad: Removing noise by adding noise. *arXiv* **2017**, arXiv:1706.03825. [CrossRef]
36. Collins, E.; Achanta, R.; Süsstrunk, S. Deep Feature Factorization for Concept Discovery. *arXiv* **2018**, arXiv:1806.10206. [CrossRef]
37. Kapishnikov, A.; Venugopalan, S.; Avci, B.; Wedin, B.; Terry, M.; Bolukbasi, T. Guided Integrated Gradients: An Adaptive Path Method for Removing Noise. In Proceedings of the IEEE/CVF Conference on Computer Vision and Pattern Recognition (CVPR), Nashville, TN, USA, 20–25 June 2021; pp. 5050–5058. [CrossRef]
38. Selvaraju, R.R.; Cogswell, M.; Das, A.; Vedantam, R.; Parikh, D.; Batra, D. Grad-cam: Visual explanations from deep networks via gradient-based localization. In Proceedings of the IEEE International Conference on Computer Vision (ICCV), Venice, Italy, 22–29 October 2017; pp. 618–626. [CrossRef]
39. Ruan, D.; Yan, Y.; Lai, S.Q.; Chai, Z.H.; Shen, C.H.; Wang, H.Z. Feature Decomposition and Reconstruction Learning for Effective Facial Expression Recognition. In Proceedings of the IEEE/CVF Conference on Computer Vision and Pattern Recognition (CVPR), Nashville, TN, USA, 20–25 June 2021; pp. 13967–13976. [CrossRef]

forests

MDPI

Article

Multiple Defect Classification Method for Green Plum Surfaces Based on Vision Transformer

Weihao Su, Yutu Yang *, Chenxin Zhou, Zilong Zhuang and Ying Liu

Jiangsu Co-Innovation Center of Efficient Processing and Utilization of Forest Resources, College of Mechanical and Electronic Engineering, Nanjing Forestry University, Nanjing 210037, China; suvi@njfu.edu.cn (W.S.); zhouchx@njfu.edu.cn (C.Z.); zzl0702@njfu.edu.cn (Z.Z.); liuying@njfu.edu.cn (Y.L.)

* Correspondence: yangyutu@njfu.edu.cn

Abstract: *Green plums* have produced significant economic benefits because of their nutritional and medicinal value. However, *green plums* are affected by factors such as plant diseases and insect pests during their growth, picking, transportation, and storage, which seriously affect the quality of *green plums* and their products, reducing their economic and nutritional value. At present, in the detection of *green plum* defects, some researchers have applied deep learning to identify their surface defects. However, the recognition rate is not high, the types of defects identified are singular, and the classification of *green plum* defects is not detailed enough. In the actual production process, *green plums* often have more than one defect, and the existing detection methods ignore minor defects. Therefore, this study used the vision transformer network model to identify all defects on the surfaces of *green plums*. The dataset was classified into multiple defects based on the four types of defects in *green plums* (scars, flaws, rain spots, and rot) and one type of feature (stem). After the permutation and combination of these defects, a total of 18 categories were obtained after the screening, combined with the actual situation. Based on the VIT model, a fine-grained defect detection link was added to the network for the analysis layer of the major defect hazard level and the detection of secondary defects. The improved network model has an average recognition accuracy rate of 96.21% for multiple defect detection of *green plums*, which is better than that of the VGG16 network, the Desnet121 network, the Resnet18 network, and the WideResNet50 network.

Keywords: vision transformer; *green plums*; deep learning; multiple defect detection

Citation: Su, W.; Yang, Y.; Zhou, C.; Zhuang, Z.; Liu, Y. Multiple Defect Classification Method for Green Plum Surfaces Based on Vision Transformer. *Forests* **2023**, *14*, 1323. https://doi.org/10.3390/f14071323

Academic Editor: Cate Macinnis-Ng

Received: 16 May 2023
Revised: 15 June 2023
Accepted: 26 June 2023
Published: 28 June 2023

1. Introduction

Green plums are widely distributed in hills and sloping forests all over the world. They are rich in a large number of amino acids, vitamins, lipids, trace elements, and other nutrients, of which a variety of natural acids are important for human metabolism and have a rich nutritional and economic value [1]. *Green plum* sarcocarp is crisp and tender; it is thick, the core is small, and the taste is sweet and sour, so it is very popular among people. Not only is it unique in flavor, healthy, and appetizing, but it is also beneficial to human health.

With the improvement in people's living standards, their demand for high-quality fruits is also increasing. Consumers are more inclined to buy fruits without defects, and fruit product manufacturers are more inclined to choose high-quality fruits as raw materials. However, *green plums* are susceptible to diseases, insect pests, and knocks during their growth and production [2], resulting in different defects. Damage to the nutritional content and appearance of the product caused by defects will affect the market and price of the product. After picking, *green plums* are not easy to preserve, and they need to be sorted and selected as soon as possible. However, in China, the sorting of *green plums* is mainly carried out manually. The efficiency of manual sorting is low and the cost is high, which makes it impossible to sort a large amount of greengage in a short time.

Moreover, the sorting experience requirements for people are very high, and the sorting accuracy cannot be guaranteed. In addition, the efficiency and accuracy of manual detection are affected by human fatigue. These are the factors that cause quality problems in the secondary processing of *green plums*. To improve the economic value and nutritional value of *green plums* and their products, it is of great significance to carry out a variety of defect detection and classification processes on *green plums* that utilize high-level automation and intelligence. The main defects of *green plums* are divided into four categories: scars, rot, flaws, and rain spots. These defects will lead to quality and nutritional problems for *green plums* and their products. Therefore, before their sale and further processing, it is necessary to carry out defect detection on *green plums*, reject unqualified *green plums*, and classify *green plums*. This research focuses on the detection of the above four types of defects in order to achieve accurate identification of the main defects and other defects.

Computer vision technology is equivalent to the role of human vision in fruit and vegetable quality inspection. It perceives images, interprets and recognizes characters electronically, and provides information for quality grading and sorting machines. By combining machine vision and image processing with the advancement of computer technology, such systems have been applied in different fields of food engineering to accurately identify product characteristic defects in real time [3]. With the development of machine learning [4–7], researchers have applied machine vision [8] and deep learning to defect detection, making the non-destructive testing of fruit processing technology more efficient and accurate. The efficiency and accuracy of defect detection have been greatly improved through machine vision and deep learning. Yao et al. [9] developed a defect detection model based on You Only Look Once (YOLOv5) and optimized the network aiming at kiwifruit defects. This model can accurately and quickly detect defects in kiwifruit. The detection accuracy rate reached 94.7%, nearly 9% higher than the original algorithm. It only takes 0.1 s to process a single image, realizing real-time high-precision detection of kiwifruit defects. R. Nithya et al. [10] developed a computer-aided grading system for mango defect detection to classify high-quality mangoes. After training and testing the system using the publicly available Mango database, an accuracy rate of 98 percent was obtained. Huang et al. [11] used a multichannel hyperspectral imaging system for non-destructive testing of apple varieties. They achieved the best overall classification accuracy of 99.4% in the near-infrared and full-region spectral ranges, whose wavelengths range from 550–1650 nm. The multichannel hyperspectral imaging system provides more spatial–spectral information, and the non-destructive testing effect is excellent. In their research on *green plum* surface defect detection, Zhou et al. [12] and Zhou et al. [13] proposed a computer vision system for *green plum* surface defect detection based on the convolutional neural networks VGG16 and WideResNet50, respectively, which can detect the main defects of *green plums*. The average accuracy rates were 93.8% and 98.95%, respectively. Although the main defects of *green plums* can be accurately identified, each *green plum* may have more than one defect. The previous detection methods for *green plum* defects could only identify and output the main defects of the recognized *green plums* but couldn't identify other defects. According to the degree of impact of defects on production, from large to small, the defects of *green plums* are characterized by rot, flaws, scars, and rain spots. The production of different *green plum* products has different requirements regarding the defects of *green plums*. For example, *green plums* should have no rot or flaw defects to produce *green plum* wine. These defects indicate that the *green plums* have become moldy and contain a large number of microorganisms in their bodies. Such defective *green plums* damage the quality of green plum wine and pose a risk to human health. However, these defective *green plums* can be used as fertilizer after fermentation [14]. *Green plums* with only small scar defects can also be used to produce *green plum* wine to improve production efficiency. However, for *green plums* whose main defect is the scar, it is impossible to know whether it has other defects if there are small-scale rot defects on the surface. Additionally, this type of *green plum* still has food safety problems and cannot be used to produce *green plum* products. *Green plums* with milder scars and rain spots only have surface problems and no internal necrosis, so they can still be used as raw materials for *green plum* wine, candied fruit, and plum powder; *green plums* without defects can be further sold or processed in the market. Therefore, it is necessary to carry out multi-defect detection on *green plums*, which can improve secondary production efficiency and the utilization rate of defective *green plums*.

Rain spot defects are the most common among the four types of defects in *green plums*. Although *green plums* with rain spot defects will not cause food safety problems, these affect the quality classification of *green plums*. The previous defect detection methods could only identify the main defects but could not judge whether other defects threatened food production safety in *green plums*. This makes it impossible to ensure that such *green plums* identified as having rain spot defects will not have safety problems and thus cannot be used as the raw material for producing *green plum* products in the next step. They can only be completely discarded due to food safety issues, which greatly reduces the economic value of *green plums*. This study used a deep learning method based on the vision transformer (VIT). Compared with the WideResNet50, Resnet18, and VGG16 models, the vision transformer network model has higher accuracy, added hazard degree analysis, and fine-grained detection abilities. Using a multi-defect detection scheme, it can identify all of the surface defects of *green plums*. This system can accurately detect major and minor defects in the output, enabling a more meticulous classification of defective *green plums*. Therefore, the precision and accuracy in identifying defects of green plums can be improved.

This study has the following innovations: (a) Aiming at the multi-defect identification problem of *green plums*, a defect identification network model based on the VIT network was proposed. (b) Compared with the single-defect classification processing of traditional data sets, this study's data set was processed using multi-defect classification. (c) After the MLP layer, the *green plum* defect risk level analysis layer and fine-grained detection link were added. The contribution of this study lies in the realization of a more detailed classification of *green plum* defect levels, the ability to accurately identify major defects and the remaining minor defects, and the output of the results of multiple defects. A new method for identifying multiple defects on a surface is proposed.

2. Materials and Methods

2.1. Data Collection and Processing

The dataset used in this study was a batch of *green plums* from Zhangzhou, Fujian, and 2799 RGB images of *green plums* were collected through visible light images. To simulate the real scene of actual production and inspection, this research transported *green plums* on a conveyor belt and collected images, as shown in Figure 1. A light gate from Yue Jiang Company (Hong Kong, China) was installed on the conveyor belt, and when the *green plums* were transported to the light gate, the conveyor belt stopped, and the acquisition device located above collected images. The acquisition system is shown in Figure 1. Viewed from the top down, the acquisition system's first equipment is the camera holder, the second equipment is the camera, the third equipment is the light source holder, the fourth equipment is the light source, the fifth equipment is the light gate, the sixth equipment is the *green plum* to be photographed, and the seventh is the conveyor belt. The entire image acquisition stage is in a closed lighting environment, and the material of the conveyor belt has a light-absorbing effect. The defective *green plum* is located on the conveyor belt, and the LED ring light source is used for supplementary light. The camera bracket can be adjusted to keep the camera at a fixed height, and the *green plum* should be rotated at random angles during the shooting process to obtain multi-angle *green plum* defect pictures.

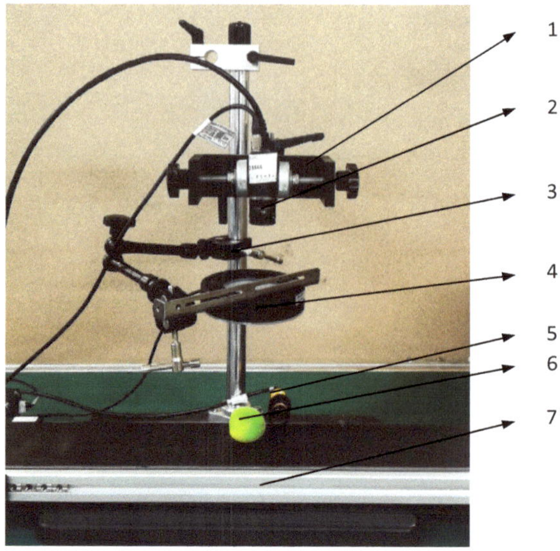

Figure 1. Collection equipment diagram: 1. camera holder; 2. camera; 3. light source holder; 4. light source; 5. light gate; 6. *green plum* sample; 7. conveyer belt.

The camera lens adopted is the M1620-MP2 industrial camera lens from Computer Company (Tokyo, Japan), whose focal length is 16 mm and minimum object distance is 20 cm. The industrial camera adopted the MER-531-20GC-P industrial camera of Beijing Daheng Image Technology Co., Ltd. (Beijing, China). A PYTHON 5000 frames exposure CMOS sensor chip was adopted. The light source used for the collection was an LED ring light source. The image collection stage was carried out in a closed lighting environment. During the shooting process, the *green plum* rotated to obtain multi-angle images of the surface defects of *green plums*.

In this study, 2799 pictures of various *green plum* defects and intact pictures were taken with a dot matrix camera, and the original pictures collected by the camera were 2592 × 2048 pixels. Due to the large size of the original image, in order to ensure the efficiency of image processing, the original images were preprocessed, and the noise in the image was removed at the same time [15]. The final image obtained had a size of 224 × 224 pixels. The defects of *green plums* were divided into four categories according to the degree of damage, from heavy to shallow: rot, flaws, scars, and rain spots. Among them, the rain spot defect had the characteristics of smallness, light color, and dispersion and occupied a small number of pixels in the image; thus, it is not easy to identify or misidentify [16]. At the same time, some plum pictures contained fruit stems from *green plums*. Although the feature of fruit stems is not a defect, it is affected by factors such as image acquisition angle, light changes, and lens distortion [17], resulting in the color and shape of fruit stems and rain spots. Consequently, the recognition of rain spots was disturbed. In Zhou H. Y.'s [12] previous *green plum* defect detection method, the algorithm (an improved VGG network model) did not yet solve the problem of misjudging fruit stems as defective rain spots. Traditional visual detection algorithms still have poor accuracy and limitations with fruit stems and rain spots [18], resulting in misjudgments of defects. Moreover, in the VIT model used by Zhang Xiao [19], compared to the recognition accuracy of other defects, the recognition error rate of rain spots was the highest, reaching 2.62%, which lowered the overall recognition accuracy. In order to avoid the misjudgment of rain spots and fruit stems and achieve higher recognition accuracy, the characteristic fruit stems were divided into one category for training. To sum up, *green plums* could be classified into the following six categories: scars, rain spots, flaws, rot, intact, and fruit stems, as shown in Figure 2.

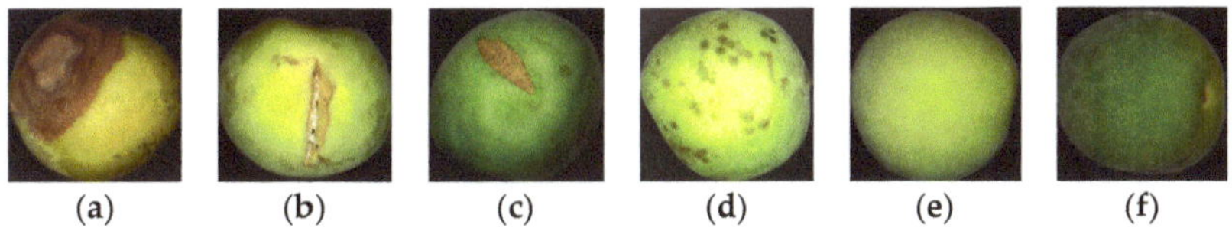

Figure 2. *Green plum* surface defect classification chart: (**a**) rot; (**b**) flaw; (**c**) scar; (**d**) spot; (**e**) intact; (**f**) stem.

2.2. Dataset Processing Methods

In terms of the classification method of the data set, the previous *green plum* defect research team chose to divide the *green plum* defects into four categories: rot, flaws, scars, and rain spots. When faced with *green plums* with multiple defects, they did not use the hazards of the defects as the classification standard. They chose the defect with the largest area as the defect class for the *green plums*. However, if a more harmful defect appeared in a small area, the final output could still be the defect in a larger area, ignoring the harm of other defects to the *green plums*. Moreover, small-area defects occupy fewer pixels, and training features may be lost after repeated convolution and pooling operations during training. This is also one of the reasons for the poor recognition effect of previous rain spot training. In contrast to the above classification methods, it was considered that in the actual detection process, multiple defects might appear on a *green plum*, as shown in Figure 3. In order to express and output the multiple defects of *green plums* more clearly, the dataset was divided more carefully. The sum of defects on each *green plum* picture was used as its defect category, as shown in Figure 3, which contains flaws, stems, and rain spots; then, this picture was used as the flaw + stem + spot category. In the actual detection, there were very few *green plums* with more than three kinds of defects, and some plum categories had only a few pictures or even none. In order to ensure the quality of the dataset, according to the hazard based on a combination of harmfulness and quantity, the following 18 theoretical types of combination classes were finally obtained: scar—1, scar + rot—2, scar + stem—3, scar + stem + spot—4, scar + spot—5, rot—6, rot + flaw—7, rot + stem—8, rot + stem + spot—9, rot + spot—10, intact—11, flaw—12, flaw + stem—13, flaw + stem + spot—14, flaw + spot—15, stem—16, stem + spot—17, spot—18, carried out with these types using order numbers 1–18 (Note: for concise expression, the category names and numbers in the following diagrams, such as the confusion matrix, correspond one-to-one). The combined images of 18 types of *green plum* defects are shown in Figure 4. They were divided into a training set, a test set, and a validation set. The images in the test set and validation set do not intersect. In addition, in order to ensure the quality of the dataset, data enhancement was performed on it, and operations such as mirroring, rotating, and adjusting the brightness and contrast of the original picture were performed. Finally, a total of 27,990 *green plum* sample pictures were obtained. The dataset was divided into a training set, a test set, and a validation set in the ratio of 8:1:1 and then enhanced. The category distribution of the dataset after image enhancement is shown in Table 1. Putting it into the VIT model, the VIT model could effectively learn various types of defect features and finally output all the defects of *green plums*. It only needed to classify them to meet the needs of improving the productivity of *green plums*. According to the degree of harmfulness of the defect, as long as the *green plums* with rot and flaws were listed as a hazard, as this type of *green plum* seriously affects food safety and the manufacturer can use it as fertilizer after fermentation, the *green plums* with scars and rain spots were listed as defective. Plums can be used as raw materials for secondary production. Fruit stems and perfect *green plums* are listed as normal plums, which can be further processed or sold directly.

Figure 3. *Green plum* map with multiple defects (flaw + stem + spot—13).

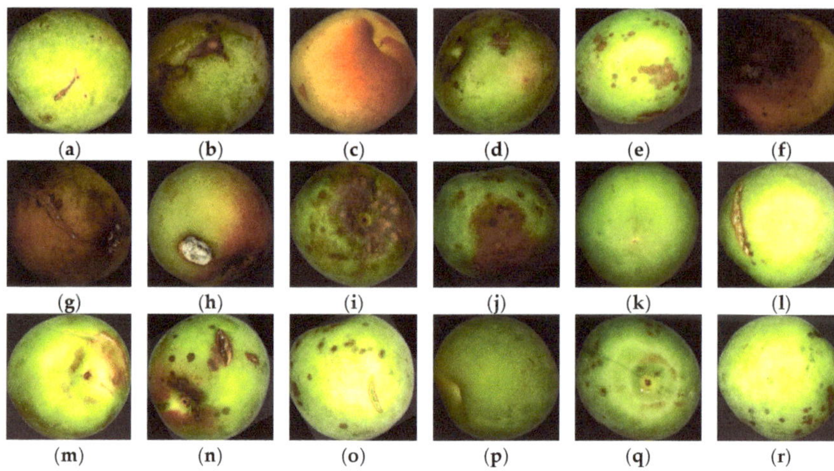

Figure 4. Shown are the 18 categories of *green plum* multi-defect classification: (**a**) scar; (**b**) scar + rot; (**c**) scar + stem; (**d**) scar + stem + spot; (**e**) scar + spot; (**f**) rot; (**g**) rot + flaw; (**h**) rot + stem; (**i**) rot + stem + spot; (**j**) rot + spot; (**k**) intact; (**l**) flaw; (**m**) flaw + stem; (**n**) flaw + stem + spot; (**o**) flaw + spot; (**p**) stem; (**q**) stem + spot; (**r**) spot.

2.3. Multiple Defect Detection Model of Green Plum Based on Vision Transformer

The vision transformer network [20] adopts self-attention and multi-head attention mechanisms, using residual connection and layer normalization techniques to accelerate training. The self-attention mechanism obtains information for each position in the input sequence. Among them, self-supervised learning reduces VIT's dependence on large-scale training [21]. Therefore, this study chose to use this network model to identify the *green plum* defects. In the field of image classification, the common convolutional neural network (CNN) [22,23] uses continuous stacking convolution layer operations to extract local features, which has certain limitations in extracting global features. As an encoder–decoder architecture based on the self-attention mechanism [24,25], the vision transformer model does not use RNN (cyclic neural network) sequential structure parallel training and can reflect complex spatial transformations and long-distance feature dependencies. Through the softmax function, the gradient is reduced. With the multiple sets of independent weights and parameter quantities added to the multi-head attention mechanism [26], the information obtained by different learning methods is combined, improving the expression ability of the network. Its global feature representation ability is stronger, and the migration effect is better.

Table 1. Distribution of dataset.

Number	Class	Original Data Set	Data Augmentation	Validation Set	Training Set	Test Set
1	scar	168	1680	168	1344	168
2	scar + rot	34	340	34	272	34
3	scar + stem	312	3120	312	2496	312
4	scar + stem + spot	37	370	37	296	37
5	scar + spot	18	180	18	144	18
6	rot	546	5460	546	4368	546
7	rot + flaw	62	620	62	496	62
8	rot + stem	130	1300	130	1040	130
9	rot + stem + spot	67	670	67	536	34
10	rot + spot	178	1780	178	1424	178
11	intact	616	6160	616	4928	616
12	flaw	114	1140	114	912	114
13	flaw + stem	62	620	62	496	62
14	flaw + stem + spot	30	300	30	240	30
15	flaw + spot	23	230	23	184	23
16	stem	54	540	54	432	54
17	stem + spot	60	600	60	480	60
18	spot	288	2880	288	2304	288

The VIT model consists of three modules: the linear projection of flattened patches (embedding layer), the transformer encoder, and the multilayer perceptron (MLP) head. The input image (224 pixels × 224 pixels) first passes through the embedding layer and is divided into 196 patches according to the size of 16 × 16. This step is realized by a convolution operation with a convolution kernel size of 16 × 16, a step size of 16, and a number of 768. The methods of adding position embedding and patch embedding can better reflect the information of the whole image. Secondly, the data enters the transformer encoder layer. The encoder contains a multi-head attention mechanism, which can represent the global features more accurately and repeatedly stack the encoder block L times. The output shape after the transformer encoder is consistent with the input shape. Finally, the defect classification result for *green plums* is obtained through the linear output in the MLP head [27]. Among them, the calculation formula for multi-head self-attention is as follows:

$$\mathrm{MultiHead}(\mathrm{Q}, \mathrm{K}, \mathrm{V}) = \mathrm{Concat}(\mathrm{head}_1, \cdots, \mathrm{head}_\mathrm{h})\mathrm{W}^\mathrm{O} \quad (1)$$

where Q, K, V, H, and W^O represent the query vector, key vector, value vector, number of heads, and output transformation matrix, respectively.

In Formula (1), the output head_i of each head can be expressed as follows:

$$\mathrm{head}_\mathrm{i} = \mathrm{Attention}\left(\mathrm{QW}_\mathrm{i}^\mathrm{Q}, \mathrm{KW}_\mathrm{i}^\mathrm{K}, \mathrm{VW}_\mathrm{i}^\mathrm{V}\right) \quad (2)$$

In Formula (2), W_i^Q, W_i^K, and W_i^V represent the query, key, and value transformation matrix of the head_i, respectively. The self-attention calculation formula is as follows:

$$\mathrm{Attention}\ (\mathrm{Q}, \mathrm{K}, \mathrm{V}) = \mathrm{Softmax}\ (\frac{\mathrm{QK}^\mathrm{T}}{\sqrt{\mathrm{d}_\mathrm{k}}})\mathrm{V} \quad (3)$$

Although the classification method in this study could facilitate the network to learn various defects and output all the defects of *green plums*, it was also necessary to output the main defects according to the degree of defect damage. This study proposed primary defect detection based on previous research, enabling the new VIT network to output the first hazard defect of the *green plum* as the major defect based on the defect area size and hazard level. A hazard level analysis layer was added after the MLP output layer to obtain the main defects more accurately. In this layer, a convolutional neural network was inserted into the network. It was used for training in the identification of the hazards of each defect. Different degrees and different types of defects have different effects on *green plums*. The network layer obtains the defect hazard factors of *green plums* and judges the degree of influence of the hazard on *green plums* through factor size analysis in order to determine the main defects of *green plums*. After analyzing the degree of harm, the entire network outputs the main defects of *green plums* more precisely. The main defect detection structure diagram is shown in Figure 5.

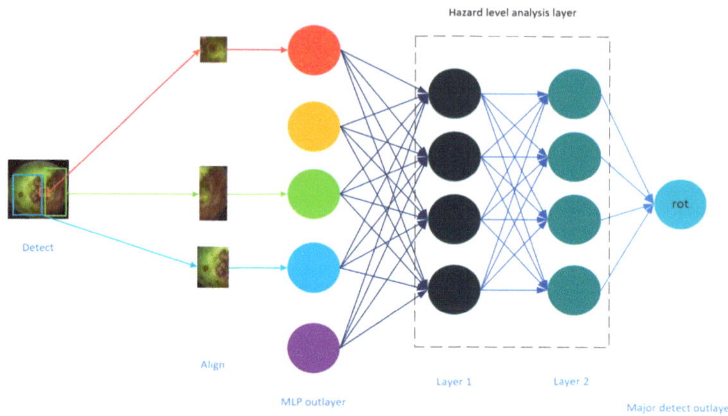

Figure 5. Major defect detection structure diagram.

The improved network could identify the most harmful defect features, but there may be multiple defects on the surface of *green plum*, causing the network to ignore the remaining defects and the harm they bring. In order to solve the multi-defect detection problem of *green plums*, the network structure was improved, and a fine-grained multi-defect detection link was added after the MLP. This link existed in parallel with the risk level analysis to identify all the defects for *green plums* and output them. In this link, multiple defects on *green plums* are first identified, the confidence of the corresponding defects in the graph is calculated, the confidence threshold is set to 0.6, and the confidence of defects higher than the threshold is output as secondary defects. If there are multiple secondary defects, the output sequence is in order of the degree of harm; finally, the network model can output all the defects in *green plums*, major defects + minor defects. The structure diagram of the whole network is shown in Figure 6.

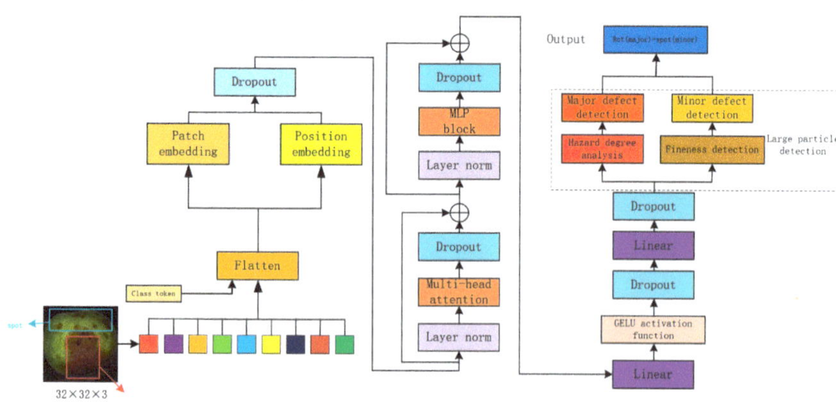

Figure 6. Network flowchart.

3. Results

The *green plum* defect classification network built in this research used the deep learning framework PyTorch to define the network calculation graph. The hardware, software, and compilation environment configurations used in this study are shown in Table 2.

Table 2. Software and hardware environment configuration.

Software and Hardware	Name
System	Windows 10 × 64
CPU	Inter I7 11700K@3.6 GHz
GPU	Nvidia GeForce RTX 3080Ti(12G)
Environment configuration	PyCharm 2022.3.3 + Pytorch 1.7.1 + Python 3.7.7 Cuda 10.2 + cudnn 7.6.5 + tensorboardX 2.1

Before training, the *green plum* defect classification network was parameterized. Batch_size was set to 64, Heads (the number of "heads" in the multi-head attention) was 2, Mlp_dim (the number of neurons in the hidden layer in the multilayer perceptron) was 64, and the learning rate parameter in the Adam optimizer was set to 0.01. After the parameters were set, the dataset was input into the model for training until the loss reached the minimum value and remained stable for 30 epochs. At this stage, the training of the *green plum* defect classification network model was completed. The *green plum* data from the test set were imported into the trained *green plum* defect classification network model. The model generated the test results for the main defects through the risk level analysis layer. As shown in Table 3, the accuracy rate of the VIT network for the classification of the main defects on the *green plum* surface reached 96.21%.

Table 3. Results of *green plum* defect classification.

Methods		Vision Transformer
Major Defect Classification Accuracy	Scar	94.02%
	Rot	98.62%
	Intact	93.89%
	Flaw	96.42%
	Spot	93.68%
Accuracy		96.21%
Loss		0.078

Inputting 2799 test set pictures into the fine-grained defect detection link for testing, the confusion matrix of multi-defect detection on the *green plum* surface obtained by the VIT network is shown in Figure 7. The VIT model had the best detection effect on scar + spot and rot + stem + spot, with an accuracy of 100%. The effects of the intact category, the scar + stem + spot category, and the stem + spot category were poor. Among the 616 intact pictures, 1 was misjudged as the scar category, 4 were misjudged as the scar + stem category, and 15 were misjudged for rain spots. Among the 37 pictures in the scar + stem + spot category, 1 was identified as a spot, 1 was misjudged as a rot + stem, 1 was misjudged as a flaw + spot, and 6 were identified as a stem + spot kind. Among the 60 pictures of stem + spot, 4 were misjudged as scar + stem + spot, 1 was misjudged as a flaw, and 7 pictures were misjudged as spot without the characteristics of the stem.

Figure 8 is a test result diagram of a part of the test set, and the colored boxes in the figure show some misjudged plum cases. Picture (11 -> 18) in the green box in the figure misjudged the intact class as a spot. This may have occurred because the fruit tip of the intact *green plum* turned yellow, the rain spot was a small target defect, and its shape and color were similar to the fruit tip, resulting in misjudgment. The pictures in the purple frame (12 -> 13) identified the flaw as a flaw + stem category, which may have been caused by the fact that the pulp at the flaw was oxidized by air and was similar in color to the fruit stems. The rain spot fruit stems in the yellow and red boxes are, respectively, identified as a rain spot and a flaw. The comprehensive analysis showed that because the rain spot defect was too dense, small, and round in shape, resulting in a misjudgment of the fruit stem, the rain spot defect was similar to a flaw when it was distributed laterally. The pictures in the blue box (1 -> 18) classified the scars as rain spots because the small scars were similar to rain spots, which led to their misjudgment.

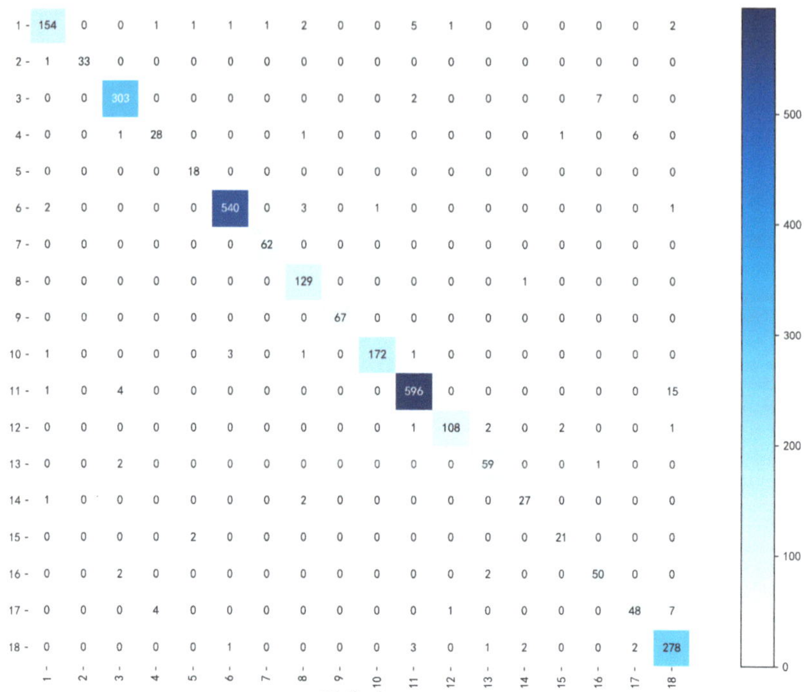

Figure 7. Confusion matrix. Type correspondence: scar—1, scar + rot—2, scar + stem—3, scar + stem + spot—4, scar + spot—5, rot—6, rot + flaw—7, rot + stem—8, rot + stem + spot—9, rot + spot—10, intact—11, flaw—12, flaw + stem—13, flaw + stem + spot—14, flaw + spot—15, stem—16, stem + spot—17, spot—18.

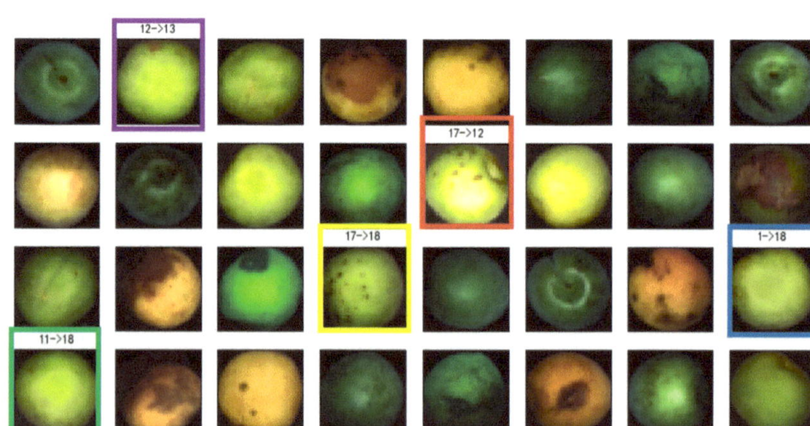

Figure 8. Results of the test.

4. Discussion

In this study, the multi-defect detection of *green plums* was a classification task. Compared with the target detection algorithm, the VIT model did not need to label each defect in each picture in the dataset. It only needed to classify different *green plum* defect pictures in the dataset. The target detection algorithm, such as the YOLO series model, needed to use labeling software to frame the defect area in each picture. Multiple defects often overlapped when labeling, resulting in repeated labeling of the frame, as shown in Figure 9. The yellow box in Figure 9 shows the stem, the blue box shows the scar, and the red box shows the rain spot. The rain spot feature and the rot feature overlapped. Furthermore, the process of manual labeling is a subjective job, after all, so there is also a certain error rate that will interfere with subsequent training. This not only consumes a lot of time but also leads to a decrease in recognition accuracy. Moreover, the purpose of this defect detection process was not to determine the exact position of the *green plum* [28], but only to identify the defect type of the *green plum*. Therefore, there was no need to label and locate defects in the dataset. Regardless of the perspective of dataset production or the final research goal, the target detection algorithm was unsuitable for this research.

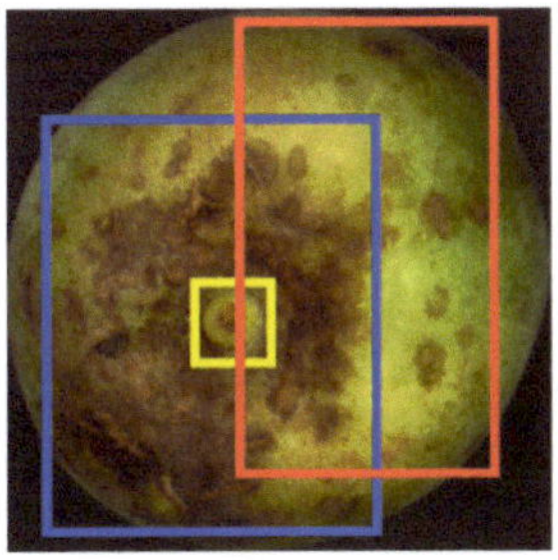

Figure 9. *Green plum* defect picture.

This study compared the performance of the *green plum* multi-defect classification network with that of other networks such as ResNet18, WideResNet50, Desnet121, and VGG16. This was conducted to validate the performance of the *green plum* network further. After the above models were fully trained, the accuracy rate of the main defect classification and the average test time were used as performance index comparisons. The test results are shown in Table 4.

Table 4. Accuracy of surface main defect classification for *green plum*.

	Accuracy of Surface Defect Classification					Accuracy	Average Test Time
Network Name	**Scar**	**Rot**	**Intact**	**Flaw**	**Spot**		
ResNet18	86.54%	90.95%	79.04%	94.18%	93.10%	89.92%	0.88 ms
WideResNet50	89.53%	89.68%	94.48%	86.03%	88.51%	91.39%	1.05 ms
Desnet121	93.83%	92.63%	96.57%	89.52%	97.70%	94.14%	1.39 ms
VGG16	92.34%	95.18%	98.06%	90.83%	97.17%	95.42%	0.96 ms
Vision Transformer	94.02%	98.62%	93.89%	96.42%	93.68%	96.21%	1.43 ms

In Table 4, in the classification of main surface defects of *green plums*, the accuracies of the VIT model for the main defects of scars, rot, intact, flaws, and rain-spotted *green plum* images reached 94.02%, 98.62%, 93.89%, 96.42%, and 93.68%, respectively. The average discrimination accuracy rate of the network was 96.21%, and the processing time of a single image was 1.43 ms. The accuracy rate of all kinds of main defect discrimination was significantly better than other models, such as WideResNet50. The VIT model could also identify other defects in *green plums*.

In terms of model accuracy, compared with ResNet18 [29] and WideResNet50, the vision transformer had a larger lead in the accuracy of the *green plum* multi-defect detection task. The vision transformer was higher than ResNet18 and WideResNet50 by 6.29% and 4.82%, respectively, and slightly ahead of the Desnet121 and VGG16 models [30] (2.07% and 0.79%). The detection accuracy for scars, rot, and flaws was higher than that of other models, and the accuracy regarding intact and rain spots was slightly lower than that of Desnet121 and VGG16 [31]. The overall effect of VIT was better. However, in terms of image processing time, since a hazard level analysis layer and a fine-grained detection link were added to the model, the VIT model took 0.55 ms longer to process a single image than the fastest ResNet18 but could obtain a high recognition rate of detection of main defects and multiple defects.

The loss curves of the vision transformer, ResNet18, WideResNet50, Desnet121, and VGG16 networks are shown in Figure 10. Although the VIT model outperformed other models' accuracy, its convergence speed during training was obviously not as good as other models. It may be that the effect of the optimizer in the VIT model was not as good as that of other models. In subsequent studies, we may consider replacing the optimizer with one that is more suitable for the VIT model in order to promote earlier convergence of the model and improve the efficiency of training.

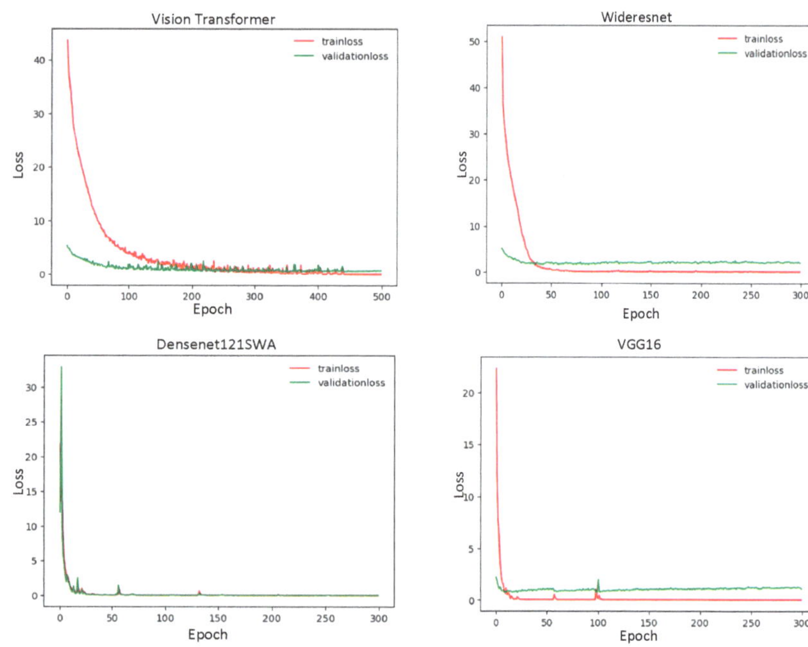

Figure 10. Loss curves of the four models.

Comprehensive analysis shows that the VIT network demonstrated excellent classification performance. To sort the quality of *green plums* based on the VIT network model and classify *green plums* with multiple flaws, the softmax function was used to reduce the gradient, and the multi-head attention mechanism was added. The overall feature representation ability is stronger, resulting in improved feature learning and migration effects. As a result, the network learns more features of defects, increasing the feature recognition rate. Consequently, the network performs better in the multi-defect classification of *green plums*. The average discrimination accuracy of the final model was 96.21%. This method not only accurately identifies the main defects of *green plums* but also classifies and outputs the defects in a more detailed manner and completes the multi-defect detection task of *green plums*. Manufacturers can consider the rational use of defective *green plums* that can ensure food safety according to the defect situation of *green plums*. They can further classify multi-category *green plums* according to their own needs, which greatly improves the utilization rate of *green plums* and can increase the profit margins of enterprises.

5. Conclusions

Some researchers have previously conducted research on the defects of *green plums* and were able to identify the main defects. However, each *green plum* may have more than one defect. Their methods could only identify a single type of defect due to the lack of a detailed classification of *green plum* defects. The previous *green plum* defect detection methods could not detect other defects in *green plums*. This study proposed a model for the detection and classification of *green plum* multi-defects based on the vision transformer, aiming at the problem of multiple defects on the surfaces of *green plums*. This vision transformer was based on the four main defects of *green plums* (scars, rot, flaws, and spots) and a class of features (stems). There were 2799 *green plum* pictures classified with multiple defects to obtain a more detailed dataset, divided into 18 categories according to the actual situation. Moreover, the training set, validation set, and test set were allocated according to a ratio of 8:1:1. Then the dataset was expanded by changing parameters such as image angle, contrast, brightness, etc. to ensure the quality of the dataset and adding a risk level analysis layer and fine-grained detection links. The model was trained with the improved network. The network realized the effective classification of the main defects and multiple defects on the *green plum* surface, and the average recognition accuracy rate reached 96.21%. The single test image processing time was 1.43 ms.

This study also compared the established network with the accuracy of various major defects and the training loss curves of the ResNet18, Desnet121, WideResNet50, and VGG16 networks. The superiority of the vision transformer network was verified in defect classification performance compared to other network methods. It completed the automatic detection of multiple types of defects on the surfaces of *green plums* and classified the defect levels of *green plums* more carefully. However, there is still room for optimization in the training speed of the model. In addition, the more detailed classification method for *green plum* surface defects used in this study can also be applied to the defect detection of other fruits. This can help manufacturers further classify defective fruits and improve the utilization of non-hazardous and minimally hazardous fruits, thereby increasing production profit.

This research was based on static *green plum* surface images, and a static *green plum* surface multi-defect classification model was constructed based on the vision transformer model, achieving good surface multi-defect classification results. However, the training efficiency of the model was not high enough. This can be improved by changing the optimizer to accelerate the convergence speed of the model. The surface defect detection method used in this study could not understand the chemical composition of *green plums*, such as sugar content, pH, soluble solids, etc. It could not identify whether there were internal defects in the *green plums*. Moreover, under static conditions, only one side of the *green plums'* defects could be identified. In actual testing, the conveyor belt can be improved to make the *green plum* rotate continuously during transportation, allowing the camera to recognize all defects on the plum. In subsequent research, we should study how to identify internal defects in *green plums* in order to achieve higher food safety rates. Additionally, high-spectral imaging technology can be used to obtain the internal chemical components of *green plums* to select high-quality *green plums*.

Author Contributions: Conceptualization, W.S. and Y.Y.; data curation, W.S.; formal analysis, C.Z.; funding acquisition, Y.L.; investigation, Y.L. and Y.Y.; methodology, W.S.; project administration, Y.L.; resources, Y.Y.; software, W.S., Y.Y. and C.Z.; supervision, Z.Z.; validation, Y.L.; visualization, W.S.; writing—original draft, W.S.; writing—review and editing, Y.Y. and Y.L. All authors have read and agreed to the published version of the manuscript.

Funding: This research was funded by the Jiangsu agricultural science and technology innovation fund project (Project#: CX(18)3071): research on key technologies of intelligent sorting for *green plums*; and the 2020 Jiangsu graduate research and innovation plan (Project#: KYCX20_0882): research on *green plum* defect sorting systems based on artificial intelligence.

Data Availability Statement: Not applicable.

Acknowledgments: In addition to the funds we received, we would also like to thank Haiyan Zhou for providing us with materials and support for the image collection.

Conflicts of Interest: The authors declare no conflict of interest.

References

1. Fyfe, S.; Schirra, H.J.; Rychlik, M.; van Doorn, A.; Tinngi, U.; Sultanbawa, Y.; Smyth, H.E. Future flavours from the past: Sensory and nutritional profiles of green plum (*Buchanania obovata*), red bush apple (*Syzygium suborbiculare*) and wild peach (*Terminalia carpentariae*) from East Arnhem Land, Australia. *Future Foods* **2022**, *5*, 100136. [CrossRef]
2. Mahanti, N.K.; Pandiselvam, R.; Kothakota, A.; Ishwarya, S.P.; Chakraborty, S.K.; Kumar, M.; Cozzolino, D. Emerging non-destructive imaging techniques for fruit damage detection: Image processing and analysis. *Trends Food Sci. Technol.* **2022**, *120*, 418–438. [CrossRef]
3. Cubero, S.; Aleixos, N.; Molto, E.; Gomez-Sanchis, J.; Blasco, J. Advances in Machine Vision Applications for Automatic Inspection and Quality Evaluation of Fruits and Vegetables. *Food Bioprocess Technol.* **2011**, *4*, 829–830. [CrossRef]
4. Bhargava, A.; Bansal, A.; Goyal, V. Detection and Grading of Different Vegetable Using Machine Learning. *Adv. Data Inf. Sci.* **2022**, *318*, 9. [CrossRef]
5. Bhargava, A.; Bansal, A.; Goyal, V. Machine Learning-Based Detection and Sorting of Multiple Vegetables and Fruits. *Food Anal. Meth.* **2022**, *15*, 15. [CrossRef]
6. Knott, M.; Perez-Cruz, F.; Defraeye, T. Facilitated machine learning for image-based fruit quality assessment. *J. Food Eng.* **2023**, *345*, 9. [CrossRef]
7. Nturambirwe, J.F.I.; Hussein, E.A.; Vaccari, M.; Thron, C.; Perold, W.J.; Opara, U.L. Feature Reduction for the Classification of Bruise Damage to Apple Fruit Using a Contactless FT-NIR Spectroscopy with Machine Learning. *Foods* **2023**, *12*, 210. [CrossRef] [PubMed]
8. Jiang, H.; Jiang, X.; Ru, Y.; Chen, Q.; Li, X.; Xu, L.; Zhou, H.; Shi, M. Rapid and non-destructive detection of natural mildew degree of postharvest *Camellia oleifera* fruit based on hyperspectral imaging. *Infrared Phys. Technol.* **2022**, *123*, 11. [CrossRef]
9. Yao, J.; Qi, J.M.; Zhang, J.; Shao, H.M.; Yang, J.; Li, X. A Real-Time Detection Algorithm for Kiwifruit Defects Based on YOLOv5. *Electronics* **2021**, *10*, 1711. [CrossRef]
10. Nithya, R.; Santhi, B.; Manikandan, R.; Rahimi, M.; Gandomi, A.H. Computer Vision System for Mango Fruit Defect Detection Using Deep Convolutional Neural Network. *Foods* **2022**, *11*, 3483. [CrossRef]
11. Huang, Y.P.; Yang, Y.T.; Sun, Y.; Zhou, H.Y.; Chen, K.J. Identification of Apple Varieties Using a Multichannel Hyperspectral Imaging System. *Sensors* **2020**, *20*, 5120. [CrossRef] [PubMed]
12. Zhou, H.Y.; Zhuang, Z.L.; Liu, Y.; Liu, Y.; Zhang, X. Defect Classification of Green Plums Based on Deep Learning. *Sensors* **2020**, *20*, 6933. [CrossRef] [PubMed]
13. Zhou, C.X.; Wang, H.H.; Liu, Y.; Ni, X.Y.; Liu, Y. Green Plums Surface Defect Detection Based on Deep Learning Methods. *IEEE Access* **2022**, *10*, 100397–100407. [CrossRef]
14. Zhu, Z.; Zhao, Y.; Guo, Y.; Zhang, R.; Pan, Y.; Zhou, T. A novel additional carbon source derived from rotten fruits: Application for the denitrification from mature landfill leachate and evaluation the economic benefits. *Bioresour. Technol.* **2021**, *334*, 125244. [CrossRef]
15. Bhargava, A.; Bansal, A. Fruits and vegetables quality evaluation using computer vision: A review. *J. King Saud Univ. Comput. Inf. Sci.* **2021**, *33*, 243–257. [CrossRef]
16. Sun, M.L.; Zhao, R.N.; Yin, X.; Xu, L.C.; Ruan, C.Z.; Jia, W.K. FBoT-Net: Focal bottleneck transformer network for small green apple detection. *Comput. Electron. Agric.* **2023**, *205*, 107609. [CrossRef]
17. Xie, W.J.; Wang, F.H.; Yang, D.Y. Research on Carrot Grading Based on Machine Vision Feature Parameters. *IFAC Pap.* **2019**, *52*, 30–35. [CrossRef]
18. Liang, C.X.; Xiong, J.T.; Zheng, Z.H.; Zhong, Z.; Li, Z.H.; Chen, S.M.; Yang, Z.G. A visual detection method for nighttime litchi fruits and fruiting stems. *Comput. Electron. Agric.* **2020**, *169*, 105192. [CrossRef]
19. Zhang, X.; Huo, L.T.; Liu, Y.; Zhuang, Z.L.; Yang, Y.T.; Gou, B.L. Research on 3D Phenotypic Reconstruction and Micro-Defect Detection of Green Plum Based on Multi-View Images. *Forests* **2023**, *14*, 218. [CrossRef]
20. Guo, Z.X.; Wang, C.S.; Yang, G.; Huang, Z.Y.; Li, G. MSFT-YOLO: Improved YOLOv5 Based on Transformer for Detecting Defects of Steel Surface. *Sensors* **2022**, *22*, 3467. [CrossRef]
21. Liu, Y.; Wang, H.H.; Fei, Y.Q.; Liu, Y.; Shen, L.X.; Zhuang, Z.L.; Zhang, X. Research on the Prediction of Green Plum Acidity Based on Improved XGBoost. *Sensors* **2021**, *21*, 930. [CrossRef]

22. Xie, W.J.; Wei, S.; Zheng, Z.H.; Yang, D.Y. A CNN-based lightweight ensemble model for detecting defective carrots. *Biosyst. Eng.* **2021**, *208*, 287–299. [CrossRef]
23. Lu, P.; Zhao, Y.; Xu, Y. A Two-Stream CNN Model with Adaptive Adjustment of Receptive Field Dedicated to Flame Region Detection. *Symmetry* **2021**, *13*, 397. [CrossRef]
24. Jannat, F.E.; Willis, A.R. *Improving Classification of Remotely Sensed Images with the Swin Transformer*; IEEE: Piscataway, NJ, USA, 2022; pp. 611–618.
25. Li, A.; Zhao, Y.; Zheng, Z. Novel Recursive BiFPN Combining with Swin Transformer for Wildland Fire Smoke Detection. *Forests* **2022**, *13*, 2032. [CrossRef]
26. Chen, R.Y.; Zhan, S.Y.; Chen, Y. *Underwater Target Detection Algorithm Based on YOLO and Swin Transformer for Sonar Images*; IEEE: Piscataway, NJ, USA, 2022.
27. Zheng, H.; Wang, G.H.; Li, X.C. Swin-MLP: A strawberry appearance quality identification method by Swin Transformer and multi-layer perceptron. *J. Food Meas. Charact.* **2022**, *16*, 2789–2800. [CrossRef]
28. Liang, X.T.; Jia, X.Y.; Huang, W.Q.; He, X.; Li, L.J.; Fan, S.X.; Li, J.B.; Zhao, C.J.; Zhang, C. Real-Time Grading of Defect Apples Using Semantic Segmentation Combination with a Pruned YOLO V4 Network. *Foods* **2022**, *11*, 3150. [CrossRef] [PubMed]
29. Yang, Y.T.; Wang, H.H.; Jiang, D.; Hu, Z.K. Surface Detection of Solid Wood Defects Based on SSD Improved with ResNet. *Forests* **2021**, *12*, 1419. [CrossRef]
30. Ding, F.L.; Zhuang, Z.L.; Liu, Y.; Jiang, D.; Yan, X.A.; Wang, Z.G. Detecting Defects on Solid Wood Panels Based on an Improved SSD Algorithm. *Sensors* **2020**, *20*, 5315. [CrossRef] [PubMed]
31. Xie, W.J.; Wei, S.; Zheng, Z.H.; Jiang, Y.; Yang, D.Y. Recognition of Defective Carrots Based on Deep Learning and Transfer Learning. *Food Bioprocess Technol.* **2021**, *14*, 1361–1374. [CrossRef]

forests

MDPI

Article

Assessment of Forest Ecological Function Levels Based on Multi-Source Data and Machine Learning

Ning Fang [1,2,3], Linyan Yao [4], Dasheng Wu [1,2,3], Xinyu Zheng [1,2,3] and Shimei Luo [5,*]

1 College of Mathematics and Computer Science, Zhejiang A&F University, Hangzhou 311300, China
2 Key Laboratory of State Forestry and Grassland Administration on Forestry Sensing Technology and Intelligent Equipment, Hangzhou 311300, China
3 Key Laboratory of Forestry Intelligent Monitoring and Information Technology of Zhejiang Province, Hangzhou 311300, China
4 Zhejiang Public Information Industry Co., Ltd., Hangzhou 310000, China
5 College of Economic and Management, Zhejiang A&F University, Hangzhou 311300, China
* Correspondence: luosm@zafu.edu.cn

Abstract: Forest ecological function is one of the key indicators reflecting the quality of forest resources. The traditional weighting method to assess forest ecological function is based on a large amount of ground survey data; it is accurate but costly and time-consuming. This study utilized three machine learning algorithms to estimate forest ecological function levels based on multi-source data, including Sentinel-2 optical remote sensing images and digital elevation model (DEM) and forest resource planning and design survey data. The experimental results showed that Random Forest (RF) was the optimal model, with *overall accuracy* of 0.82, *recall* of 0.66, and *F1* of 0.62, followed by CatBoost (*overall accuracy* = 0.82, *recall* = 0.62, *F1* = 0.58) and LightGBM (*overall accuracy* = 0.76, *recall* = 0.61, *F1* = 0.58). Except for the indicators from remote sensing images and DEM data, the five ground survey indicators of forest origin (QI_YUAN), tree age group (LING_ZU), forest category (LIN_ZHONG), dominant species (YOU_SHI_SZ), and tree age (NL) were used in the modeling and prediction. Compared to the traditional methods, the proposed algorithm has lower cost and stronger timeliness.

Keywords: multi-source data; machine learning; forest ecological function level; forest ecological function index

Citation: Fang, N.; Yao, L.; Wu, D.; Zheng, X.; Luo, S. Assessment of Forest Ecological Function Levels Based on Multi-Source Data and Machine Learning. *Forests* **2023**, *14*, 1630. https://doi.org/10.3390/f14081630

Academic Editor: Russell G. Congalton

Received: 26 June 2023
Revised: 24 July 2023
Accepted: 10 August 2023
Published: 12 August 2023

1. Introduction

As one of the most important components of the ecosystem, forests provide the basic natural resource foundation for the sustainable development of human beings [1]. The forest ecological function index can comprehensively reflect the structure and ecological benefits of forests. Therefore, establishing a scientific and dynamic comprehensive system to evaluate forest ecological functions plays an important role in accurately addressing ecological and economic development [2]. As research progresses, a single analysis of a particular forest characteristic no longer meets the current requirements. Therefore, it is important to assess forests' ecological functions by integrating the synergistic effects of multiple factors [3,4]. Many scholars have attempted to present concepts or tools for forest ecological function analysis [5–7], greatly helping to measure and quantify species' actions.

Currently, some progress in the quantitative estimation of forest ecological functions has been made in China. In terms of research methods, Li Ma et al. established the Beijing Forest Ecosystem Health Evaluation Index System (EIS-BFEH) to evaluate the health function of forest ecosystems and used the hierarchical analysis process (AHP) to obtain a comprehensive index (CI) representing the health status of forest ecosystems [8]. Fang Xiaomin et al. used a comprehensive index method and a statistical grouping method to evaluate forest functions and calculated the forest ecological function index to compare forest ecological functions for different age groups, species structure, and origin [9]. Du Qun et al. proposed that the three factors of forest quantity, quality, and spatial distribution should be used for forest ecological function evaluation [10]. Although there has been some progress in research on computational methods and evaluation factors, the current studies are mainly focused on simple statistical methods, and there is a lack of attempts of using machine learning methods; this procedure cannot meet the requirements of accurately evaluating and monitoring forest ecological functions based on multi-data [11].

Among machine learning algorithms, ensemble learning algorithms can combine multiple classifiers together to improve the accuracy and generalization ability of classifiers. Overall, based on the presence or absence of dependencies between base classifiers, ensemble learning algorithms are divided into two types: boosting and bagging. Boosting algorithms have a strong dependency between base classifiers, and a series of base classifiers needs to be generated serially, represented by AdaBoost and GBDT. In fact, GBDT is more suitable for multi-category classification, and LightGBM and CatBoost are two important improved algorithms based on GBDT. In contrast, bagging algorithms do not have a strong dependency between base classifiers, and a series of base classifiers can be generated in parallel, represented by Random Forest. Regarding the research data, Wang Daling et al. evaluated the forest ecological function of arboreal forests based on the data of subcompartments from the 2018 forest resources planning and design survey of the Sanchazi Forestry Bureau [12]. Jun Yang et al. conducted a qualitative analysis of the health and ecological status of Chinese forests in 2009 based on the data of the seventh national forest resource inventory [13]. Liu Lixia et al. evaluated the ecological functions of forests based on the data of the forest resources planning and design survey [14]. Zhang Xianwu et al. used the results of the three continuous forest inventories from 2004, 2009, and 2014 to analyze the current situation, changes, and reasons of changes in the comprehensive index of forest ecological functions in Shanghai using nine indicators including forest stock, forest naturalness, and proportion of forest land area to national land area [15]. Even though forest resource inventory data has high accuracy, there are still issues of high cost and difficulty in data acquisition. Forest managers, decision makers, and politicians need to be able to make data-driven rapid decisions based on short-term and long-term monitoring information, complex modeling, and analysis approaches [11]. Thus, researchers are increasingly considering incorporating lower-cost data such as remote-sensing images and lower-cost strategies into our study.

It is necessary to attempt to integrate multi-source data to evaluate the forest ecological function. In recent years, the application of multi-source data fusion in the forestry field has become increasingly common, indicating that multi-source data have broad application prospects. For instance, Wang et al. used multi-source remote sensing data (Gaofen 1, Sentinel-2, Landsat 9, and Gaofen 3) to classify mangrove species in urban areas of Leizhou City, Guangdong Province [16]; Abd Rahman Kassim et al. used hyperspectral images and airborne LiDAR data to evaluate the ecological status of the FRIM campus forest ecosystem [17]. Some of the studies on multi-source data fusion in the field of forestry have achieved good results, but there is still more room to explore its application in forest ecological function level assessment.

This paper aimed to combine multi-source data with machine learning algorithms to assess the forest ecological function levels based on the unit of forest subcompartment. Spectral features and topographic features of the study area were collected from remotely sensed images and DEM. Some ground data with low acquisition cost, such as forest origin, tree age group, forest category, dominant species, and tree age, were provided by the forest resource planning and design survey. Based on the multi-source data scheme, three classic machine learning algorithms, I.e., Random Forest, LightGBM, and CatBoost, were involved in the study.

2. Materials and Methods

2.1. Schematic Framework of Materials and Methods

The schematic framework of the Materials and Methods of this study is shown in Figure 1, including multi-source data, preprocessing methods, models to predict forest ecological function levels.

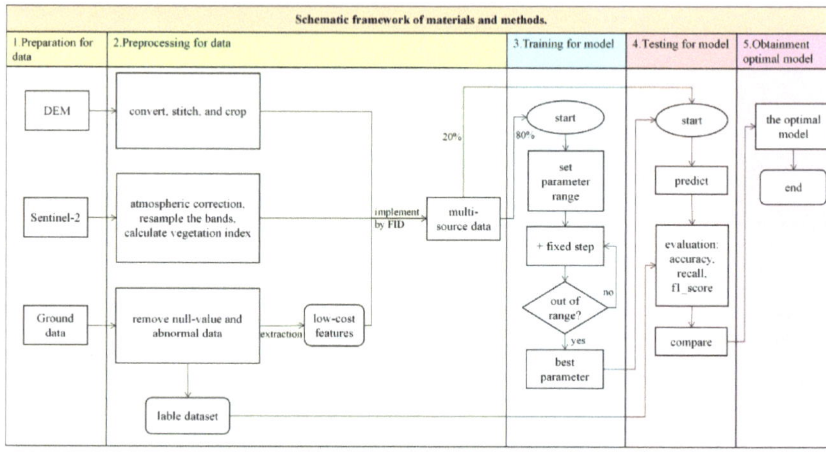

Figure 1. Schematic framework of Materials and Methods.

2.2. Overview of the Study Area

Lin'an District (118°51′~119°52′ E, 29°56′~30°23′ N) is located in the northwest of Zhejiang Province, with a total area of 3126.8 square kilometers, shown in Figure 2. It is in the central subtropical monsoon climate zone, rich in plant resources. As a typical southern forest city, the forestry land area of Lin'an is up to 263,868.79 hm^2, with 1603.88 hm^3 of forest standing stock and 81.99% of forest canopy density.

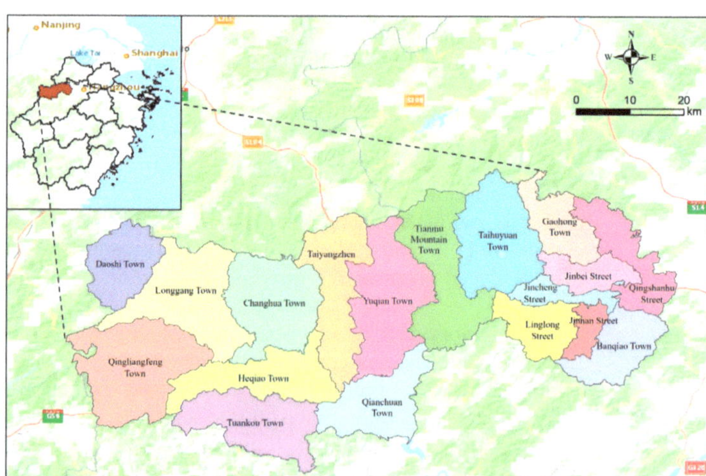

Figure 2. Location map of Lin'an District.

2.3. Processing of Label Dataset

According to the Technical Operation Rules for forest resources planning and design survey of Zhejiang Province of 2014 [18] and the Technical Regulations for the Continuous

Inventory of Forest Resources of 2020 (GB/T 38590-2020) [19], there are eight investigation factors (Table 1), i.e., forest biomass, forest naturalness, forest community structure, tree species structure, vegetation coverage, forest canopy density, mean tree height, and thickness of dead leaves, involved into the traditional algorithm to evaluate a forest ecological function levels. Their weights were determined according to their importance to the evaluation of forest ecological functions (Table 1).

Table 1. Factors to evaluate forest ecological function levels.

Code	Factors	Classification Standards			Weight	References
		I	II	III		
1	Forest biomass (t/hm^2)	≥150	50~149	<50	0.20	[19]
2	Forest naturalness	1, 2	3, 4	5	0.15	
3	Forest community structure	1	2	3	0.15	
4	Tree species structure	6, 7	3, 4, 5	1, 2	0.15	
5	Vegetation coverage (%)	≥70	50~69	<50	0.10	
6	Canopy density	≥0.70	0.40~0.69	0.20~0.39	0.10	
7	Mean tree height/m	≥15.0	5.0~14.9	<5.0	0.10	
8	Thickness of dead leaves	1	2	3	0.05	

To facilitate the data standardization, the values of the above eight evaluation factors were uniformly classified into three classes, I.e., I, II, and III. According to the data type and the distribution of the values, the above eight evaluation factors were divided into three categories. The first category, including vegetation coverage, canopy density, mean tree height, forest community structure, and thickness of dead leaves, was directly classified into types I, II, and III (as shown in Table 1). The second category. consisting of forest naturalness and tree species structure, was classified first by the division standards of Table 2 for naturalness and Table 3 for tree species structure, and then into types of I, II, and III. The third category merely included forest biomass, which was firstly calculated according to the amount of forest volume for various dominant species by equations (shown in Table 4) and then classified into types I, II, and III.

Table 2. Criteria and codes for the classification of naturalness in the continuous inventory of forest resources.

Naturalness	Division Standard	Code	References
I	Forest types are pristine or in a largely untouched state, with little human influence.	1	[19]
II	Natural forest types with obvious human interference or secondary forest types in the later stage of succession, mainly consisting of tree species with high adaptability at the top level of zonality.	2	
III	A secondary forest type with great human disturbance, in the late stage of secondary succession. In addition to pioneer species, top-level species can also be seen.	3	
IV	Highly disturbed by humans, succession retrograde, is in an extremely fragile secondary forest stage.	4	
V	Highly and continuously disturbed by humans, with the destruction of almost all zonal forest types, in the late stage of difficult-to-recover retrograde succession.	5	

Table 3. Criteria and codes for the classification of forest species structure.

Tree Species Structure Type	Division Standard	Code	References
I	Pure coniferous forests, where the volume of individual coniferous species is greater than or equal to 90% of the total volume.	1	[19]
II	Pure broadleaved forests, where the volume of individual broadleaved species is greater than or equal to 90% of the total volume.	2	
III	Relatively pure coniferous forest, where the volume of individual coniferous species is greater than or equal to 65% and less than 90% of the total volume.	3	
IV	Relatively pure broad-leaved forests, where the volume of individual broad-leaved species is greater than or equal to 65% and less than 90% of the total volume.	4	
V	Mixed coniferous forests, where the volume of total coniferous species is greater than or equal to 65% of the total volume.	5	
VI	Mixed coniferous and broad-leaved forests, where the volume of total coniferous species or total broad-leaved species is greater than or equal to 35% and less than 65% of the total volume.	6	
VII	Broad-leaved mixed forests, where the volume of total broad-leaved species is greater than or equal to 65% of the total volume.	7	

Table 4. Biomass models of major tree species and vegetation types.

Code	Tree Species/Vegetation Type	Biomass Model	References
1	Cunninghamia lanceolata	W = 0.3999 V + 22.5410	[20]
2	P. massoniana	W = 0.5101 V + 1.0451	
3	Other pine and conifer tree species (besides P. massoniana, Tsuga, Cryptomeria, and Keteleeria), coniferous mixed forest	W = 0.5168 V + 33.2378	
4	Cypress	W = 0.6129 V + 46.1451	
5	Mixed conifer and deciduous forests	W = 0.8019 V + 12.2799	
6	Betula	W = 0.9644 V + 0.8485	
7	Deciduous oaks	W = 1.3288 V − 3.8999	
8	Eucalyptus	W = 1.0357 V + 8.0591	
9	Mixed deciduous and Sassafras	W = 0.6255 V + 91.0013	
10	Tsuga, Cryptomeria, Keteleeria	W = 0.4158 V + 41.3318	

Note: W is the biomass of forest stand measured in t/hm^2, V is the forest volume per hectare measured in m^3/hm^2.

2.4. Data Sources and Pre-Processing

2.4.1. Data Sources

The data sources mainly included remote sensing images and Digital Elevation Model and ground survey data. The remote sensing images from the satellite Sentinel-2 (with 13 Bands, a spatial resolution of 10 m, 20 m, and 60 m) and DEM (with a spatial resolution of 30 m) with a format of ASTER GDEM, were all downloaded from the International Science and Technology Data Mirror of the Computer Network Information Centre of the Chinese Academy of Sciences (www.gscloud.cn) on 27 September 2021. The ground data were obtained from the forest resources planning and design survey provided by the Lin'an District Forestry Bureau in 2019.

2.4.2. Data Pre-Processing

(1) Forest Resources Planning and Design Survey Data

The original data of forest resource planning and design survey consist of 119,792 sub-compartments. To eliminate erroneous data, two steps were taken. Firstly, the data with non-forest, zero volume, and null value of forest ecological function level were removed. Secondly, according to the Pauta criterion [21], the abnormal data exceeding the mean (u) $\pm$ three times the standard deviation (3σ) were also removed. Consequently, 47,596 valid samples were retained (Table 5), consisting of 26 dominant tree species, that is, broad-leaved mixed forest, horsetail pine, fir, coniferous mixed forest, coniferous mixed forest, oak, other hard broad-leaved forest, maple, camphor, yellow pine, etc.

Table 5. Number of experimental samples.

Number of Original Samples	Number of Valid Samples	Number of Tree Species
119,792	47,596	26

(2) Extraction and processing of characteristic factors based on images from remote sensing and DEM

For the Sentinel-2 remote sensing images, Sen2Cor® (v2.8, European Space Agency, Paris, France) was used for atmospheric correction, and SNAP® (v6.0, European Space Agency, Paris, France) was exploited to resample the bands (band1, band5, band6, band7, band8A, band9, band10, band11, and band12) to fuse the lower-resolution images of 20 m and 60 m with higher-resolution images of 10 m by the nearest neighbor method. After that, by the operations of image mosaic and clipping in ArcGIS® (v10.8, Environmental Systems Research Institute, Inc., Redlands, CA, USA), a valid and complete remote sensing image of the Lin'an District (Figure 3) was produced.

Figure 3. Remote sensing image from Sentinel-2 of Lin'an District.

The characteristic factors extracted from the Sentinel-2 optical remote sensing images consisted of two main types: original factors and derived factors. The former consisted of 13 bands [22–24]. There were three bands, namely, Band1 for the coastal band, Band9 for the water vapor band, and Band10 for the cirrus band, which were not relevant to the experiment and were removed. Therefore, the remaining ten original bands (as shown in Table 6) and eleven vegetation indices-derived factors (as shown in Table 7), that is, a total of 21 spectral feature factors, were used as independent variables.

Table 6. Vegetation index formula.

Code	Vegetation Index	Formula
1	Atmospherically resistant vegetation index (ARVI)	ARVI = (NIR – (2 * R) + B)/(NIR + (2 * R) + B)
2	Enhanced vegetation index (EVI)	EVI = 2.5 × (NIR − R)/(NIR + 6 × R − 7.5 × B + 1)
3	Differential environmental vegetation index (DVI)	DVI = NIR − R
4	Normalized vegetation index (NDVI)	NDVI = (NIR − R)/(NIR + R)
5	Ratio red-edge vegetation index (RVIre)	RVIre = NIR/Re
6	Inverted red-edge chlorophyll index (IRECI)	IRECI = (Re3 − R)/(Re1 − Re2)
7	Normalized red-edge vegetation index1 (NDVIre1)	NDVIre1 = (NIR − Re1)/(NIR + Re1)
8	Normalized red-edge vegetation index2 (NDVIre2)	NDVIre2 = (NIR − Re2)/(NIR + Re2)
9	Non-linear red-edge index (NLIre)	NLIre = ((NIR * NIR) − Re1)/((NIR * NIR) + Re1)
10	Improved normalized red-edge vegetation index (mNDVIre)	mNDVIre = (NIR − Re1)/(NIR + Re1 − 2 * B)
11	Red-edge chlorophyll index (CIre)	CIre = (NIR/Re1) − 1

Note: R represents the red band, B represents the blue band, NIR represents the near-infrared band, and Re represents the red-edge band.

Table 7. Evaluation factors.

No.	Factor Name	Explanation	Source of Data
1	Band 2	Bule	Sentinel-2
2	Band 3	Green	
3	Band 4	Red	
4	Band 5	VNIR1	
5	Band 6	VNIR2	
6	Band 7	VNIR3	
7	Band 8	NIR	
8	Band 8A	Narrow NIR	
9	Band 11	SWIR 1	
10	Band 12	SWIR 2	
11	HAI_BA	Elevation	DEM
12	PO_DU	Slope	
13	PO_XIANG	Aspect	
14	LIN_ZHONG	Forest category	Forest Resources Planning and Design Survey Data
15	QI_YUAN	Forest origin	
16	YOU_SHI_SZ	Dominant species	
17	NL	Tree age	
18	LING_ZU	Tree age group	
19–29	Refer to Table 6		Vegetation indices generated from optical remote sensing images

In this study, the remote sensing images and DEM were preprocessed by converting, stitching, and cropping in ArcGIS. The three topographic factors—elevation, slope, and aspect—were obtained from DEM images (Figure 4), with a spatial resolution of 30 m.

2.5. Extraction of Feature Factors from Ground Survey Data

Even if the forest Resources Planning and Design Survey data have authenticity and reliability, many factors, such as forest biomass, vegetation coverage, forest canopy density, and mean tree height, with very high investigation costs, are not suitable for evaluating forest ecological function levels. Therefore, the eight factors (shown in Table 1) which were traditionally used to compute forest ecological function levels were removed, and unrelated factors such as county code and county name were also removed. After that, 36 variables remained. To further reduce the dimensionality of the features and improve the efficiency of the model, the feature importance was ranked (as shown in Figure 5). Furthermore, five feature factors, that is, QI_YUAN, LING_ZU, NL, YOU_SHI_SZ, and LIN_ZHON, were selected as independent variables of the model from top to bottom.

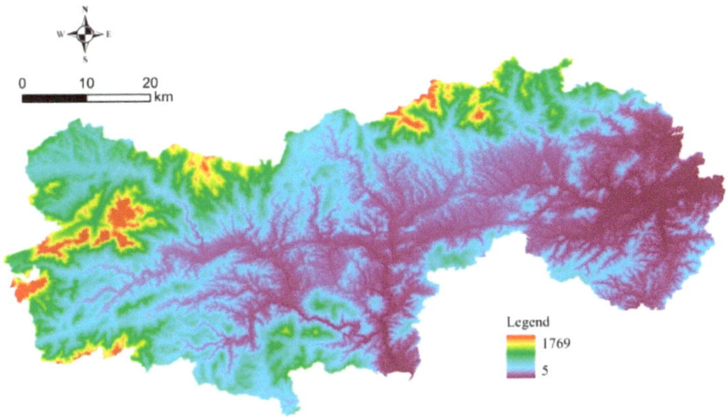

Figure 4. DEM images of Lin'an District.

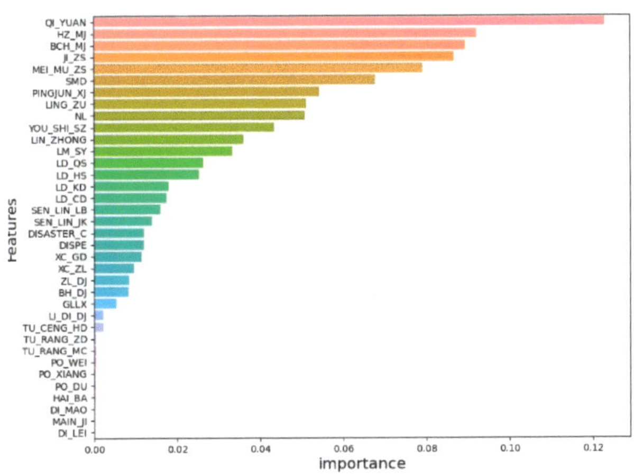

Figure 5. Ranking the feature importance of 36 variables.

2.6. Multi-Source Data Integration

A total of 29 evaluation factors (as shown in Table 7) were involved in the study, including 10 spectral bands, 11 vegetation indices, 5 ground survey factors, and 3 topographic factors. The integration of the multi-source data was implemented according to the FID.

Finally, the dataset was randomly divided into a training set with 80% of the samples for modeling and a test set with 20% of the samples for testing.

2.7. Methods

2.7.1. Grid SearchCV

To prevent overfitting and underfitting, a hyper-parameter optimization method—grid search (Grid SearchCV [25])—was used to select the optimal hyper-parameter values for the three models. Grid SearchCV allows performing hyperparameter tuning in order to determine the optimal values for a given model. Specifically, based on a specified parameter range and a validation dataset, the parameters are gradually adjusted based on a pre-set step size, and finally the optimal parameter value was selected with the highest accuracy.

2.7.2. Random Forest (RF)

Random forest [26,27] is a collection of multiple decision tree algorithms with random sampling, which is a combination of Breiman's "bagging" idea and a random selection of features. The procedure consists in making a precise prediction by taking the average or mode of the output of multiple decision trees (shown in Figure 6). Typically, the greater the decision trees' number, the more precise the output and the greater the overhead. In our study, to balance accuracy and overhead, the default number of decision trees was set to 100, and the parameter range of the grid search was set to 100–500 with a step size of 5. The final experimental results showed that when the number of subtrees reached 200, the increase in the number of subtrees had a minimal effect on the model enhancement. Therefore, the optimal parameter value was finally set to 200. Also, the maximum number of features was set to 195, which was the square root of the number of training samples (47,596 * 0.8).

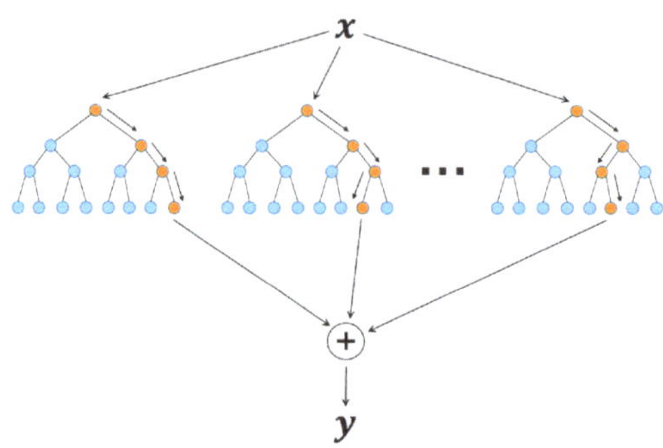

Figure 6. Random forest schematic.

2.7.3. Light Gradient Boosting Machine (LightGBM)

The Light Gradient Boosting Machine, XGBoost, and Catboost are lifting algorithms [28]. For the LightGBM, during the training process, the decision tree algorithm of Histogram was adopted, which greatly reduced the calculation amount of the model. Meanwhile, the leaf-wise growth strategy (shown in Figure 7) was introduced into the growth process of the subtree, which reduced the splitting of invalid nodes.

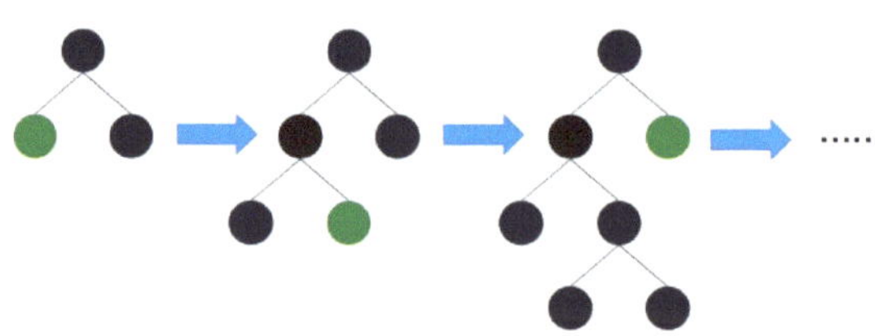

Figure 7. Leaf-wise growth strategy.

There are three hyperparameters in LightGBM that need to be determined through the grid search method, namely, n_estimators, max_depth, and learning_rate. Here, n_estimators is the maximum number of base learners, max_depth is the maximum depth of the tree, and learning_rate indicates the magnitude of each parameter update. Correspondingly, the parameter ranges were set to [100, 300], [2, 10], and [0.05, 0.2], with step sizes of 5, 1, and 0.01, respectively. Finally, the optimal values of the three parameters were 200 for n_estimators, 3 for max_depth, and 0.1 for learning_rate, respectively.

2.7.4. CatBoost

CatBoost [29,30] is a GBDT algorithm based on a symmetric binary tree, which can automatically process category-based features and effectively solve gradient bias and prediction shift problems and has excellent accuracy and generalization capabilities. In addition, CatBoost combines multiple categorical features by adding a priori distributed modified Greedy TS approach to reduce the effect of noise and low-frequency categorical data on the data distribution (as calculated in Equation (1)).

$$X_{\sigma_p,k} = \frac{\sum_{p-1}^{j=1}\left[X_{\sigma_j,k} = X_{\sigma_p,k}\right]Y_{\sigma_j} + a \times p}{\sum_{p-1}^{j=1}\left[X_{\sigma_j,k} = X_{\sigma_p,k}\right] + a} \quad (1)$$

where a is a weighting factor greater than 0; p is the priori.

The default value of n_estimators in CatBoost package we used was 500, which was already large enough. For the other two parameters, tree depth and learning rate, their ranges were set to [5, 12] and [0.05, 0.2], with step sizes of 1 and 0.01, respectively. Consequently, the optimal tree depth was 11, and the learning rate was 0.05.

2.7.5. Performance Metrics

The performance metrics of the model are generally calculated based on a confusion matrix [31], as shown in Table 8. Here, a_{ij} denotes the number of samples, the measured value is denoted by I, and the predicted value is denoted by j, N is the total number of samples, k is the number of target categories, and $a_{I+} = \sum_j a_{ij}$, $a_{+j} = \sum_I a_{ij}$.

Table 8. Confusion matrix for multi-classification models.

Confusion Matrix		Predicted Value			
		Category 1	**Category 2**	**Category k**	**Total**
Measured value	Category 1	a_{11}	a_{21}	a_{1k}	a_{1+}
	Category 2	a_{21}	a_{22}	a_{1k}	a_{2+}
	Category k	a_{k1}	a_{k2}	a_{kk}	a_{3+}
	Total	a_{+1}	a_{+2}	a_{+3}	N

Furthermore, the performance of the predicted results was evaluated by *accuracy* (Formulas (3)), *recall* (Formulas (5)), and *F1_score* (Formulas (8)).

$$accuracy_I = \frac{a_{II}}{N} \quad (2)$$

$$accuracy = \sum_{I=1}^{k} accuracy_I \quad (3)$$

$$recall_I = \frac{a_{II}}{a_{+I}} \quad (4)$$

$$recall = \sum_{I=1}^{k} recall_I \quad (5)$$

$$precision_I = \frac{a_{II}}{a_{I+}} \quad (6)$$

$$F1\_score_I = 2 * \frac{precision_I * recall_I}{precision_I + recall_I} \quad (7)$$

$$F1\_score = \sum_{I=1}^{k} F1\_score_I \quad (8)$$

3. Results

3.1. Labeling of the Data

The quantification scores for Type I, II, and III in Table 1 were assigned to 1, 2, and 3, respectively [32], and the composite score of forest ecological functions were calculated according to Formula (9) [19]

$$Y = \sum_{I=1}^{8} W_I X_I \tag{9}$$

where Y is the composite score of the forest, X_I is the standardized score of the I-th evaluation factor, and W_I is the weight of the I-th evaluation factor.

Hence, the ecological function index was calculated according to the composite score and is represented by Equation (10)

$$K = \frac{1}{Y} \tag{10}$$

where K is the the ecological function index with a value less than or equal to 1. The larger the value of K, the better the forest's ecological function.

Further, according to the values of K, the ecological function levels were divided into three groups, I.e., Good, Medium, and Poor (as shown in Table 9).

Table 9. Criteria and codes for rating forest ecological functions.

Ecological Function Level	Comprehensive Score Value (Y)	Forest Ecological Function Index (K)	Code	References
Good	<1.5	>0.6667	1	[19]
Medium	1.5~2.4	0.6667~0.4167	2	
Poor	≥2.5	≤0.4	3	

Consequently, the classification results based on Equation (10) acted as the labeled data and were plotted in Figure 8. As shown, 10,844 forest subcompartments were graded as "good", 36,365 forest subcompartments were graded as "medium", and 386 forest subcompartments were graded as "poor". In terms of area, the ecological function levels of "good", "medium", and "poor" were 59,094.3 hectares, 92,011.9 hectares, and 487.6 hectares, respectively, accounting for 38.98%, 60.70%, and 0.32% of the total area of forest land, respectively. The average forest ecological function index of forested land in Lin'an was 0.63, which was only 0.04 lower than the good ecological function index of 0.67, and the overall ecological function was at a medium, tending to good, level.

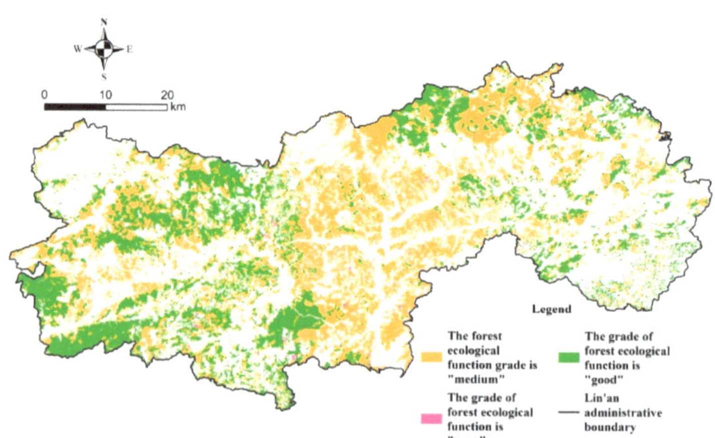

Figure 8. Distribution of forest ecological function levels in Lin'an.

There was an upward tendency for forest ecological function levels from east to west, which was consistent with the zoning of construction land in Lin'an District, with more urban construction land to the east and more ecological forest areas to the west. The extent to which the experimental results were consistent with the actual situation from the cross-reference map of the focus areas (Figure 9) was determined. For example, the two national nature reserves in Figure 9a,b corresponded to areas where the ecological function of the forest was rated as "good" on a large scale and, partly, as "medium", with no areas rated as "poor". Closer to urban areas, there was a greater chance of a "poor" forest ecological function level (as shown in Figure 9c). To some extent, this reflected the impact of human activities on the levels of forest ecological function. The regions with lower human activity often had higher levels of forest ecological function, while the regions with higher human activity had lower levels of forest ecological function.

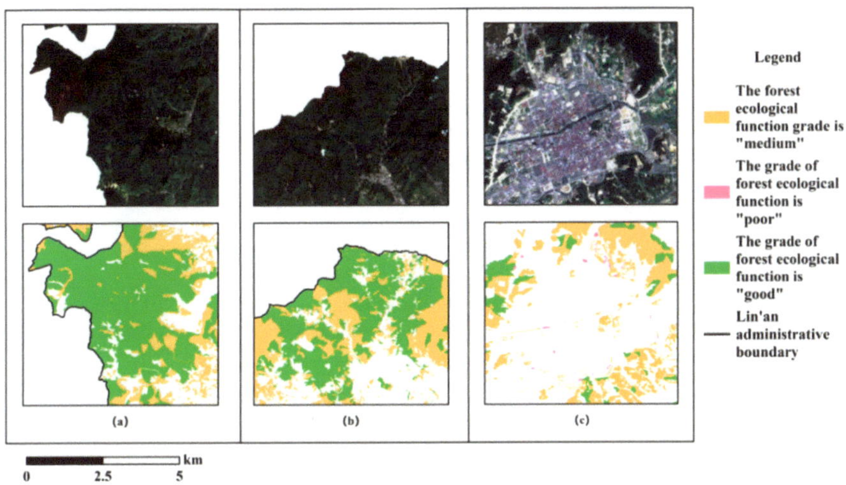

Figure 9. Cross-reference map of the focus areas. (**a**) Qingliang Peak National Nature Reserve; (**b**) Tianmu Mountain National Nature Reserve; (**c**) Lin'an urban area.

3.2. Design of the Data Scheme

Four data combination schemes (as shown in Table 10)—A, B, C, and D—were designed according to the three data sources.

Table 10. Data combination schemes.

Data Combination Scheme	Data Source
A	Sentinel-2
B	Sentinel-2, DEM
C	Sentinel-2, forest resource planning and design survey data
D	Sentinel-2, DEM, forest resource planning and design survey data

3.3. Testing Results

The results of the optimal hyperparameters for three models were obtained by grid search (Section 2.7 for details), as shown in Table 11. Furthermore, the four different data combination schemes (shown in Table 10) were modeled and analyzed using the random forest, LightGBM, and CatBoost algorithms, and the experimental results are shown in Table 12.

Table 11. Combinations of optimal hyperparameters for three models.

Model	Optimal Values of Hyperparameters
RF	n_estimators = 200, max_features = 195
LightGBM	n_estimators = 200, max_depth = 3, learning_rate = 0.1
CatBoost	n_estimators = 500, depth = 11, learning_rate = 0.05

Table 12. Performance of the three models based on the four data combination schemes.

Program	Overall Accuracy Rate	Category Accuracy Rate			Recall	F1 Score
		Good	Medium	Poor		
RF-A	0.46	0.57	0.80	0.35	0.39	0.39
RF-B	0.47	0.62	0.80	0.40	0.40	0.41
RF-C	0.82	0.73	0.89	0.80	0.54	0.57
RF-D	0.82	0.76	0.89	0.83	0.66	0.62
LightGBM-A	0.47	0.61	0.80	0.32	0.41	0.40
LightGBM-B	0.47	0.62	0.80	0.33	0.40	0.41
LightGBM-C	0.73	0.71	0.90	0.58	0.52	0.55
LightGBM-D	0.76	0.73	0.90	0.64	0.61	0.58
CatBoost-A	0.46	0.59	0.80	0.35	0.42	0.41
CatBoost-B	0.48	0.62	0.81	0.42	0.42	0.43
CatBoost-C	0.73	0.73	0.90	0.57	0.55	0.56
CatBoost-D	0.82	0.75	0.90	0.80	0.63	0.58

Comparing the performance metrics of the four data schemes in the three models in Table 12, the data scheme A (single-data combination scheme) had the worst performance, with an *overall accuracy* rate of only 0.46~0.47 and a *classification accuracy* of 0.32~0.35 for the "poor" category samples. However, the data scheme D (multi-source-data combination scheme) performed the best, with an *overall accuracy* rate of 0.76~0.82. The accuracy of the "good", "medium", and "poor" categories reached 0.73~0.76, 0.89~0.90, and 0.64~0.80, respectively, and the *F1* score was 0.58~0.62. When comparing the data scheme B (after adding the DEM data to scheme A) with A, it was found that the DEM had an insignificant contribution to the model, with an *overall accuracy* improvement of only 0.01~0.02, and the accuracy of the "good", "medium", and "poor" categories, respectively, improved by 0.01~0.16, 0~0.01, and 0.01~0.07. However, when comparing the data scheme C (after adding the ground survey data to scheme A) with A, it was found that the addition of the forest resource planning and design survey data made a significant positive contribution to the model, with an *overall accuracy* improvement of 0.26~0.36, significantly improving the performance metrics of "good", "medium", and "poor" categories, respectively.

Ultimately, the RF-D executed the optimal program with *overall accuracy* of 0.82, *recall* of 0.66, and *F1* score of 0.62, and the classification accuracy was significantly improved, especially for the small sample category of "poor".

3.4. Ranking of Features' Importance

Based on the optimal data scheme D, the performance metrics (as shown in Table 12) were calculated by the three machine learning algorithms of RF, LightGBM, and CatBoost, and the ranking of the feature importance was obtained and shown in Figure 10.

As shown in Figure 10, there were five factors from the ground survey data that led to a higher ranking of feature importance and played an important role in the model. Nevertheless, there were three factors from the DEM data with a lower ranking of feature importance that played an unimportant role in the model, which is consistent with the results in Table 11. In the optical remote sensing data, the factors of b12, NDVIre2, EVI, IRECI, b2, b11, and b4 ranked relatively high in the model.

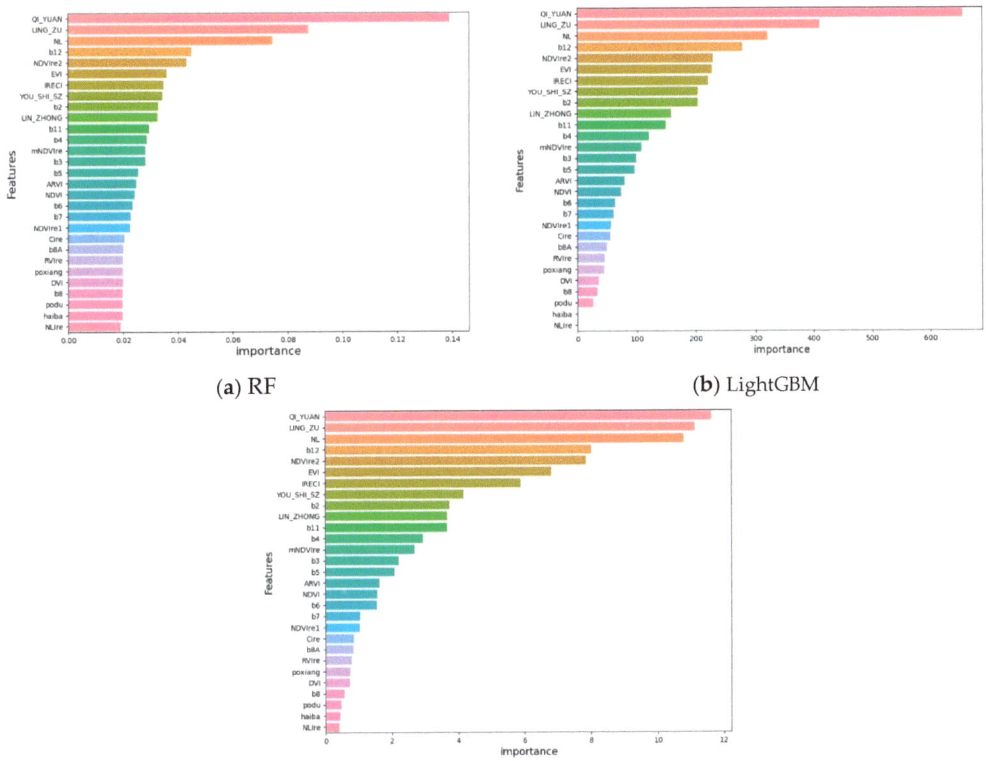

(**a**) RF (**b**) LightGBM

(**c**) CatBoost

Figure 10. Ranking the importance of the features based on the multi-source data scheme D. (**a**) Ranking the importance of the features by RF; (**b**) ranking the importance of the features by LightGBM; (**c**) ranking the importance of the features by CatBoost.

4. Discussion

4.1. Performance Metrics

Even if many studies have been aimed to the forest ecological function rating methodology, some deficiencies still exist, as reported below.

(1) The evaluation indicators are heavily influenced by foresters' experience

In the evaluation of forest ecological functions, due to many qualitative indicators such as forest naturalness, tree species structure, and thickness of dead leaves, the original way of obtaining data is heavily influenced by foresters' experience, which makes it difficult to establish an objective indicator system and may lead to inconsistent forest ecological function levels evaluated by different foresters for the same forest stand [33,34].

(2) High cost of data acquisition for evaluation indicators

The data of the eight evaluation factors (forest naturalness, forest community structure, tree species structure, vegetation coverage, forest canopy density, mean tree height, and thickness of dead leaves) used to calculate the forest ecological function levels are obtained from ground surveys, which leads to high acquisition costs and time consumption [35].

The proposed evaluation method combined the advantages of multiple sources of data and machine learning algorithms to effectively reduce the human influence on the evaluation system and save data acquisition expenses and time. Specifically, the addition of remote sensing data effectively reduced the influence of human subjective experiences and increased the frequency of data acquisition. The costs of acquiring remote sensing data are also much lower than that of ground surveys [36], especially in areas that are difficult to access by humans, such as deep forests and cliffs.

4.2. Complementarity of Multi-Source Data

As machine learning algorithms are data-driven, variations in the data could greatly affect the accuracy of a classification [37]. When providing insufficient data, such as for data scheme A (merely optical remote sensing data from Sentinel-2), poor results were obtained (*overall accuracy* of 0.46~0.47 for scheme A in Table 11). Due to influences by environmental conditions such as different angles and intensities of sunlight, topography, water content, and other factors, it is possible that the same object may have different spectrums, and different objects may have the same spectrum [38,39]. For instance, the spectral information for the same vegetation on sunny and shady slopes could be different, and the height of lower vegetation is easily obscured by shadows. This tends to increase the errors during training. Complementarity of multiple sources of data is generally used to address this problem.

Compared to the data scheme A, the data scheme D (addition of DEM data and some ground survey data) significantly increased the *overall accuracy* by 0.29~0.36, with the accuracy rates for the "good", "medium", and "poor" categories increasing by 0.12~0.19, 0.09~0.1, and 0.32~0.48, respectively. The addition of DEM data complements vertical structure parameters which are lacking in optical remote sensing data [40], allowing areas of deciduous trees to be distinguished from areas of vegetation with similar spectral characteristics (e.g., high-density grassland) [41], which in turn has an impact on the accuracy of the results. The addition of ground survey data further complements the growth status information for the vegetation—such as the age of trees that can lead to changes in their growth rate—thereby improving the accuracy of the model.

4.3. The Feasibility of Machine Learning

Different from the traditional statistical methods used by Huafu Liu et al. [42], Hailong Yin et al. [43], and Kassim et al. [17], this paper exploited a machine learning algorithm to develop a comprehensive model to evaluate forest ecological function levels, which has a higher flexibility and faster processing speed for high-dimensional data with more complex relationships among feature factors. Most indicators are non-linearly related to forest ecological function levels, including ground survey factors and spectral characteristics. For instance, with the increasing NL, the forest ecological function levels increase first and then decrease. Machine learning algorithms are non-linear approximations to an objective function, different from than the traditional comprehensive evaluation methods bound to a linear function, such as the scoring method, principal component analysis, etc. Therefore, their powerful fitting ability could make the predicted results closer to the reality and improve the evaluation accuracy of the model.

The Random Forest showed the best performance among the above three models, with an *overall accuracy* of 0.82 and an *F1* score of 0.62. As a non-linear, parametric classifier, Random Forest is robust with non-equilibrium data and can randomly generate multiple decision trees to form a forest, effectively avoiding overfitting [44,45]. It allows the fusion of high-dimensional data from multiple sources [46] and has a high tolerance for missing values and outliers, so that it can effectively reduce the interference of noise in the data. In addition, it can automatically determine the importance of variables, which in turn improves its accuracy and usability.

4.4. Limitations of this Study

This study, based on the multi-source data of Sentinel-2 remote sensing images and DEM and partial data from the forest resource planning and design survey, as well as the three machine learning algorithms RF, LightGBM, and CatBoost, evaluated the forest ecological function levels. Overall, our results may promote research on the evaluation of forest ecological function levels. However, the 10 m resolution of the Sentinel-2 images may limit a further improvement of the performance metrics. If higher-resolution images can be acquired in the future, such as remote sensing images from Gaofen series satellites or UAV images, it will be possible to further increase the performance of our model and even further reduce the participation of ground survey indicators. On the other hand, from the perspective of research methods, deep learning algorithms such as YOLO are also worth a try in the future.

5. Conclusions

Optical remote sensing data, DEM data, and forest resource planning and design survey data were used in this study to evaluate the forest ecological function levels of Lin'an District using three machine learning algorithms, I.e., RF, LightGBM, and CatBoost.

In the three models, Random Forest was the best-performing model, with an *overall accuracy* rate of 0.82 (the *accuracy* rates for the "good", "medium", and "poor" categories being 0.76, 0.89, and 0.83, respectively) and with an *F1* score of 0.62.

The multi-source data significantly improved the performance metrics. Furthermore, the acquisition of ground survey data such as QI_YUAN, LING_ZU, LIN_ZHONG, YOU_SHI_SZ, and NL, was achieved at lower costs than those required for the traditional eight indicators of forest biomass, forest naturalness, forest community structure, tree species structure, vegetation coverage, forest canopy density, mean tree-height, and thickness of dead leaves.

If more data sources are used, such as higher-resolution remote sensing images, LiDAR remote sensing images, etc., the estimation performance might further improve in the future.

Author Contributions: Conceptualization, D.W.; Formal analysis, S.L.; data curation, L.Y.; Funding acquisition, D.W. and X.Z.; Methodology, X.Z.; Resources, L.Y.; Writing—original draft, N.F.; writing—review and editing, L.Y. All authors have read and agreed to the published version of the manuscript.

Funding: This research was financially supported by the Zhejiang Forestry Science and Technology Project (2023SY08), the National Natural Science Foundation of China (Grant No. 42001354), and the Natural Science Foundation of Zhejiang Province (Grant No. LQ19D010011).

Data Availability Statement: The remote sensing data can be found here: [https://scihub.copernicus.eu/ (accessed on 27 September 2021)]. The DEM data can be found here: [www.gscloud.cn (accessed on 13 December 2021)]. The ground survey data are not publicly available (for policy reasons, these data are kept confidential).

Conflicts of Interest: The authors declare no conflict of interest.

References

1. Liu, Y.; Wang, S.; Chen, Z.; Tu, S. Research on the Response of Ecosystem Service Function to Landscape Pattern Changes Caused by Land Use Transition: A Case Study of the Guangxi Zhuang Autonomous Region, China. *Land* **2022**, *11*, 752. [CrossRef]
2. Tang, Y.; Shao, Q.; Liu, J.; Zhang, H.; Yang, F.; Cao, W.; Wu, D.; Gong, G. Did Ecological Restoration Hit Its Mark? Monitoring and Assessing Ecological Changes in the Grain for Green Program Region Using Multi-source Satellite Images. *Remote Sens.* **2019**, *11*, 358. [CrossRef]
3. Holling, C.S. Simplifying the complex: The paradigms of ecological function and structure. *Eur. J. Oper. Res.* **1987**, *30*, 139–146. [CrossRef]
4. Aerts, R.; Honnay, O. Forest restoration, biodiversity and ecosystem functioning. *BMC Ecol.* **2011**, *11*, 29. [CrossRef]
5. Brodie, J.F.; Redford, K.H.; Doak, D.F. Ecological Function Analysis: Incorporating Species Roles into Conservation. *Trends Ecol. Evol.* **2018**, *33*, 840–850. [CrossRef]
6. Shen, W.; Li, M.; Huang, C.; Wei, A. Quantifying Live Aboveground Biomass and Forest Disturbance of Mountainous Natural and Plantation Forests in Northern Guangdong, China, Based on Multi-Temporal Landsat, PALSAR and Field Plot Data. *Remote Sens.* **2016**, *8*, 595. [CrossRef]
7. Rizky, P.M.; Safe'I, R.; Kaskoyo, H.; Febryano, I.G. Forestry Value for Health Status: An Ecological Review. *IOP Conf. Ser. Earth Environ. Sci.* **2022**, *995*, 012002. [CrossRef]
8. Ma, L.; Han, H.R. Evaluation of the forest ecosystem health in Beijing area. *For. Stud. China* **2007**, *2*, 157–163. [CrossRef]
9. Fang, X.M. Evaluation of ecological functions of forests in the Aojiang River Basin. *For. Surv. Des.* **2022**, *1*, 36–39+61.

10. Du, Q.; Ji, B.Y.; Xu, J. Research on forest ecological function evaluation based on forest quantity, quality and spatial distribution. *Zhejiang For. Sci. Technol.* **2013**, *33*, 46–50.
11. Lausch, A.; Borg, E.; Bumberger, J.; Dietrich, P.; Heurich, M.; Huth, A.; Jung, A.; Klenke, R.; Knapp, S.; Mollenhauer, H.; et al. Understanding Forest Health with Remote Sensing, Part III: Requirements for a Scalable Multi-Source Forest Health Monitoring Network Based on Data Science Approaches. *Remote Sens.* **2018**, *10*, 1120. [CrossRef]
12. Wang, D.L.; Zhang, X.; Yang, Q.; Zhang, X.Y.; Gong, J.Q.; Dong, H. Evaluation of forest ecological functions in Sanjiuzi Forestry Bureau. *Jilin For. Sci. Technol.* **2022**, *51*, 34–37.
13. Yang, J.; Dai, G.; Wang, S. China's National Monitoring Program on Ecological Functions of Forests: An Analysis of the Protocol and Initial Results. *Forests* **2015**, *6*, 809–826. [CrossRef]
14. Liu, L.X.; Zhu, S.W. Evaluation of forest ecological functions in Beijing Jingxi Forestry Management Office. *Green Technol.* **2022**, *24*, 152–155.
15. Zhang, X.W.; Zheng, C.M.; Tang, X.J.; Zhang, W.D. Comprehensive index assessment of forest ecological functions in Shanghai. *Subtrop. Soil Water Conserv.* **2015**, *27*, 34–37.
16. Wang, X.Z.; Tan, L.L.; Fan, J.C. Performance Evaluation of Mangrove Species Classification Based on Multi-Source Remote Sensing Data Using Extremely Randomized Trees in Fucheng Town, Leizhou City, Guangdong Province. *Remote Sens.* **2023**, *15*, 1386. [CrossRef]
17. Abd Rahman Kassim, M.A.M.; Faidi, M.A.; Omar, H. A tool for assessing ecological status of forest ecosystem. *IOP Conf. Ser. Earth Environ. Sci.* **2016**, *37*, 012026.
18. Relevant Attachments to the Technical Regulations for Category 2 Investigation in Zhejiang Province in 2014. Available online: https://wenku.baidu.com/view/441978e3b80d6c85ec3a87c24028915f804d848a.html (accessed on 28 September 2021).
19. *GB/T 38590-2020*; Technical Regulations for Continuous Inventory of Forest Resources. China National Standard Publishing House: Beijing, China, 2020; pp. 1–48.
20. Fang, J.Y.; Wang, G.G.; Liu, G.H.; Xu, S.L. Forest biomass of China: An estimate based on the biomass-volume relationship. *Ecol. Appl.* **1998**, *8*, 1084–1091.
21. Liu, Y.; Fan, Z.P.; You, T.H.; Zhang, W.Y. Large group decision-making (LGDM) with the participators from multiple subgroups of stakeholders: A method considering both the collective evaluation and the fairness of the alternative. *Comput. Ind. Eng.* **2018**, *122*, 262–272. [CrossRef]
22. Keshta, A.E.; Riter, J.C.A.; Shaltout, K.H.; Baldwin, A.H.; Kearney, M.; Sharaf El-Din, A.; Eid, E.M. Loss of Coastal Wetlands in Lake Burullus, Egypt: A GIS and Remote-Sensing Study. *Sustainability* **2022**, *14*, 4980. [CrossRef]
23. Ebodé, V.B. Land Surface Temperature Variation in Response to Land Use Modes Changes: The Case of Mefou River Sub-Basin (Southern Cameroon). *Sustainability* **2023**, *15*, 864. [CrossRef]
24. Fernandez-Beltran, R.; Baidar, T.; Kang, J.; Pla, F. Rice-Yield Prediction with Multi-Temporal Sentinel-2 Data and 3D CNN: A Case Study in Nepal. *Remote Sens.* **2021**, *13*, 1391. [CrossRef]
25. Sumathi, B. Grid search tuning of hyperparameters in random forest classifier for customer feedback sentiment prediction. *Int. J. Adv. Comput. Sci. Appl.* **2020**, *11*, 173–178.
26. Yoshida, Y.; Yuda, E. Workout Detection by Wearable Device Data Using Machine Learning. *Appl. Sci.* **2023**, *13*, 4280. [CrossRef]
27. Xiao, S.; Laurie, A.C. Classification of Australian Native Forest Species Using Hyperspectral Remote Sensing and Machine-Learning Classification Algorithms. *IEEE J. Sel. Top. Appl. Earth Obs. Remote Sens.* **2014**, *7*, 2481–2489.
28. Han, P.; Zhai, Y.; Liu, W.; Lin, H.; An, Q.; Zhang, Q.; Ding, S.; Zhang, D.; Pan, Z.; Nie, X. Dissection of Hyperspectral Reflectance to Estimate Photosynthetic Characteristics in Upland Cotton (*Gossypium hirsutum* L.) under Different Nitrogen Fertilizer Application Based on Machine Learning Algorithms. *Plants* **2023**, *12*, 455. [CrossRef]
29. Chang, W.; Wang, X.; Yang, J.; Qin, T. An Improved CatBoost-Based Classification Model for Ecological Suitability of Blueberries. *Sensors* **2023**, *23*, 1811. [CrossRef]
30. Li, H.; Zhang, G.; Zhong, Q.; Xing, L.; Du, H. Prediction of Urban Forest Aboveground Carbon Using Machine Learning Based on Landsat 8 and Sentinel-2: A Case Study of Shanghai, China. *Remote Sens.* **2023**, *15*, 284. [CrossRef]
31. Foody, G.M. Global and Local Assessment of Image Classification Quality on an Overall and Per-Class Basis without Ground Reference Data. *Remote Sens.* **2022**, *14*, 5380. [CrossRef]
32. Yuan, Y.; Liu, Z.G.; Dong, L.B. GIS-based evaluation and analysis of forest ecological function level of Pangu forestry field in Daxinganling. *J. Cent. South Univ. For. Sci. Technol.* **2016**, *36*, 108–114.
33. Stewart, D.H. Data capture in forestry research. *J. R. Stat. Soc.* **1968**, *18*, 377–411. [CrossRef]
34. Olga, M.; Maria, T. The effect of expertise on the quality of forest standards implementation: The case of FSC forest certification in Russia. *For. Policy Econ.* **2009**, *11*, 422–428.
35. Dube, T.; Onisimo, M.; Cletah, S.; Sam, A.A.; Tsitsi, B. Remote sensing of aboveground forest biomass: A review. *Trop. Ecol.* **2016**, *57*, 125–132.
36. White, J.C.; Coops, N.C.; Wulder, M.A.; Vastaranta, M.; Hilker, T.; Tompalski, P. Remote sensing technologies for enhancing forest inventories: A review. *Can. J. Remote Sens.* **2016**, *42*, 619–641. [CrossRef]
37. Hou, Z.S.; Wang, Z. From model-based control to data-driven control: Survey, classification and perspective. *Inf. Sci.* **2013**, *235*, 3–35. [CrossRef]

38. Liu, Y.; Chen, X.; Qiu, Y.; Hao, J.; Yang, J.; Li, L. Mapping snow avalanche debris by object-based classification in mountainous regions from Sentinel-1 images and causative indices. *Catena* **2021**, *206*, 105559. [CrossRef]
39. Yi, Z.; Jia, L.; Chen, Q. Crop Classification Using Multi-Temporal Sentinel-2 Data in the Shiyang River Basin of China. *Remote Sens.* **2020**, *12*, 4052. [CrossRef]
40. Wu, F.; Ren, Y.F.; Wang, X.K. Application of Multi-Source Data for Mapping Plantation Based on Random Forest Algorithm in North China. *Remote Sens.* **2022**, *14*, 4946. [CrossRef]
41. Masayasu, M.; Toshiharu, K.; Tsuyoshi, A. Potential of ASTER DEM for Forest Ecological Research. *J. Remote Sens. Soc. Jpn.* **2007**, *27*, 39–45.
42. Liu, H.F.; Qin, O.C.; Meng, H.L.; Lan, H.B. Evaluation of forest ecological functions in Maolan Reserve. *Inn. Mong. For. Surv. Des.* **2021**, *44*, 59–63.
43. Yin, H.L.; Zhang, Z.W.; Song, W.; Zhang, Z.W.; Wu, L.Q.; Deng, T.; Zhang, Q. Evaluation of forest ecological function level in Sehanba North Mandian Forest. *J. Beijing For. Univ.* **2023**, *45*, 89–98.
44. Pes, B. Learning from High-Dimensional and Class-Imbalanced Datasets Using Random Forests. *Information* **2021**, *12*, 286. [CrossRef]
45. Orynbaikyzy, A.; Gessner, U.; Mack, B.; Conrad, C. Crop Type Classification Using Fusion of Sentinel-1 and Sentinel-2 Data: Assessing the Impact of Feature Selection, Optical Data Availability, and Parcel Sizes on the Accuracies. *Remote Sens.* **2020**, *12*, 2779. [CrossRef]
46. Shaeela, A.; Muhammad, K.H.; Ramzan, T. Overview and comparative study of dimensionality reduction techniques for high dimensional data. *Inf. Fusion* **2020**, *59*, 44–58.

forests

MDPI

Article

Future Reductions in Suitable Habitat for Key Tree Species Result in Declining Boreal Forest Aboveground Biomass Carbon in China

Bin Zhu [1], Zengxin Zhang [1,2,*], Rui Kong [3], Meiquan Wang [1], Guangshuai Li [1], Xiran Sui [1] and Hui Tao [2]

1 Joint Innovation Center for Modern Forestry Studies, College of Forestry, Nanjing Forestry University, Nanjing 210037, China; binzhu@njfu.edu.cn (B.Z.); xiaocuipi@njfu.edu.cn (M.W.); liguangshuai@iga.ac.cn (G.L.); suixr@njfu.edu.cn (X.S.)

2 Xinjiang Institute of Ecology and Geography, Chinese Academy of Sciences, Urumqi 830011, China; taohui@ms.xjb.ac.cn

3 State Key Laboratory of Hydrology-Water Resources and Hydraulics Engineering, Hohai University, Nanjing 210098, China; kongrui@hhu.edu.cn

* Correspondence: nfuzhang@163.com

Abstract: China's forest ecosystem plays a crucial role in carbon sequestration, serving as a cornerstone in China's journey toward achieving carbon neutrality by 2060. Yet, previous research primarily emphasized climate change's influence on forest carbon sequestration, neglecting tree species' suitable area changes. This study combinates the Lund–Potsdam–Jena model (LPJ) and the maximum entropy model (MaxENT) to reveal the coupling impacts of climate and tree species' suitable area changes on forest aboveground biomass carbon (ABC) in China. Key findings include the following: (1) China's forests are distributed unevenly, with the northeastern (temperate coniferous broad-leaved mixed forest, TCBMF), southwestern, and southeastern regions (subtropical evergreen broad-leaved forest, SEBF) as primary hubs. Notably, forest ABC rates in TCBMF exhibited a worrisome decline, whereas those in SEBF showed an increasing trend from 1993 to 2012 based on satellite observation and LPJ simulation. (2) Under different future scenarios, the forest ABC in TCBMF is projected to decline steadily from 2015 to 2060, with the SSP5-8.5 scenario recording the greatest decline (−4.6 Mg/ha/10a). Conversely, the forest ABC in SEBF is expected to increase under all scenarios (2015–2060), peaking at 1.3 Mg/ha/10a in SSP5-8.5. (3) Changes in forest ABC are highly attributed to climate and changes in tree species' highly suitable area. By 2060, the suitable area for *Larix gmelinii* in TCBMF will significantly reduce to a peak of 65.71×10^4 km^2 under SSP5-8.5, while *Schima superba Gardner & Champ* and *Camphora officinarum* in SEBF will expand to peaks of 94.07×10^4 km^2 and 104.22×10^4 km^2, respectively. The geographic detector's results indicated that the climate and tree species' suitable area changes showed bi-variate and nonlinear enhanced effects on forest ABC change. These findings would offer effective strategies for achieving carbon neutrality.

Keywords: forest aboveground biomass carbon; climate change; tree species' suitable area; Lund–Potsdam–Jena model; maximum entropy model; China

Citation: Zhu, B.; Zhang, Z.; Kong, R.; Wang, M.; Li, G.; Sui, X.; Tao, H. Future Reductions in Suitable Habitat for Key Tree Species Result in Declining Boreal Forest Aboveground Biomass Carbon in China. *Forests* **2023**, *14*, 2053. https://doi.org/10.3390/f14102053

Academic Editor: Matthias Peichl

Received: 10 September 2023
Revised: 29 September 2023
Accepted: 6 October 2023
Published: 13 October 2023

1. Introduction

Forests play a pivotal role in global carbon sequestration, significantly contributing to climate change mitigation and sustainability efforts [1,2]. It is noteworthy that forest biomass carbon stock alone constitutes more than 90% of the global vegetation biomass carbon reservoir [3,4]. However, the accurate estimation of forest carbon sequestration remains a challenging scientific issue. Considerable debate persists concerning whether forthcoming forest ecosystems will enhance carbon sequestration within the context of climate variation [5]. Consequently, precise evaluations of forest carbon sequestration hold great importance in appraising strategies for ecosystem management and mitigation policies [6,7].

Estimating China's forest carbon sequestration potential has remained a focal point of research, given that forest coverage in China reaches 22.96% [8,9]. However, preceding studies have focused on forecasting carbon sink trends and examining the correlation with climate alterations, often lacking in-depth mechanistic investigations [8]. For example, Tong et al. (2018) [10] explored the trend of vegetation aboveground biomass carbon (ABC) and its response to natural drivers in China's karst region from 1982 to 2015 through integrating satellite data with the Lund–Potsdam–Jena model (LPJ). Meanwhile, leveraging the LPJ model and datasets of global climate models (GCMs), Kong et al. (2022) [2] demonstrated that China's vegetation ABC is projected to experience an upward trajectory from 2041 to 2060 under the SSP1-2.6 and SSP2-4.5 scenarios, with temperature and precipitation identified as primary influencing factors. While the LPJ model comprehensively considers the dynamic characteristics of the carbon cycle using the theory of vegetation physiology and ecology [2,11], its applicability is limited in explaining key ecological functional outcomes at the species level due to its typical focus on temporal and spatial variations of general vegetation functional types [12].

Climate change drives changes in forest vegetation growth and suitable zones, thereby influencing the spatial pattern of forest carbon sequestration capacity. Nevertheless, current research generally lacks an in-depth understanding of the connection between shifts in tree species' suitable areas and forest carbon sequestration under changing conditions. For instance, under the background of climate warming, the distribution patterns of the dominant conifer tree species may exhibit a shift towards higher latitudes and altitudes [13–15]. Li et al. (2011) and Yu et al. (2011) [16,17] demonstrated that rising temperatures would lead to a decline in spruce growth within the Changbai Mountain Natural Reserve, while other prevalent tree species would experience benefits from the warming climate. Although recent studies have started to consider the impacts of forest type and age on carbon sequestration capacity [18], such investigations are often constrained by experimental scales and data collection limitations. Presently, statistical models rooted in the maximum entropy theory (MaxENT) have gained prominence due to their small sample requirements. These models accurately project species point data to suitable distribution regions on a regional scale through establishing relationships between species data and environmental variables. For instance, Wang and Guan (2023) [19] employed the MaxENT model to investigate the impact of climate change on the suitability of conformation trees. Nevertheless, these studies fall short of delving into the underlying variations in forest carbon sequestration resulting from changes in tree species' suitable areas. Consequently, the mechanism of climate variation's effects on species migration and forest carbon sequestration potential remains unclear.

This study aims to bridge the gap in understanding the combined impacts of climate variation and changes in species' suitable areas on forest carbon sequestration capacity. Through integrating the LPJ model and the MaxEnt model, our research seeks to (1) investigate the temporal and spatial variations in forest ABC across China and (2) explore the coupling effects of climate and tree species' suitable area changes on forest carbon sequestration potential in China. Through conducting case studies in different forest regions of China, this research aims to provide empirical evidence and enhance our understanding of the complex interplay between climate variation, tree species' suitable area changes, and the carbon sequestration potential of forests. Ultimately, this study will contribute to the development of targeted strategies for sustainable forest management and carbon neutrality in China.

2. Materials and Methods

2.1. Study Area and Data Sources

China's complex topography, coupled with variations in temperature, precipitation, and vegetation types, creates a rich tapestry of ecosystems that respond differently to climate drivers. Figure 1 illustrates that China is divided into seven distinct geographical zones. The response of terrestrial ecosystems to climate change varies across these regions.

China's diverse ecology, ranging from the tropical forests in the south to the alpine meadows in the north, offer an ideal setting for investigating the impacts of climate change on ABC [2]. Forests in China occupy approximately 23.57% of the total land area, mainly distributed across the eastern, central, southern, southwestern, and northeastern regions. The typical natural vegetation zone in the northeastern region is temperate coniferous broad-leaved mixed forest (TCBMF), while the southern region is characterized as subtropical evergreen broad-leaved forest (SEBF).

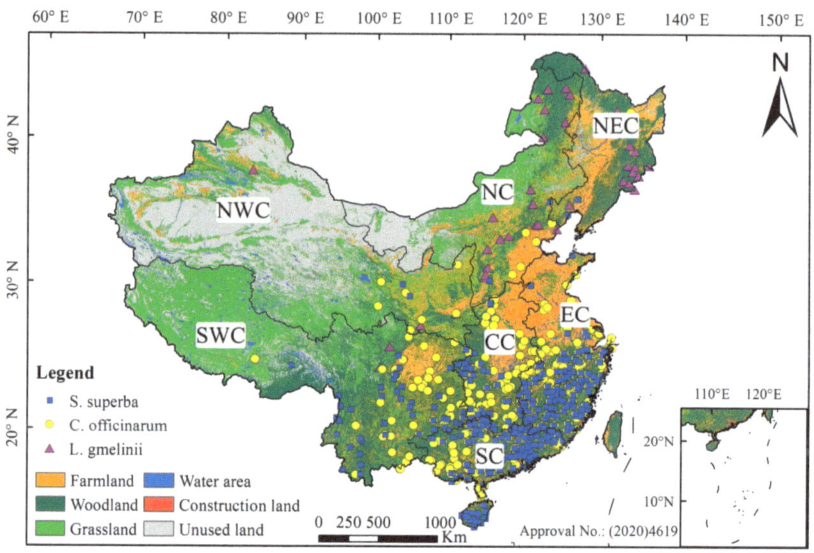

Figure 1. The land use, and land cover, and distribution points of *L. gmelinii*, *S. superba*, and *C. officinarum* in China. EC: East China; NEC: Northeast China; NC: North China; CC: Central China; SC: Southern China; NWC: Northwest China; SWC: Southwest China.

The forest ABC dataset was obtained from a VOD dataset collected by global satellites monthly, with a resolution of 0.25°, covering the years 1993–2012 (http://www.wenfo.org/wald/globalbiomass, accessed on 22 August 2019). The meteorological observation data were collected from 2472 stations with a spatial resolution of 0.5° × 0.5° in China from 1982 to 2014 (https://data.cma.cn/site/index.html, accessed on 31 January 2020). The data on wet days and cloud cover were obtained from the Climate Research Unit (CRU) version TS4.04 datasets spanning the period between 1982 and 2014. These datasets were all provided at a resolution of 0.5° × 0.5° and can be accessed at (https://sites.uea.ac.uk/cru/, accessed on 2 April 2022). The carbon dioxide (CO_2) data were sourced from the CO_2 Program of the Scripps Institution of Oceanography (SIO), accessible through (https://scrippsco2.ucsd.edu/, accessed on 2 April 2022).

Larix gmelinii (*L. gmelinii*) is one of the commonly encountered species in the boreal forests of TCBMF in Northeast China, playing a significant role in the high carbon storage capacity of the forest [20,21]. In addition, *L. gmelinii* habitats provide a diverse range of support for various organisms, including endemic species, and any alterations to the habitat would impact their distribution [22]. Similarly, *Schima superba Gardner & Champ* (*S. superba*) and *Camphora officinarum* (*C. officinarum*) are dominant evergreen broad-leaved tree species widely distributed in China's SEBF, with a high capacity for CO_2 absorption [23–25]. Therefore, *L. gmelinii* serve as the typical tree species of TCBMF, while *S. superba* and *C. officinarum* represent typical tree species for SEBF in this study. The initial geographic points of *L. gmelinii*, *S. superba*, and *C. officinarum* within China were acquired from the Global Biodiversity Information Facility (GBIF, https//www.gbif.org, accessed on 12 May 2023). These data primarily originated from field observations and preserved specimens.

Duplicate entries and original records lacking geographic coordinates (longitude or latitude) were excluded. Additionally, only one occurrence point per 1 km × 1 km grid cell was retained. This selection was primarily due to the spatial resolution of bioclimatic variables being 30 arcseconds (equivalent to approximately 1 km), which aimed to mitigate potential clustering effects predicted using the model [19]. Adhering to the input data format of the MaxEnt model, the distribution data for the target species were organized into CSV files, ordered by species name, longitude, and latitude (refer to Figure 1).

The environmental information considered in this research encompassed 19 bioclimatic datasets, elevation data, and 34 soil-related variables. The bioclimatic data, spanning the years 1970–2000, were computed as averages and possessed a spatial resolution of 30 arcseconds. These bioclimatic data and elevation information were sourced from the dataset provided by WorldClim (http://www.worldclim.org, accessed on 28 May 2021). This study took into account the impact of soil attributes on species distribution modeling (SDM) predictions, where a comprehensive set of 34 soil variables was procured from the Harmonized World Soil Database (HWSD v. 1.21). All data were clipped to China for model simulation.

In this study, forthcoming climate datasets (precipitation and temperature) of China were extracted from the BCC-CSM2-MR, MRI-ESM2-0, and CanESM5 global climate models, considering three shared socio-economic pathways (SSP1-2.6, SSP2-4.5, and SSP5-8.5). These three sets of GCMs data, bias-corrected through the Equidistant Cumulative Distribution Functions (EDCDF) [2,26], accurately captured the spatial-temporal variation characteristics of temperature and precipitation in China (Figures S1 and S2). All the data points were averaged over 20-year intervals (from 2041 to 2060). The anticipated future distribution for each species was determined based on the average results generated by the MaxEnt model, utilizing inputs from the three specified GCMs.

To establish an efficient model using a reduced set of variables, all the variables were initially fed into the model for a preliminary run. Variables that collectively contributed to over 90% of the model's simulation were then retained. Subsequently, a Pearson correlation analysis was conducted on the retained variables. If the correlation between the two variables exceeded 0.8, the variable with a greater contribution rate during the preliminary model run was preserved [27]. The variable selection results of *L. gmelinii*, *S. superba*, and *C. officinarum* are presented in Table 1.

Table 1. Variable selection results of *L. gmelinii*, *S. superba*, and *C. officinarum*.

Variable	Variable Description	*L. gmelinii*	*S. superba*	*C. officinarum*
Bio2	Mean diurnal range (mean of monthly (max temp–min temp))	√	√	
Bio3	Isothermality (bio2/bio7) (×100)	√		
Bio4	Temperature seasonality (standard deviation × 100)	√		√
Bio5	Max temperature of the warmest month		√	√
Bio6	Temperature annual range (bio5–bio6)		√	
Bio8	Mean temperature of wettest quarter	√		
Bio12	Annual precipitation		√	
Bio13	Precipitation of the wettest month	√		
Bio14	Precipitation of the driest month		√	√
Bio15	Precipitation seasonality (coefficient of variation)		√	
Bio17	Precipitation of the driest quarter	√		
elev	Elevation	√		
t_ph_h2o	Topsoil pH (H_2O)		√	
t_bs	Topsoil Base Saturation	√		
s_ece	Subsoil Salinity (Elco)	√		
s_cec_soil	Subsoil CEC (soil)	√		
s_teb	Subsoil TEB (Total exchangeable bases)	√		

2.2. Methods

2.2.1. LPJ-DGVM

The Lund–Potsdam–Jena Dynamic Global Vegetation Model (LPJ-DGVM) serves as a widely utilized computational tool devised for simulating the intricate dynamics of global vegetation under the influence of environmental factors and climate shifts. This model, at the forefront of technology, amalgamates insights from ecology, biogeography, and climatology to replicate the intricate interplays among the atmosphere, land surface, and terrestrial ecosystems [11]. The LPJ-DGVM relied on a process-driven methodology, embodying essential ecological mechanisms like photosynthesis, respiration, carbon allocation, and vegetation evolution. The model effectively captures the multifaceted feedback loops between vegetation and the natural environment through encompassing meticulous depictions of plant physiology, the carbon cycle, and land surface procedures [12]. The embedded physical mechanisms and causalities allow LPJ to provide valuable insights into the potential future trajectories of global vegetation and the associated feedback with climate change.

In the LPJ model, ten distinct plant functional types (PFTs) are featured, each governed by constraining factors that regulate the effective influences of light, temperature, and moisture [28]. Every simulated PFT represents the collective characteristics of the entire population (e.g., tree height and vegetation carbon pools; population-based method) [29]. The simulation process for the LPJ model involves employing climate data spanning 1981 to 2010, allowing for 1000 model years spin-up to establish ecosystem equilibrium. Following the attainment of equilibrium, a dynamic simulation was executed, employing climate data from both 1981 to 2014 and 1981 to 2060 [2]. To support the LPJ model simulations from 1981 to 2014, observed meteorological data such as temperature and precipitation, along with the CRUTS 4.04 climate dataset containing wet days and cloud cover information, were incorporated. Furthermore, the annual average atmospheric CO_2 concentration was integrated.

2.2.2. MaxEnt Model

Using MaxEnt (version 3.4), the training dataset encompassed 75% of the occurrence data, while the remaining 25% was allocated for testing purposes [19,30]. The algorithm underwent 1000 iterations of these steps. This process was iterated 10 times to yield the mean simulation outcomes of tree species' geographical distributions [31]. Subsequently, these results were consolidated within ArcGIS 10.5 and transformed into a raster format for subsequent analysis.

Model performance was assessed through the utilization of the receiver operating characteristic (ROC) curve and the corresponding area under the ROC curve (AUC), a metric that varied from 0 to 1 [27,31]. Evaluation of the model's performance was typically categorized into five grades as follows: failure (0.5–0.6), moderately accurate (0.6–0.7), reasonably accurate (0.7–0.8), highly accurate (0.8–0.9), and exceptionally accurate (0.9–1) [32]. According to the AUC results generated from the MaxEnt model, the average AUC values of 10 repetitions for *L. gmelinii*, *S. superba*, and *C. officinarum* were 0.819, 0.928, and 0.914, respectively, indicating that the model yielded highly reliable results in predicting the potential distribution (Figure S3).

The MaxEnt model's output for species distribution was presented in ASCII format, with values ranging from 0 to 1 [27]. Employing the natural breaks (Jenks) methodology within ArcGIS, the suitability levels for *L. gmelinii*, *S. superba*, and *C. officinarum* were divided into five categories: unsuitable habitat (<0.10; <0.08; <0.10), poorly suitable habitat (0.10–0.28; 0.08–0.27; 0.10–0.27), moderately suitable habitat (0.28–0.50; 0.27–0.48; 0.27–0.48), and highly suitable habitat (0.50–1.00; 0.48–1.00; 0.48–1.00), respectively.

2.2.3. Statistical Methods

The Theil–Sen trend analysis approach was applied to compute long-term trends on a pixel-by-pixel basis. The Theil–Sen trend analysis method is a robust non-parametric estimation algorithm that effectively mitigates the impact of outliers in extended time-series analyses, representing an enhancement over the least squares linear regression method [33]. The calculation formula is provided below:

$$\text{slope} = \text{median}\left(\frac{x_j - x_i}{j - i}\right), i < j \le n \tag{1}$$

where "slope" signifies the rate of change for the particular factor; i and j denote the respective years; n refers to the length of the time series; and x_i and x_j represent the values associated with the corresponding years. Ordinarily, a slope of 0 indicates the absence of change in the variable. However, in practice, instances of a slope precisely equaling 0 may be rare. Consequently, to account for this, a very small slope value is considered, implying negligible change. Thus, we establish a threshold interval of [−0.0005–0.0005] to signify no change. An affirmative slope value (>0.0005) signifies an ascending trend, while a negative value (<−0.0005) signifies a descending trend.

The Theil–Sen trend analysis is frequently paired with the Mann–Kendall test, a nonparametric statistical method, to evaluate long-term data trends. This approach was initially introduced by Mann in 1945 and subsequently improved by Kendall and Sneyers. This method does not require normal distribution or a linear trend and is not affected by missing values or outliers. This approach had found broad application in assessing trends in extended time-series data [33,34]. The test statistic S was computed using the following formula:

$$\mathrm{S} = \sum_{i=1}^{n-1}\sum_{j=i+1}^{n} sign(x_j - x_i) \tag{2}$$

$$\text{sign}(v) = \begin{cases} 1 & for \quad v > 0 \\ 0 & for \quad v = 0 \\ -1 & for \quad v < 0 \end{cases} \tag{3}$$

The variance of S was:

$$\text{var}(\mathrm{S}) = \frac{n(n-1)(2n+5)}{18} \tag{4}$$

Statistic Z was defined as:

$$\mathrm{Z} = \begin{cases} \frac{S-1}{\sqrt{var}} & for \quad S > 0 \\ 0 & for \quad S = 0 \\ \frac{S+1}{\sqrt{var}} & for \quad S < 0 \end{cases} \tag{5}$$

where $|\mathrm{Z}| \ge 1.96$ signifies that the trend has achieved statistical significance at the 95% confidence level.

Geodetector has proven to be an effective technique for the comprehensive analysis of the interplay between independent and cooperative variables concerning dependent variables [35,36]. The Geodetector model encompasses components such as the factor detector, interaction detector, risk detector, and ecological detector. In this study, we employed the factor detector and interaction detector to investigate the impact of climate factors and tree species' suitable area changes on forest ABC changes in China.

(1) The factor detector systematically measures the degree to which driving factor "X" elucidates the spatial disparities in forest ABC via the q-statistic value:

$$q = 1 - \frac{\sum_{h=1}^{L} N_h \delta_h^2}{N\delta^2} \tag{6}$$

where q symbolizes the explanatory potency of the influencing factor concerning the temporal and spatial shifts in ABC, h signifies the categorization levels of a driving factor, L represents the sample size of the impact factor, N_h and N stand for the counts of units within layer h and the entire region, respectively, and δ_h^2 and δ^2 denote the variances of h and the overall region, respectively.

(2) The interaction detector was designed to identify the way two driving factors interact and affect the ABC. Firstly, we calculated the q values of two driving factors that affect ABC, namely q(X1) and q(X2). Subsequently, we calculated the q values of the interactive effect (q(X1∩X2)) and compared them with q(X1) and q(X2) to identify the interaction type between the two driving factors (Table 2).

Table 2. Definition of the different interaction types in the Geodetector model.

Interaction Relationship	Interaction Types
q(Xi∩Xj) < Min(q(Xi), q(Xj))	Nonlinear-weaken
Min(q(Xi), q(Xj)) < q(Xi∩Xj) < Max(q(Xi), q(Xj))	Uni-variable weaken
q(Xi∩Xj) = q(Xi) + q(Xj)	Independent
Max(q(Xi), q(Xj)) < q(Xi∩Xj) < q(Xi) + q(Xj)	Bi-variable enhanced
q(Xi∩Xj) > q(Xi) + q(Xj)	Nonlinear-enhanced

3. Results

3.1. Satellite-Observed and Simulated Forest ABC in China

Satellite observations of the spatial distribution of aboveground biomass carbon in Chinese forests reveal that from 1993 to 2012, regions with high carbon density were located in the SEBF and TCBMF. Among these regions, North, Northeast, and the southern part of Southwest China exhibited carbon densities exceeding 70 Mg/ha. The LPJ model driven by observed data, including temperature and precipitation, successfully captured the spatial distribution of aboveground carbon stocks in forests. However, overestimation was observed in the Northeast region, as well as in Central, South, East, and West China. The forest carbon stocks simulated using CMIP6-driven models showed good consistency with those driven by observed data, but underestimated carbon stocks were observed in North China and the southeastern coastal areas (Figure 2).

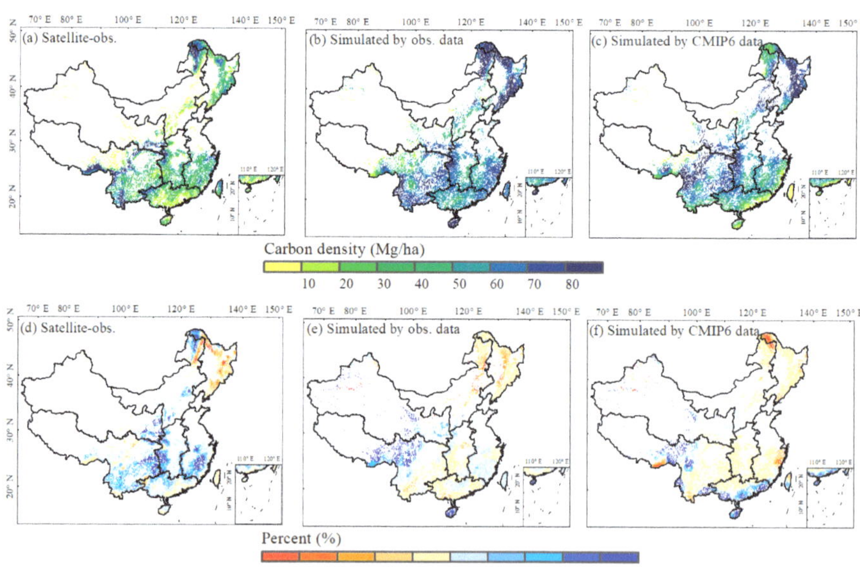

Figure 2. Spatial distribution of the annual mean forest ABC during 1993–2012 and the percentage change in the annual average forest ABC from 2001 to 2012 compared with 1993–2000 in China: (**a**,**d**) satellite-observed; (**b**,**e**) simulated, driven by observed meteorological data; (**c**,**f**) simulated, driven by CMIP6 data.

The LPJ model simulation results showed that the areas with high forest ABC in China from 1982 to 2014 were mainly distributed in the SEBFs and TCBMFs of China (Figure 3a). However, these forest areas with high ABC experienced a decreasing trend, and the area with a significant decreasing trend accounted for 11.56%, mainly distributed in Northeast China and Southwest China. Approximately 40.15% of the forest ABC showed a significant increasing trend, mainly in southwest, central, east, and northern China (Figure 3b). The results of the average annual forest ABC in China clearly show that the overall ABC in the study area increased at a growth rate of 0.19 Mg/ha/a before 1992 and then changed to a decreasing trend (−0.03 Mg/ha/a). The overall trend in TCBMF was similar to that in the study area, and the decline rate is more obvious (−0.17 Mg/ha/a) after 1992. Meanwhile, the forest ABC in SEBF showed a slower growth rate (before 2000: 0.12 Mg/ha/a; after 2000: 0.01 Mg/ha/a).

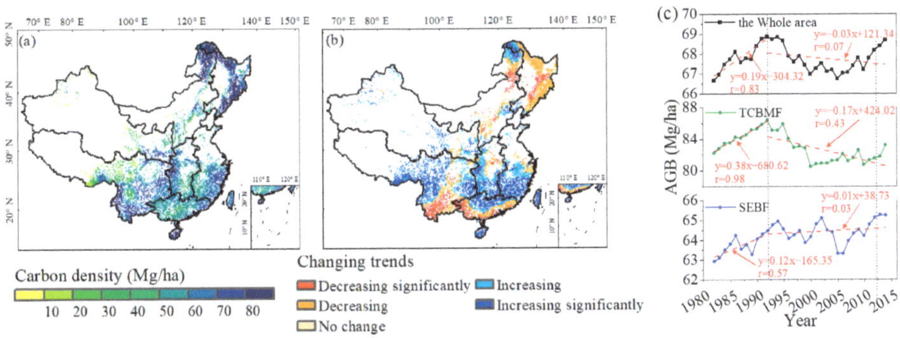

Figure 3. Spatial distribution of carbon density (**a**), changing trends (**b**), and (**c**) the annual mean forest ABC during 1982–2014.

The results of simulations using three different models and a multiple-model ensemble (MME) to drive the LPJ model to calculate the ABC trends in China from 2015 to 2060 under different SSP scenarios are shown in Figure 4. These simulations consistently indicate an overall increasing trend in forest ABC from 2015 to 2060. Especially in the SSP5-8.5 scenario, all models and the MME predicted a higher proportion of pixels with significantly increased forest ABC compared to other scenarios. The forest ABC of the TCBMF will significantly decrease in all scenarios, while forest ABC of the SEBF generally exhibit a significant increasing trend, except in the SSP1-2.6 scenario. The MME results show that under the SSP1-2.6, SSP2-4.5, and SSP5-8.5 scenarios, the proportion of pixels with a significantly increasing trend in forest ABC will be 46.16%, 67.18%, and 75.53%, respectively. Conversely, the proportion of pixels with a significantly decreasing trend will be 10.49%, 12.93%, and 16.09%, respectively. Under the SSP5-8.5 scenario, the BCC model results exhibit the highest proportion of pixels with a significantly increasing trend in forest ABC at 80.34%, while the proportion with a significantly decreasing trend will be lowest at 5.90%. The Can model results show that under the SSP1-2.6 scenario, the smallest proportion of forest ABC significantly increasing will be 29.53%, whereas, under the SSP5-8.5 scenario, the largest proportion of significantly decreasing forest ABC will be 19.63%. Overall, from 2015 to 2060, the forest ABC trajectory in TCBMF is expected to experience a continuous decrease, in stark contrast to SEBF, which is projected to undergo a significant increase.

Under different scenarios, the statistical results of annual forest ABC in different regions of China from 2015 to 2060 are shown in Figure 5. Overall, the rates of change in forest ABC in China are expected to remain stable. However, there will be notable differences in various regions across the country. In the TCBMF area, the forest ABC change rate will be consistently declining across various scenarios. The most significant rate of decreae is projected to be −0.46 Mg/ha/a under the SSP5-8.5 scenario. In contrast, the rate of change in forest ABC in the SEBF region is expected to increase in all scenarios. The maximum change rate will be 0.13 Mg/ha/year in the SSP5-8.5 scenario (Figure 5).

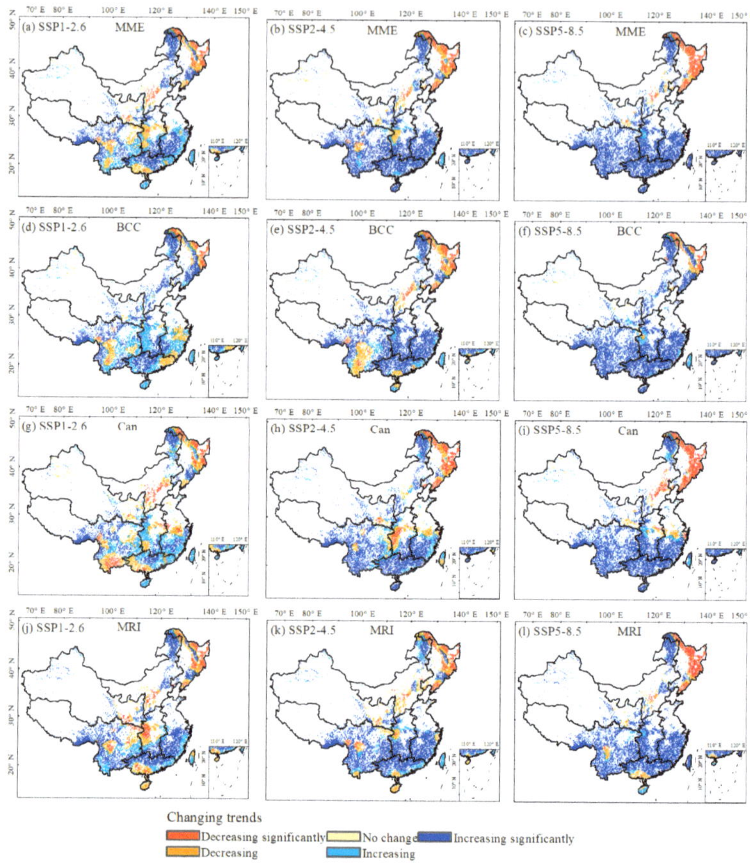

Figure 4. Changing trends of the annual mean forest ABC from 2015 to 2060 in China under scenarios of SSP1-2.6, SSP2-4.5, and SSP5-8.5.

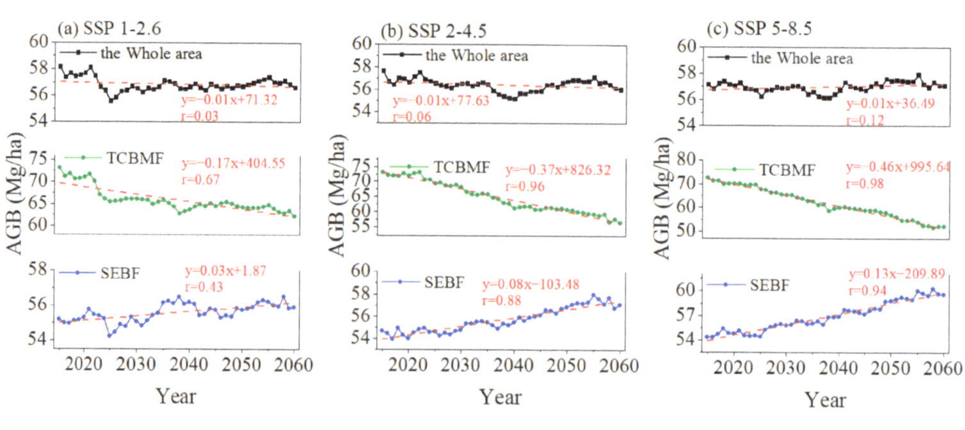

Figure 5. The annual mean forest ABC from 2015 to 2060 in China under scenarios of SSP1-2.6, SSP2-4.5, and SSP5-8.5.

3.2. Potential Suitable Area Shifts of Typical Tree Species in Response to Climate Change

Figure 6 illustrates the current potential geographic distribution of the three typical species. The statistical results showed that the high-, moderate-, and low-suitability areas for *L. gmelinii* are currently estimated to cover an area of 133.47 × 10^4 km^2, 145.31 × 10^4 km^2, and 223.58 × 10^4 km^2, respectively. The high-suitability areas for *L. gmelinii* are mainly distributed in the northern part of North China and the Northeast region. Under different scenarios, the highly suitable area of *L. gmelinii* is projected to decrease, with a reduction to 50.56 × 10^4 km^2 by 2060 under the SSP5-8.5 scenario. The total suitable area is expected to decrease from 475.25 × 10^4 km^2 in the current period to 375.29 × 10^4 km^2 in 2060 under the SSP5-8.5 scenario (Figure 6a–d, Table 3).

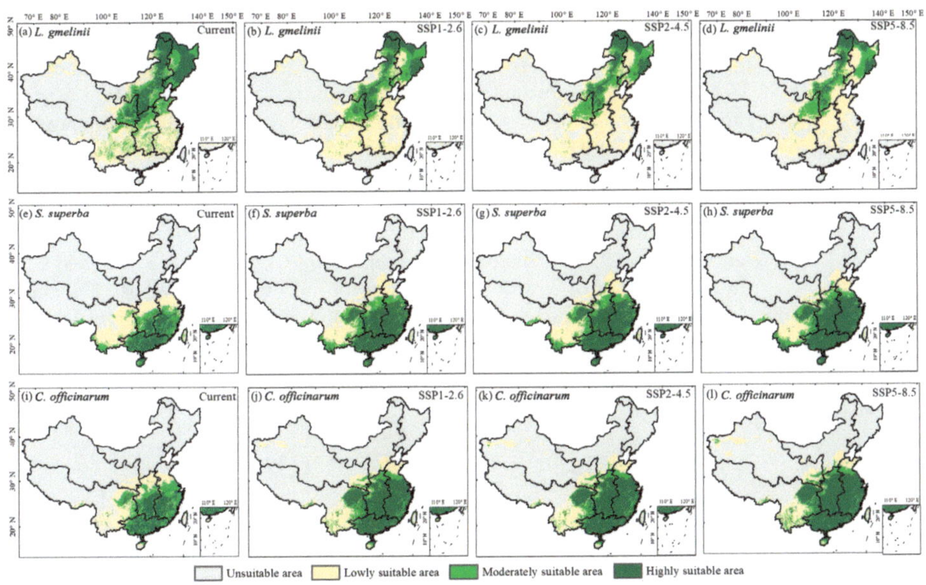

Figure 6. Distribution of current and future potential suitable areas of three typical tree species: **(a–d)** *L. gmelinii*; **(e–h)** *S. superba*; **(i–l)** *C. officinarum*.

Table 3. Statistical results of potentially suitable area (unit: 10^4 km^2).

Species	Periods	Scenarios	Low	Moderate	High	Total
L. gmelinii	Current		223.58	145.31	133.47	475.25
	2041–2060	SSP1-2.6	246.86	102.60	67.81	417.27
		SSP2-4.5	241.41	96.62	58.45	396.48
		SSP5-8.5	234.24	90.49	50.56	375.29
S. superba	Current		98.20	63.55	49.96	211.71
	2041–2060	SSP1-2.6	81.85	62.68	114.47	259.00
		SSP2-4.5	88.57	66.48	113.25	268.30
		SSP5-8.5	79.09	59.57	136.97	275.63
C. officinarum	Current		93.37	76.79	60.41	230.57
	2041–2060	SSP1-2.6	84.85	55.75	141.78	282.38
		SSP2-4.5	91.86	56.63	140.98	289.47
		SSP5-8.5	92.86	52.13	157.90	302.89

In the current period, it was estimated that the high-, moderate-, and low-suitability areas for *S. superba* are 49.96 × 10^4 km^2, 63.55 × 10^4 km^2, and 98.20 × 10^4 km^2, respectively.

The high-suitability area is primarily located in the southern region, with expansions expected in the future to cover all of Central China, East China, and the eastern part of Southwest China. Under the scenarios of SSP1-2.6, SSP2-4.5, and SSP5-8.5, the high-suitability area for *S. superba* will expand in 2060. Under the scenario of SSP5-8.5, the expansion of the high-suitability area will be the largest, with an expected expansion of 136.97×10^4 km^2 in 2060 (Figure 6e–h, Table 3).

For *C. officinarum*, the currently suitable areas were estimated to be 60.41×10^4 km^2, 76.79×10^4 km^2, and 93.37×10^4 km^2 for the high-, moderate-, and low-suitability areas, respectively. The high- and medium-potential suitable zones for *C. officinarum* are mainly distributed in Northeast China and North China. The high- and moderate-suitability areas for *C. officinarum* are primarily distributed in Central China, East China, and South China. By 2060, the area of highly suitable areas for *C. officinarum* in the SSP1-2.6 and SSP2-4.5 scenarios will increase to 141.78×10^4 km^2 and 140.98×10^4 km^2, respectively, with a maximum of 157.90×10^4 km^2 in the SSP5-8.5 scenario (Figure 6i–l, Table 3).

We further examined the potential changes in the distribution of suitable areas for different tree species based on the distribution of high-suitability areas in both current and future scenarios. (Figure 7). The results indicate that the future distribution of the high-suitability zone for *L. gmelinii* is projected to shrink primarily in North China and the Northeast region, with the greatest reduction in area under the SSP5-8.5 scenario in 2060, reaching up to 65.71×10^4 km^2. As for *S. superba* and *C. officinarum*, they will both tend to expand towards the central regions of Central China and East China, as well as the eastern part of Southwest China, under all future scenarios. The expansion is most significant under the SSP5-8.5 scenario, with an increase in area of 94.07×10^4 km^2 for *S. superba* and 104.22×10^4 km^2 for *C. officinarum* (Figure 7, Table 4).

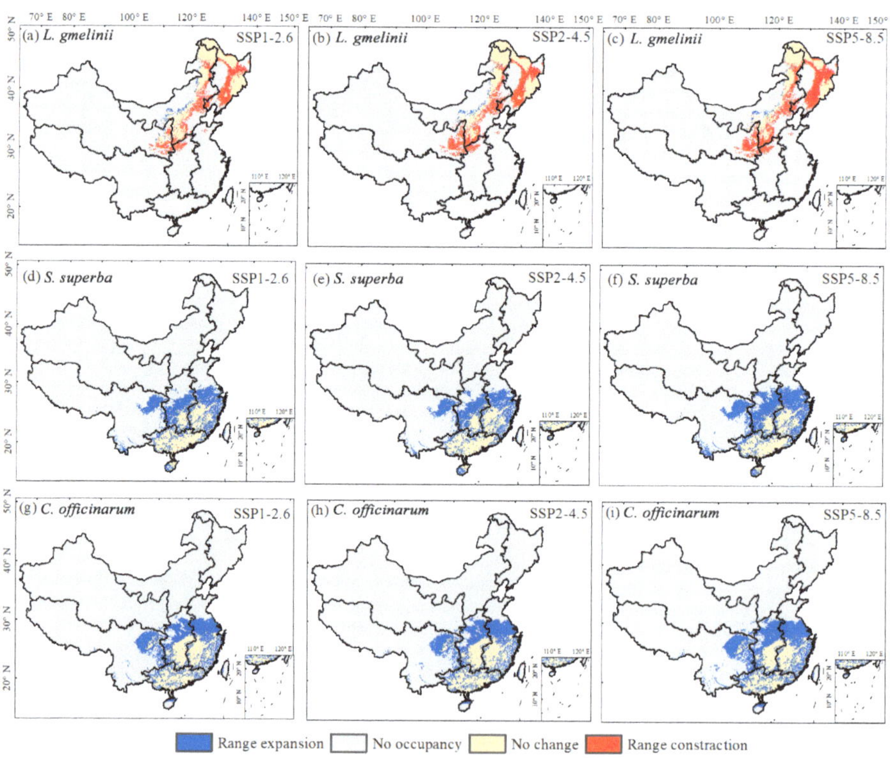

Figure 7. Future changes in the potential geographical distribution of three typical tree species: (**a–c**) *L. gmelinii*; (**d–f**) *S. superba*; (**g–i**) *C. officinarum*.

Table 4. Statistical results of area change of potential highly suitable areas (unit: 10^4 km^2).

Species	Scenarios	Range Expansion	No Occupancy	No Change	Range Contraction
L. gmelinii	SSP1-2.6	3.97	823.86	68.86	49.05
	SSP2-4.5	2.80	825.03	60.29	57.61
	SSP5-8.5	2.28	825.54	52.20	65.71
S. superba	SSP1-2.6	70.12	819.28	55.42	0.18
	SSP2-4.5	68.67	820.73	55.44	0.16
	SSP5-8.5	94.07	795.33	55.46	0.14
C. officinarum	SSP1-2.6	87.00	793.95	66.65	0.03
	SSP2-4.5	86.04	794.91	66.60	0.09
	SSP5-8.5	104.22	776.73	66.66	0.02

3.3. Coupling Impacts of Climate and Species' Suitable Area Change on Forest ABC

The results of the superposition of the ABC changes simulated using the LPJ model and the changes in the suitable areas of dominant tree species simulated using the Maxent model from 2015 to 2060 are shown in Figure 8. Under SSP1-2.6 scenario, high-suitability areas of *L. gmelinii* will decrease in some TCBMF areas in Northeast China, but forest ABC will still increase, accounting for 20.46% of the total area. Meanwhile, the increase in forest ABC in most SEBF areas will be consistent with the single or co-increase in the high-suitability areas of *S. superba* and *C. officinarum*, which account for 12.66%, 13.54%, and 24.55%, respectively (Figure 8a,d). Under the scenarios of SSP2-4.5 and SSP5-8.5, the main theme in the TCBMF region will be more evident in the reduction in forest ABC and the shrinking of highly suitable areas for dominant tree species, with an area proportion of 18.23% and 21.96%, respectively. The synergistic trend between the increase in forest ABC and the expansion of highly suitable habitat for dominant tree species in the southern SEBF region will be even more pronounced, with the proportion of this trending area accounting for 63.24% and 65.10% under the SSP2-4.5 and SSP5-8.5 scenarios, respectively (Figure 8b–d).

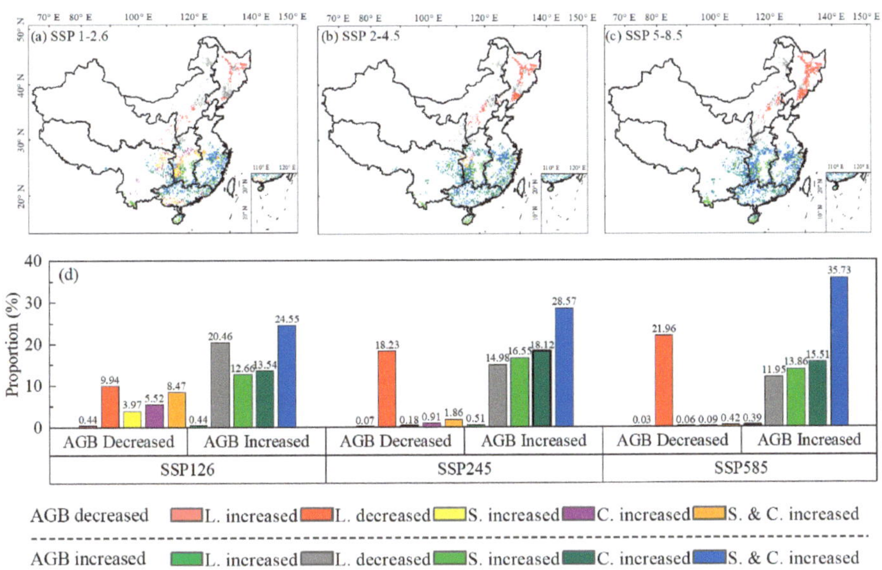

Figure 8. Relationship between ABC and tree species' suitable area changes under SSP1-2.6 (**a**), SSP2-4.5 (**b**), SSP5-8.5 (**c**), and their area proportion statistics (**d**); L.: *L. gmelinii*; S.: *S. superba*; C.: *C. officinarum*.

The individual effects of the driving factors behind forest ABC variations were analyzed using factor detectors (Figure 9). For Northeast China, climate factors have a relatively strong ability to explain forest ABC, especially temperature, followed by precipitation and the distribution of *L. gmelinii*. The *q* value of temperature interpretation will reach 0.224 in the scenario of SSP1-2.6. For the southern parts of China, the temperature will still be the driving factor with the highest explanatory power (0.094) for forest ABC in the SSP1-2.6 scenario, while in the SSP2-4.5 scenario, the *q* value of the explanatory power for forest ABC shows the order of *C. officinarum*, precipitation, *S. superba*, and temperature (Figure 9b).

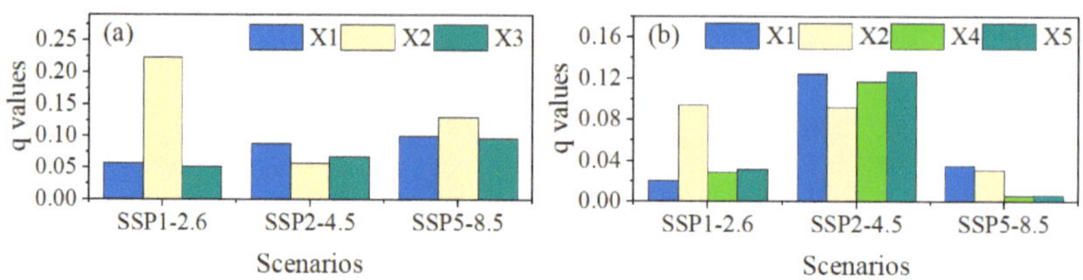

Figure 9. The *q* values of driving factors for forest ABC variations: (**a**) TCBMF; (**b**) SEBF. X1: precipitation; X2: temperature; X3: *L. gmelinii*; X4: *S. superba*; X5: *C. officinarum*.

The results of synergistic detection among the factors showed that the effects of the interaction between temperature and precipitation and *L. gmelinii* on forest ABC in northeast China will be bi-enhanced and nonlinear enhanced in the SSP1-2.6 and SSP2-4.5 scenarios, respectively (Figure 10a,b). In the SSP5-8.5 scenario, precipitation has an effect on the ABC of northeastern forests through the interaction of bi-enhanced, with temperature, and nonlinear enhanced interaction with *L. gmelinii*, respectively. Temperature and *L. gmelinii* also showed a bi-enhanced interaction affecting the northeastern forest ABC (Figure 10c).

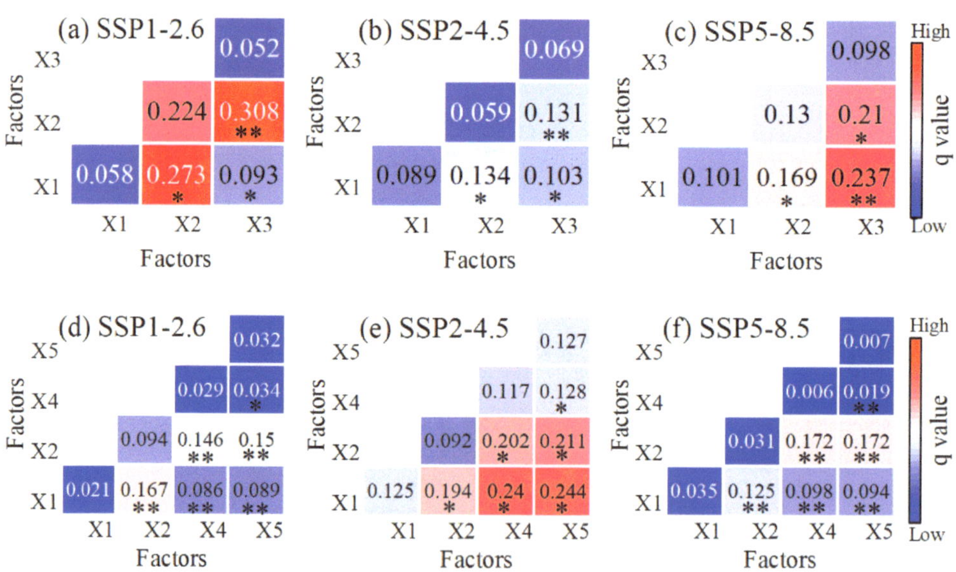

Figure 10. The *q* values of interaction between factors. Note: * and ** represent the interaction types of bi-variable enhanced and nonlinear-enhanced, respectively. (**a**–**c**) TCBMF; (**d**–**f**) SEBF. X1: precipitation; X2: temperature; X3: *L. gmelinii*; X4: *S. superba*; X5: *C. officinarum*.

In the SSP1-2.6 and SSP5-8.5 scenarios, temperature, precipitation, *S. superba*, and *C. officinarum* will affect the forest ABC in southern China through the synergistic effect of nonlinear enhancement between each two (Figure 10d,f). In the SSP2-4.5 scenario, the bi-enhanced synergistic effect between pairwise factors will have a stronger effect on forest ABC (Figure 10e).

4. Discussion

The satellite dataset showed that high values of forest ABC in China from 1993 to 2012 were mainly distributed in North China, Northeast China, and Southwest China, especially in northeastern and southwestern China (>70 Mg/ha), and forest ABC increased at a rate of approximately 0.37%yr^{-1}, which is aligned with the results of Chen et al. [6] and Tang et al. [1], who integrated various types of remote sensing observations and field measurements to analyze the spatial distribution and temporal trends of forest ABC using regression analysis and machine learning methods. The mean value of satellite observation of forest ABC (37.58 Mg/ha) was lower than Masson pine with 18-year-old stands (86.0 Mg/ha of total biomass carbon) [37]. Meanwhile, simulated forest ABC (62.87 Mg/ha) was greater than Masson pine with 28-year-old stands (112.70 Mg/ha of total biomass carbon). The main reasons for the differences in results are primarily attributed to variations in carbon sequestration capabilities among different species, disparities within the same species at different ages, and human-induced factors of interference [37]. To enhance the accuracy of remote sensing retrieval or model simulation of forest biomass carbon, it is essential to incorporate considerations for vegetation species diversity, stand age, and human-induced disturbances.

The forest ABC of TCBMF in Northeast China showed a significant downward trend from 1993 to 2012, while the ABC of SEBF in South China showed a continuous upward trend. Combining the LPJ model and GCM data, the future forest ABC indicated a significant overall increase in ABC in SEBF regions under all SSP1-2.6, SSP2-4.5, and SSP5-8.5 conditions from 2015 to 2060, which was consistent with the results obtained by Kong et al. (2022) [2]. This phenomenon would be attributed to the higher precipitation and temperature conditions in SEBF regions, which can meet the requirements for vegetation growth [2]. Liu et al. (2015) [38] believed that tropical deforestation declines in China had also contributed to the increase in forest ABC in the region. Furthermore, the forest ABC in TCBMF of China will experience a continuing significant downward trend from 2015 to 2060. Sun et al. (2022) [39] also obtained similar results regarding the changes in forest carbon stocks in the Northeast region of China using the INVEST model based on land use and land cover changes. The main reasons for these changes are associated with the conversion of forest land to agricultural land and fluctuations in the economic timber forests of the Lesser Khingan Mountains. In addition, common forest pests and diseases in temperate forests, wildfires, and the degradation of old-growth forests may also lead to a decline in the forest's carbon sequestration potential in this region [38].

Understanding the impact of climate change on forests' carbon sequestration potential needs to be linked to changes in areas suitable for tree species. Our results showed that the future distribution of the highly suitable area for coniferous species (*L. gmelinii*) is projected to shrink primarily in Northeast China, while evergreen broad-leaved tree species (*S. superba* and *C. officinarum*) will tend to expand towards the central regions of China from 2015 to 2060 based on the MaxEnt model. Similar conclusions have also been reported, for example, by Bai et al. (2016) and Chen et al. (2022) [40,41], who found, through research on the relationship between tree rings and temperature changes, that rapid warming will lead to the gradual decline of *L. gmelinii* in Northeast China. Du et al. (2022) [42] found that the future suitable habitat area of eight species of tree species in the Northeast region, including *L. gmelinii*, will decrease by 10%–30%. Wu et al. (2012) [43] found that climate warming caused a significant decrease in the growth of *L. gmelinii* in semi-northern coniferous forests during the last two decades. These findings are consistent with ours, but such studies rarely explore the impact of changes in species' suitable areas on forest ABC. Combining climate change, species' suitable area changes and forest ABC changes can enhance our understanding of the carbon sequestration potential of forests.

Previous research has primarily revealed changes in the carbon sequestration potential of China, and emphasized climate change's influence on them, but ignored the effects of changes in tree species' suitable area [43–45]. Overlay analysis of the LPJ and MaxENT models showed that the future trends of forest ABC reduction in TCBMF and increase in the SEBF region of China are spatially consistent with the shrinking of highly suitable areas for *L. gmelinii* in TCBMF and the expansion of highly suitable areas for *S. superba* and *C. officinarum* in the SEBF region. The outcomes of the Geodetector model found that climate and tree species' highly suitable area changes showed a bi-variable and nonlinear synergistic enhancement of forest ABC change [46]. Generally, warming and humidification of the climate are conducive to the growth of vegetation, but this has species and space differences [47]. Larch pollen is negatively correlated with temperature [48]. For the Northeast region, which is dominated by coniferous species such as Larch, a reduction in Larch pollen leads to an insufficient supply of young species under warming conditions, coupled with the degradation of old-aged coniferous forests in the region, resulting in a decline in forest ABC. In addition, the thawing of permafrost in Northeast China due to climate warming may provide more water and space for plants in the short term [40,49], but as warming intensifies in the upcoming decades (0.36 °C per decade), the thawing of permafrost will lead to a long-term reduction in the coniferous forest ecosystem [50], leading to forest fragmentation and forest loss and reducing the potential of forests to sequester carbon [51]. Notably, the reduction in the carbon sequestration potential of forests in the Northeast may be short-term. Temperate broad-leaved species are expanding into northern forest patches, causing the Northeastern coniferous forest to retreat and be replaced by deciduous broad-leaved forests [40,48,52,53]. The carbon sequestration potential of forests in the Northeast remains uncertain in the long term, as some relatively uncommon species will become more common under climate change. Therefore, the long-term carbon sequestration potential of forests in the Northeast would depend on newly established tree species.

Despite this study innovatively combines embedded physics-based models with statistical models, revealing changes in China's future forest carbon sequestration potential and its relationship with climate change and tree species' suitable area shifts, limitations still exist within this study. Firstly, while this study considered the impact of climate and vegetation on carbon sequestration potential, it did not account for factors like forest fires, pests, diseases, stand age, and the effects of carbon dioxide fertilization [54]. Secondly, due to model limitations, this research did not include forest belowground biomass carbon, a crucial component of the forest ecosystem's carbon composition [4,55]. Additionally, apart from the inherent uncertainties of the LPJ and MaxEnt models, the accuracy of future meteorological data, which served as input for both models, as well as the correction of biases, requires further improvement. Despite these limitations, this study introduces a novel approach to assessing the relationship between forest carbon sequestration potential, climate change, and shifts in tree species' suitable distribution areas.

5. Conclusions

Based on GCM model data coupled with the LPJ model and MaxENT model, this study predicted the distribution of forest ABC and suitable areas for typical tree species in China and explored the impact of climate variations and changes in the suitable area for tree species on forest ABC in China from 2015 to 2060. The main results are as follows:

(1) The geographical distribution of forests in China has spatial differences, with the northeast (TCBMF), southwest, and southeast (SEBF) being the predominant hubs of distribution. However, the forest ABC generated from the LPJ model clearly showed that the overall ABC in the study area increased at a growth rate of 0.19 Mg/ha/a before 1992, and then changed to a decreasing trend (−0.03 Mg/ha/a). Meanwhile, a clear and concerning downward trajectory was noticeable in the forest ABC in TCBMF (before 2000: 0.38 Mg/ha/a, after 2000: −0.17 Mg/ha/a), while that of SEBF showed an increasing

trend (before 2000: 0.12 Mg/ha/a, after 2000: 0.01 Mg/ha/a) after 2000 based on satellite observation and LPJ simulation.

(2) The trajectory of forest ABC in TCBMF is set to undergo a sustained decline with a maximum decline rate of −0.46 Mg/ha/a under the SSP5-8.5 scenario. In stark contrast, the forest ABC in SEBF is anticipated to experience a notable upswing at a rate of 0.13 Mg/ha/a under SSP5-8.5. Under the SSP5-8.5 scenario, the proportion of pixels with significantly increased or decreased forest ABC is higher compared to other scenarios. Among them, the BCC model results will have the highest proportion of pixels with a significantly increasing trend in forest ABC at 80.34%, whereas the Can will have the largest proportion of significantly decreasing forest ABC at 19.63%.

(3) The variations in forest ABC are highly attributed to climate and changes in tree species' highly suitable area. By 2060, the suitable area for *L. gmelinii* in TCBMF will be significantly reduced to a maximum of 65.71×10^4 km^2 under SSP5-8.5, while *S. superba* and *C. officinarum* in SEBF will expand to peaks of 94.07×10^4 km^2 and 104.22×10^4 km^2, respectively. Spatially, the decrease in forest ABC in TCBMF will be consistent with the reduction in highly suitable areas of *L. gmelinii*, while the increase in forest ABC in SEBF regions will be consistent with the expansion of *S. superba and C. officinarum*. Climate and tree species' suitable area changes showed bi-variable and nonlinear synergistic enhancement of forest ABC change based on the geographic detector's results. The findings of this work can provide valuable insights for developing effective sustainability management strategies under a carbon neutrality context.

Supplementary Materials: The following supporting information can be downloaded at: https://www.mdpi.com/article/10.3390/f14102053/s1, Figure S1: Spatial distribution map and Taylor chart of observed annual mean temperature/precipitation and simulated values (MME) in China during 1982–2014; Figure S2. Annual mean temperature and precipitation anomalies in China from 1982 to 2060. Shaded areas and grey bars represent the mean ± 1 standard deviation ranges for the three CMIP6 models; Figure S3. Reliability test of the MaxEnt model created for: (a) *Larix gmelinii*; *Schima superba Gardner* & Champ and (c) *Camphora officinarum*.

Author Contributions: B.Z.: data curation, methodology, and writing—original draft preparation. R.K.: data curation and methodology. M.W.: writing—review and editing. G.L. and X.S.: data curation and software. H.T.: investigation. Z.Z.: validation and supervision. All authors have read and agreed to the published version of the manuscript.

Funding: This paper is financially supported by the Key Research and Development Program of Xinjiang Uygur Autonomous Region, China (grant number 2022B03030), National Natural Science Foundation of China (grant number 41971025), West Light Foundation of the Chinese Academy of Sciences (grant numbers 2019-XBYJRC001, 2019-XBQNXZ-B-004). The project is supported by the Flexible Talent Introduction Project of Xinjiang Uygur Autonomous Region and the Priority Academic Program Development of Jiangsu Higher Education Institutions (PAPD).

Data Availability Statement: Data will be made available on request.

Acknowledgments: Acknowledgement for the data support from China Meteorological Data Service Centre (http://data.cma.cn/en, accessed on 31 January 2020). In addition, we sincerely thank the editor and anonymous reviewers for their valuable comments and suggestions to improve the quality of this paper.

Conflicts of Interest: The authors declare no conflict of interest.

References

1. Tang, X.; Zhao, X.; Bai, Y.; Tang, Z.; Wang, W.; Zhao, Y.; Wan, H.; Xie, Z.; Shi, X.; Wu, B.; et al. Carbon Pools in China's Terrestrial Ecosystems: New Estimates Based on an Intensive Field Survey. *Proc. Natl. Acad. Sci. USA* **2018**, *115*, 4021–4026. [CrossRef] [PubMed]
2. Kong, R.; Zhang, Z.; Huang, R.; Tian, J.; Feng, R.; Chen, X. Projected Global Warming-Induced Terrestrial Ecosystem Carbon across China under SSP Scenarios. *Ecol. Indic.* **2022**, *139*, 108963. [CrossRef]
3. Cairns, M.A.; Brown, S.; Helmer, E.H.; Baumgardner, G.A. Root Biomass Allocation in the World's Upland Forests. *Oecologia* **1997**, *111*, 1–11. [CrossRef] [PubMed]

4. Ma, H.; Mo, L.; Crowther, T.W.; Maynard, D.S.; Van Den Hoogen, J.; Stocker, B.D.; Terrer, C.; Zohner, C.M. The Global Distribution and Environmental Drivers of Aboveground versus Belowground Plant Biomass. *Nat. Ecol. Evol.* **2021**, *5*, 1110–1122. [CrossRef]
5. Green, J.K.; Keenan, T.F. The Limits of Forest Carbon Sequestration. *Science* **2022**, *376*, 692–693. [CrossRef]
6. Chen, Y.; Feng, X.; Fu, B.; Ma, H.; Zohner, C.M.; Crowther, T.W.; Huang, Y.; Wu, X.; Wei, F. Maps with 1km Resolution Reveal Increases in Above- and Belowground Forest Biomass Carbon Pools in China over the Past 20 Years. *Earth Syst. Sci. Data* **2023**, *15*, 897–910. [CrossRef]
7. Kumar, L.; Mutanga, O. Remote Sensing of Above-Ground Biomass. *Remote Sens.* **2017**, *9*, 935. [CrossRef]
8. He, H.; Wang, S.; Zhang, L.; Wang, J.; Ren, X.; Zhou, L.; Piao, S.; Yan, H.; Ju, W.; Gu, F.; et al. Altered Trends in Carbon Uptake in China's Terrestrial Ecosystems under the Enhanced Summer Monsoon and Warming Hiatus. *Natl. Sci. Rev.* **2019**, *6*, 505–514. [CrossRef]
9. Zhang, D.; Zhao, Y.; Wu, J. Assessment of Carbon Balance Attribution and Carbon Storage Potential in China's Terrestrial Ecosystem. *Resour. Conserv. Recycl.* **2023**, *189*, 106748. [CrossRef]
10. Tong, X.; Brandt, M.; Yue, Y.; Horion, S.; Wang, K.; Keersmaecker, W.D.; Tian, F.; Schurgers, G.; Xiao, X.; Luo, Y.; et al. Increased Vegetation Growth and Carbon Stock in China Karst via Ecological Engineering. *Nat. Sustain.* **2018**, *1*, 44–50. [CrossRef]
11. Sitch, S.; Smith, B.; Prentice, I.C.; Arneth, A.; Bondeau, A.; Cramer, W.; Kaplan, J.O.; Levis, S.; Lucht, W.; Sykes, M.T.; et al. Evaluation of Ecosystem Dynamics, Plant Geography and Terrestrial Carbon Cycling in the LPJ Dynamic Global Vegetation Model: Lpj Dynamic Global Vegetation Model. *Glob. Change Biol.* **2003**, *9*, 161–185. [CrossRef]
12. Bondeau, A.; Smith, P.C.; Zaehle, S.; Schaphoff, S.; Lucht, W.; Cramer, W.; Gerten, D.; Lotze-Campen, H.; Mueller, C.; Reichstein, M.; et al. Modelling the Role of Agriculture for the 20th Century Global Terrestrial Carbon Balance. *Glob. Change Biol.* **2007**, *13*, 679–706. [CrossRef]
13. Zhao, S.; Yan, X.; Yang, S.; Tao, D.; Dai, L. Simulating Responses of Northeastern China Forests to Potential Climate Change. *J. For. Res.* **1998**, *9*, 166–172. [CrossRef]
14. Shao, G. Sensitivities of Species Compositions of the Mixed Forest in Eastern Eurasian Continent to Climate Change. *Glob. Planet. Chang.* **2003**, *37*, 307–313. [CrossRef]
15. He, H.S.; Hao, Z.; Mladenoff, D.J.; Shao, G.; Hu, Y.; Chang, Y. Simulating Forest Ecosystem Response to Climate Warming Incorporating Spatial Effects in North-Eastern China. *J. Biogeogr.* **2005**, *32*, 2043–2056. [CrossRef]
16. Li, G.-Q.; Fan, B.A.I.; Wei-Guo, S. Different Responses of Radial Growth to Climate Warming in *Pinus Koraiensis* and *Picea Jezoensis* Var. *Komarovii* at Their Upper Elevational Limits in Changbai Mountain, China. *Chin. J. Plant Ecol.* **2011**, *35*, 500. [CrossRef]
17. Yu, D.; Wang, Q.; Wang, Y.; Zhou, W.; Ding, H.; Fang, X.; Jiang, S.; Dai, L. Climatic Effects on Radial Growth of Major Tree Species on Changbai Mountain. *Ann. For. Sci.* **2011**, *68*, 921. [CrossRef]
18. Dai, L.; Jia, J.; Yu, D.; Lewis, B.J.; Zhou, L.; Zhou, W.; Zhao, W.; Jiang, L. Effects of Climate Change on Biomass Carbon Sequestration in Old-Growth Forest Ecosystems on Changbai Mountain in Northeast China. *For. Ecol. Manag.* **2013**, *300*, 106–116. [CrossRef]
19. Wang, M.; Guan, Q. Prediction of Potential Suitable Areas for *Broussonetia Papyrifera* in China Using the MaxEnt Model and CIMP6 Data. *J. Plant Ecol.* **2023**, *16*, rtad006. [CrossRef]
20. Fang, J.; Chen, A.; Peng, C.; Zhao, S.; Ci, L. Changes in Forest Biomass Carbon Storage in China Between 1949 and 1998. *Science* **2001**, *292*, 2320–2322. [CrossRef]
21. He, X.; Burgess, K.S.; Gao, L.-M.; Li, D.-Z. Distributional Responses to Climate Change for Alpine Species of *Cyananthus* and *Primula* Endemic to the Himalaya-Hengduan Mountains. *Plant Divers.* **2019**, *41*, 26–32. [CrossRef]
22. Rivas, C.A.; Navarro-Cerillo, R.M.; Johnston, J.C.; Guerrero-Casado, J. Dry Forest Is More Threatened but Less Protected than Evergreen Forest in Ecuador's Coastal Region. *Environ. Conserv.* **2020**, *47*, 79–83. [CrossRef]
23. Zhang, H.; Xiao, Y.; Huang, F. Estimated Carbon Storage and Economic Value on the Mostly Forest Ecosystem in Hunan Province. In *The Dynamics of Urban Agglomeration in China: Preferences of Energy-Saving and Environment-Friendly Society*; Zhang, H., Zhang, H., Eds.; Aussino Academic Publishing House: Marrickville, Australia, 2009; p. 137.
24. Zhang, K.; Gong, Y.; Fa, H.; Zhao, M. CO_2 Flux Characteristics of Different Plant Communities in a Subtropical Urban Ecosystem. *Sustainability* **2019**, *11*, 4879. [CrossRef]
25. Jiang, S.; Guo, X.; Zhao, P.; Liang, H. Radial Growth–Climate Relationship Varies with Spatial Distribution of *Schima Superba* Stands in Southeast China's Subtropical Forests. *Forests* **2023**, *14*, 1291. [CrossRef]
26. Tian, J.; Zhang, Z.; Ahmed, Z.; Zhang, L.; Su, B.; Tao, H.; Jiang, T. Projections of Precipitation over China Based on CMIP6 Models. *Stoch. Environ. Res. Risk Assess.* **2021**, *35*, 831–848. [CrossRef]
27. Merow, C.; Smith, M.J.; Silander Jr, J.A. A Practical Guide to MaxEnt for Modeling Species' Distributions: What It Does, and Why Inputs and Settings Matter. *Ecography* **2013**, *36*, 1058–1069. [CrossRef]
28. Zaehle, S.; Sitch, S.; Prentice, I.C.; Liski, J.; Cramer, W.; Erhard, M.; Hickler, T.; Smith, B. The Importance of Age-Related Decline in Forest NPP for Modeling Regional Carbon Balances. *Ecol. Appl.* **2006**, *16*, 1555–1574. [CrossRef]
29. Pappas, C.; Fatichi, S.; Rimkus, S.; Burlando, P.; Huber, M.O. The Role of Local-Scale Heterogeneities in Terrestrial Ecosystem Modeling. *J. Geophys. Res. Biogeosci.* **2015**, *120*, 341–360. [CrossRef]
30. Phillips, S.J.; Dudik, M. Modeling of Species Distributions with Maxent: New Extensions and a Comprehensive Evaluation. *Ecography* **2008**, *31*, 161–175. [CrossRef]

31. Phillips, S.J.; Anderson, R.P.; Schapire, R.E. Maximum Entropy Modeling of Species Geographic Distributions. *Ecol. Model.* **2006**, *190*, 231–259. [CrossRef]
32. Zhou, Y.; Zhang, Z.; Zhu, B.; Cheng, X.; Yang, L.; Gao, M.; Kong, R. MaxEnt Modeling Based on CMIP6 Models to Project Potential Suitable Zones for Cunninghamia Lanceolata in China. *Forests* **2021**, *12*, 752. [CrossRef]
33. Fernandes, R.G.; Leblanc, S. Parametric (Modified Least Squares) and Non-Parametric (Theil–Sen) Linear Regressions for Predicting Biophysical Parameters in the Presence of Measurement Errors. *Remote Sens. Environ.* **2005**, *95*, 303–316. [CrossRef]
34. Mann, H.B. Nonparametric Tests Against Trend. *Econometrica* **1945**, *13*, 245–259. [CrossRef]
35. Kong, R.; Zhang, Z.; Zhang, Y.; Wang, Y.; Peng, Z.; Chen, X.; Xu, C.-Y. Detection and Attribution of Changes in Terrestrial Water Storage across China: Climate Change versus Vegetation Greening. *Remote Sens.* **2023**, *15*, 3104. [CrossRef]
36. Zhang, J.; Chu, L.; Zhang, Z.; Zhu, B.; Liu, X.; Yang, Q. Evolution of Small and Micro Wetlands and Their Driving Factors in the Yangtze River Delta—A Case Study of Wuxi Area. *Remote Sens.* **2023**, *15*, 1152. [CrossRef]
37. Ali, A.; Ahmad, A.; Akhtar, K.; Teng, M.; Zeng, W.; Yan, Z.; Zhou, Z. Patterns of Biomass, Carbon, and Soil Properties in Masson Pine (*Pinus Massoniana* Lamb) Plantations with Different Stand Ages and Management Practices. *Forests* **2019**, *10*, 645. [CrossRef]
38. Liu, Y.Y.; Dijk, A.I.J.M.V.; Jeu, R.A.M.D.; Canadell, J.G.; McCabe, M.F.; Evans, J.P.; Wang, G. Recent Reversal in Loss of Global Terrestrial Biomass. *Nat. Clim. Chang.* **2015**, *5*, 470–474. [CrossRef]
39. Sun, J.; Zhang, Y.; Qin, W.; Chai, G. Estimation and Simulation of Forest Carbon Stock in Northeast China Forestry Based on Future Climate Change and LUCC. *Remote Sens.* **2022**, *14*, 3653. [CrossRef]
40. Bai, X.; Chang, Y.; Zhang, X.; Yan-jun, M.A.; Tao, W.U.; Jun-xia, L.I.; Zhen-ju, C. Impacts of Rapid Warming on Radial Growth of *Larix Gmelinii* on Two Typical Micro-Topographies in the Recent 30 Years. *J. Appl. Ecol.* **2016**, *27*, 3853. [CrossRef]
41. Chen, C.; Zhang, X.; Wan, J.; Gao, F.; Yuan, S.; Sun, T.; Ni, Z.; Yu, J. Predicting the Distribution of Plant Associations under Climate Change: A Case Study on *Larix Gmelinii* in China. *Ecol. Evol.* **2022**, *12*, e9374. [CrossRef]
42. Du, Q.; Wei, C.; Liang, C.; Yu, C.; Wang, H.; Wang, W. Future climatic adaption of 12 dominant tree species in Northeast China under 3 climatic scenarios by using MaxEnt modeling. *Acta Ecol. Sin.* **2022**, *42*, 9712–9725.
43. Wu, X.; Liu, H.; Guo, D.; Anenkhonov, O.A.; Badmaeva, N.K.; Sandanov, D.V. Growth Decline Linked to Warming-Induced Water Limitation in Hemi-Boreal Forests. *PLoS ONE* **2012**, *7*, e42619. [CrossRef]
44. Goetz, S.J.; Baccini, A.; Laporte, N.T.; Johns, T.; Walker, W.; Kellndorfer, J.; Houghton, R.A.; Sun, M. Mapping and Monitoring Carbon Stocks with Satellite Observations: A Comparison of Methods. *Carbon. Balance Manag.* **2009**, *4*, 2. [CrossRef] [PubMed]
45. Xu, W.; Jin, X.; Liu, J.; Yang, X.; Ren, J.; Zhou, Y. Analysis of Spatio-Temporal Changes in Forest Biomass in China. *J. For. Res.* **2022**, *33*, 261–278. [CrossRef]
46. Strassburg, B.B.N.; Kelly, A.; Balmford, A.; Davies, R.G.; Gibbs, H.K.; Lovett, A.; Miles, L.; Orme, C.D.L.; Price, J.; Turner, R.K.; et al. Global Congruence of Carbon Storage and Biodiversity in Terrestrial Ecosystems. *Conserv. Lett.* **2010**, *3*, 98–105. [CrossRef]
47. Wen, R.; Xiao, J.; Chang, Z.; Zhai, D.; Xu, Q.; Li, Y.; Itoh, S.; Lomtatidze, Z. Holocene Climate Changes in the Mid-High-Latitude-Monsoon Margin Reflected by the Pollen Record from Hulun Lake, Northeastern Inner Mongolia. *Quat. Res.* **2010**, *73*, 293–303. [CrossRef]
48. Zhu, C.; Liu, H.; Wang, H.; Feng, S.; Han, Y. Vegetation Change at the Southern Boreal Forest Margin in Northeast China over the Last Millennium: The Role of Permafrost Dynamics. *Palaeogeogr. Palaeoclimatol. Palaeoecol.* **2020**, *558*, 109959. [CrossRef]
49. Piao, S.; Ciais, P.; Huang, Y.; Shen, Z.; Peng, S.; Li, J.; Zhou, L.; Liu, H.; Ma, Y.; Ding, Y.; et al. The Impacts of Climate Change on Water Resources and Agriculture in China. *Nature* **2010**, *467*, 43–51. [CrossRef]
50. Jones, M.C.; Booth, R.K.; Yu, Z.; Ferry, P. A 2200-Year Record of Permafrost Dynamics and Carbon Cycling in a Collapse-Scar Bog, Interior Alaska. *Ecosystems* **2013**, *16*, 1–19. [CrossRef]
51. Baltzer, J.L.; Veness, T.; Chasmer, L.E.; Sniderhan, A.E.; Quinton, W.L. Forests on Thawing Permafrost: Fragmentation, Edge Effects, and Net Forest Loss. *Glob. Chang. Biol.* **2014**, *20*, 824–834. [CrossRef]
52. Fisichelli, N.A.; Frelich, L.E.; Reich, P.B. Temperate Tree Expansion into Adjacent Boreal Forest Patches Facilitated by Warmer Temperatures. *Ecography* **2014**, *37*, 152–161. [CrossRef]
53. Li, F.; Zhou, G.; Cao, M. Responses of *Larix Gmelinii* Geographical Distribution to Future Climate Change: A Simulation Study. *J. Appl. Ecol.* **2006**, *17*, 2255–2260.
54. Wang, Y.; Zhang, Z.; Chen, X. The Dominant Driving Force of Forest Change in the Yangtze River Basin, China: Climate Variation or Anthropogenic Activities? *Forests* **2022**, *13*, 82. [CrossRef]
55. Huang, Y.; Ciais, P.; Santoro, M.; Makowski, D.; Chave, J.; Schepaschenko, D.; Abramoff, R.Z.; Goll, D.S.; Yang, H.; Chen, Y.; et al. A Global Map of Root Biomass across the World's Forests. *Earth Syst. Sci. Data* **2021**, *13*, 4263–4274. [CrossRef]

forests

MDPI

Article

FireViT: An Adaptive Lightweight Backbone Network for Fire Detection

Pengfei Shen [1], Ning Sun [1,2,*], Kai Hu [1], Xiaoling Ye [1], Pingping Wang [3], Qingfeng Xia [2] and Chen Wei [1]

1 Jiangsu Collaborative Innovation Center of Atmospheric Environment and Equipment Technology, Information and Systems Science Institute, Nanjing University of Information Science and Technology, Nanjing 210044, China; 20211249578@nuist.edu.cn (P.S.); 001600@nuist.edu.cn (K.H.); 000510@nuist.edu.cn (X.Y.); 202312490377@nuist.edu.cn (C.W.)
2 School of Automation, Wuxi University, Wuxi 214105, China; xqf@cwxu.edu.cn
3 Fire Research Institute, Shanghai 200030, China; wangpingping@shfri.cn
* Correspondence: 001764@cwxu.edu.cn

Abstract: Fire incidents pose a significant threat to human life and property security. Accurate fire detection plays a crucial role in promptly responding to fire outbreaks and ensuring the smooth execution of subsequent firefighting efforts. Fixed-size convolutions struggle to capture the irregular variations in smoke and flames that occur during fire incidents. In this paper, we introduce FireViT, an adaptive lightweight backbone network that combines a convolutional neural network (CNN) and transformer for fire detection. The FireViT we propose is an improved backbone network based on MobileViT. We name the lightweight module that combines deformable convolution with a transformer as th DeformViT block and compare multiple builds of this module. We introduce deformable convolution in order to better adapt to the irregularly varying smoke and flame in fire scenarios. In addition, we introduce an improved adaptive GELU activation function, AdaptGELU, to further enhance the performance of the network model. FireViT is compared with mainstream lightweight backbone networks in fire detection experiments on our self-made labeled fire natural light dataset and fire infrared dataset, and the experimental results show the advantages of FireViT as a backbone network for fire detection. On the fire natural light dataset, FireViT outperforms the PP-LCNet lightweight network backbone for fire target detection, with a 1.85% increase in mean Average Precision (mAP) and a 0.9 M reduction in the number of parameters. Additionally, compared to the lightweight network backbone MobileViT-XS, which similarly combines a CNN and transformer, FireViT achieves a 1.2% higher mAP while reducing the Giga-Floating Point Operations (GFLOPs) by 1.3. FireViT additionally demonstrates strong detection performance on the fire infrared dataset.

Keywords: CNN and transformer; lightweight; fire detection

Citation: Shen, P.; Sun, N.; Hu, K.; Ye, X.; Wang, P.; Xia, Q.; Wei, C. FireViT: An Adaptive Lightweight Backbone Network for Fire Detection. *Forests* **2023**, *14*, 2158. https://doi.org/10.3390/f14112158

Academic Editor: José Aranha

Received: 29 September 2023
Revised: 25 October 2023
Accepted: 27 October 2023
Published: 30 October 2023

1. Introduction

Hazards caused by fire are a serious threat to human life and property. According to data from the Global Disaster Database, from 2013 to 2022 the average number of deaths and missing persons due to forest and grassland fires alone reached 904,000 people. Data released by Global Forest Watch (GFW) and the World Resources Institute (WRI) indicate that on a global scale the forest area destroyed by wildfires is now double what it was at the beginning of this century. According to satellite data, the annual forest area destroyed by wildfires has increased by approximately 3 million hectares compared to the year 2001. In addition to forest fires, the increasing impact of other types of fires, such as electrical fires, is becoming more severe as society continues to progress and develop. The frequency of these incidents is on the rise. Real-time monitoring of fire-prone areas, timely fire alarms, and rapid localization of fire incidents are of paramount importance for safeguarding human life, property, and industrial safety.

Traditional fire detection methods primarily involve contact-based fire detectors, such as carbon monoxide sensors, temperature sensors, smoke detectors, etc. Rachman, F. et al. [1] proposed a fuzzy logic-based early fire detection system using KY-026 (fire detection), MQ-9 (smoke detection), and DS18b20 (temperature detection) sensors. Huang Ye et al. [2] proposed a wireless fire detection node design method based on multi-source sensor data fusion and provided a complete hardware selection and software data fusion processing method. Solorzano Soria, A.M. et al. [3] proposed a gas sensor-based array to speed up fire alarm response. Li Yafei et al. [4] developed a mid-infrared carbon monoxide (CO) and carbon dioxide (CO_2) dual gas sensor system for early fire detection. Liu Xiaojiang et al. [5] proposed a sensor optimization strategy and an intelligent fire detection method based on the combination of particle swarm optimization algorithm. Although contact fire detectors are commonly used in various public scenes, their detection range is limited to small indoor spaces, and it is difficult to apply them to large indoor spaces and outdoor open spaces where open flames are strictly prohibited. Moreover, traditional contact fire detectors are prone to age-related failures and require a lot of manpower and resources for maintenance and management.

Compared to contact fire detection using sensors, non-contact video fire detection technology has the advantages of no additional hardware, intuitive and comprehensive fire information, and large detection range. While real-time monitoring of fire appears to be a binary classification problem for images, in practice it requires further detection of fire images based on the classification. The use of fire image detection is justified due to the extensive coverage of video surveillance systems. Early fire incidents are often challenging to detect, and sometimes fires can escalate to an uncontrollable stage within a short timeframe. Therefore, relying solely on image classification is insufficient for accurately pinpointing the actual location of a fire. This limitation could undoubtedly hinder the timely response and effective management of fire incidents. Traditional fire target detection algorithms include region selection, feature extraction, and classifier design. Qiu, T. et al. [6] proposed an adaptive canny edge detection algorithm for fire image processing. Ji-neng, O. et al. [7] proposed an early flame detection method based on edge gradient features. Khalil, A. et al. [8] proposed a fire detection method based on multi-color space and background modeling. However, in traditional fire target detection algorithms, manually designed features lack strong generalization and exhibit limited robustness. The emergence of CNNs has gradually replaced traditional handcrafted feature methods, offering superior generalization and robustness compared to traditional fire detection approaches. Majid, S. et al. [9] proposed an attention-based CNN model for the detection and localization of fires. Chen, G. et al. [10] proposed a lightweight model for forest fire smoke detection based on YOLOv7. Dogan, S. et al. [11] proposed an automated accurate fire detection system using ensemble pretrained residual network. In recent years, a new generation of transformer-based deep learning network architectures has gradually started to shine. Li, A. et al. [12] proposed a combination of BiFPN and Swin transformer for the detection of smoke from forest fires. Huang, J. et al. [13] proposed a small target smoke detection method based on a deformable transformer. Although these transformer-based network architectures have achieved good results in fire detection, they tend to be more complex (i.e., the number of parameters reaches about 20 M or 30 M) and not very lightweight. In order to ensure a lightweight transformer architecture-based model, scholars have started to research solutions such as EfficientViT [14], EfficientFormerV2 [15], MobileViT [16], etc., although these network models have not become widely used in fire detection to date.

Most of the methods that utilize deep learning for fire target detection use rectangular convolution of fixed shapes for feature extraction of smoke and flame in fires; however, it is well known that smoke and flame features in a fire situation are scattered and irregular, which is undoubtedly a very challenging task in fire target detection. Therefore, in this paper we propose a lightweight adaptive backbone network called FireViT using deformable convolution [17] combined with transformer for fire detection to better adapt to the irregularly varying smoke and flames in fire scenarios. The adaptive lightweight backbone network, FireViT that we propose here is capable of meeting the requirements of various other application scenarios.

The contributions of our paper are as follows:

- An adaptive lightweight backbone network consisting of deformable convolution combined with a transformer, which we name FireViT, is proposed for smoke and flame detection in fire. Our proposed DeformViT block is the main module in FireViT.
- An improved adaptive activation function, AdaptGELU, is proposed to increase the nonlinear representation of the model and further enhance the accuracy of the network.
- Considering the relatively small number of publicly available labeled fire datasets, we collected and built one of the richest labeled fire datasets with the largest number of fire scenes and fire images to evaluate our model. Our labeled fire dataset contains a fire natural light dataset and fire infrared dataset.

The rest of this paper is organized as follows. Section 2 presents previous works on the backbone network and the target detection header in the target detection algorithm that are related to this paper. Section 3 presents FireViT, a lightweight backbone network consisting of deformable convolution combined with a transformer that can be used for fire detection, along with AdaptGELU, an improved adaptive activation function. Section 4 discusses the process of determining the FireViT backbone network using the AdaptGELU activation function proposed in Section 3 and verifies the validity of the model by comparing it to other mainstream lightweight backbone networks for fire detection experiments on self-made labeled fire datasets. Finally, in Section 5, we summarize and conclude the paper.

2. Related Work

In this section, we present previous research work related to this paper on lightweight backbone networks and target detection headers in target detection algorithms.

2.1. *MobileViT*

The emergence of ViT [18] has led people to realize the tremendous potential of transformers in the field of computer vision. The transformer architecture has become a new neural network paradigm in the field of computer vision, following the advent of CNNs, with and more researchers starting to use networks with the transformer architecture. However, although transformers are powerful, they have a number of problems; the pure transformer model structures are usually bulky and not very lightweight; furthermore, inductive bias is a form of prior knowledge, and unlike CNNs, transformers do not have the same kind of induction bias, meaning that transformers require a substantial amount of data to learn such prior information.

The inductive bias of CNNs can generally be categorized into two types. The first is locality; CNNs convolve input feature maps using a sliding window approach, which means that objects that are closer together exhibit stronger correlations. Locality helps to control the complexity of the model. The second type is translation equivariance; regardless of whether object features in an image are first convolved and then translated, or first translated and then convolved, the resulting features are the same. Translation equivariance enhances the model's generalization capabilities.

However, CNNs are not without imperfections. The spatial features they extract are inherently local in nature, which to a certain extent constrains the model's performance, whereas transformers can obtain global information through their self-attention mechanism.

MobileViT is a lightweight network that combines the strengths of both CNN and ViT. MobileViT is available in three versions depending on model size: MobileViT-S, MobileViT-XS, and MobileViT-XXS. MobileViT primarily consists of standard convolutions, inverted residual blocks from MobileNetV2 (MV2), MobileViT blocks, global pooling, and fully connected layers. The network architecture is illustrated in Figure 1. In this paper, we have replaced the MobileViT block in the MobileViT-XS network with our proposed DeformViT block. Additionally, we have removed the Conv(1×1), global pooling, and fully connected layers at the bottom of MobileViT-XS to create the FireViT network, which serves as the network backbone in our fire detection approach.

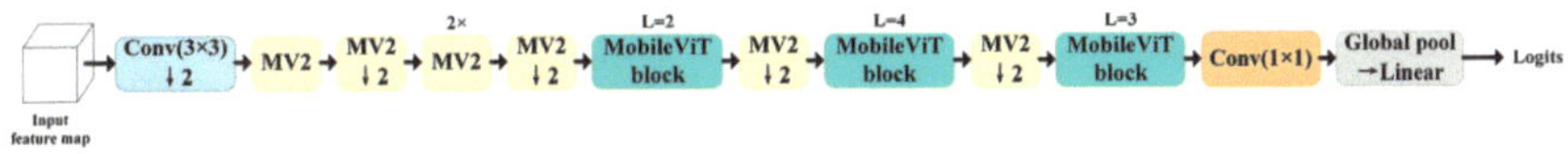

Figure 1. Structure of the MobileViT network; "(3×3)" and "(1×1)" denote the size of the convolution kernel, "MV2" denotes the inverted residual block in MobileNetV2, "↓2" denotes the downsampling operation, and "L" denotes the number of layers in the transformer block.

2.2. Prediction Head

Target detection is the ability to accurately localize objects of interest in an image; the prediction head (prediction part or output part) is required to perform classification and regression tasks. There are two types of detection heads, namely, coupled detection heads and decoupled detection heads. A coupled detection head means that the classification and regression tasks share a significant portion of their input parameters. In [19], the authors pointed out that the features of interest for classification and regression tasks during the learning of input parameters are different. These two subtasks are coupled, leading to spatial misalignment issues that can significantly impact network convergence speed. Decoupled detection heads address this issue. A decoupled detection head separately processes the input parameters for the classification and regression tasks, which enhances the detection accuracy and convergence speed of the detection network. The YOLOX [20] network uses decoupled heads for target detection, showing a 1.1% improvement in terms of mAP compared to YOLOv3 [21]. As a result, most target detection networks including PPYOLO-E [22], YOLOv6 [23], and YOLOv8 [24], have begun to adopt this paradigm. To ensure the fairness of our designed adaptive lightweight FireViT backbone feature extraction network in comparison with other backbone network models for fire target detection, all the models studied in this paper use three YOLOv8 decoupled heads at the bottom of their respective networks for fire target detection.

The prediction head of YOLOv8 uses decoupled classification and regression branches; the detailed structure is shown in Figure 2. The classification branch uses the Binary Cross-Entropy (BCE) loss. The regression branch employs the Complete Intersection over Union (CIoU) loss [25], and utilizes a novel loss function called the Distribution Focal Loss (DFL) [26]. The DFL is designed to optimize the probabilities of the left (y_i) and right (y_{i+1}) positions that are closest to the label y in a manner similar to cross-entropy. This allows the network to quickly focus on the distribution in the vicinity of the target location. The formula is represented as Equation (1):

$$DFL(S_i, S_{i+1}) = -((y_{i+1} - y)\log(S_i) + (y - y_i)\log(S_{i+1})). \tag{1}$$

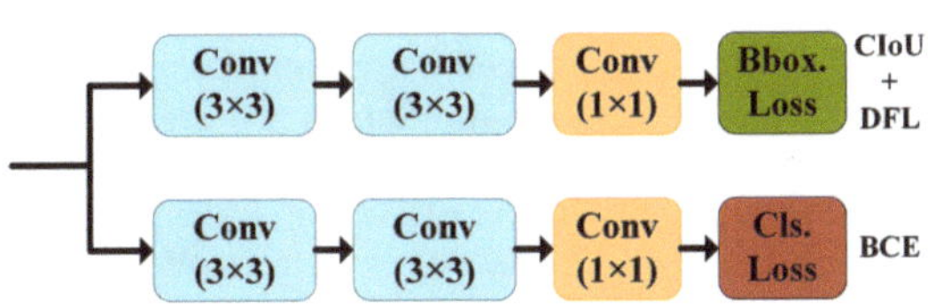

Figure 2. Structure of the prediction head of YOLOv8; CIOU is the complete intersection over union loss, DFL is the distribution focal loss, and BCE is the binary cross-entropy loss.

3. Methods

In this section, we discuss an adaptive lightweight backbone network called FireViT, which is an improvement based on the MobileViT-XS network. FireViT is designed for the detection of smoke and flame targets in fire scenarios; the overall structure is illustrated in Figure 3. The Deformable Vision Transformer (DeformViT) block is an adaptive lightweight

CNN combined with a transformer module, which plays a major role in FireViT for feature extraction. The DeformViT block is able to better extract the irregularly varying smoke and flame features in a fire situation, and can capture the flame and smoke in a fire situation locally and holistically better than the traditional convolutional module, thereby improving the accuracy of fire detection. The improved adaptive activation function proposed in our model, called AdaptGELU, can increase the model's nonlinear expressive power and further enhance the accuracy of fire detection.

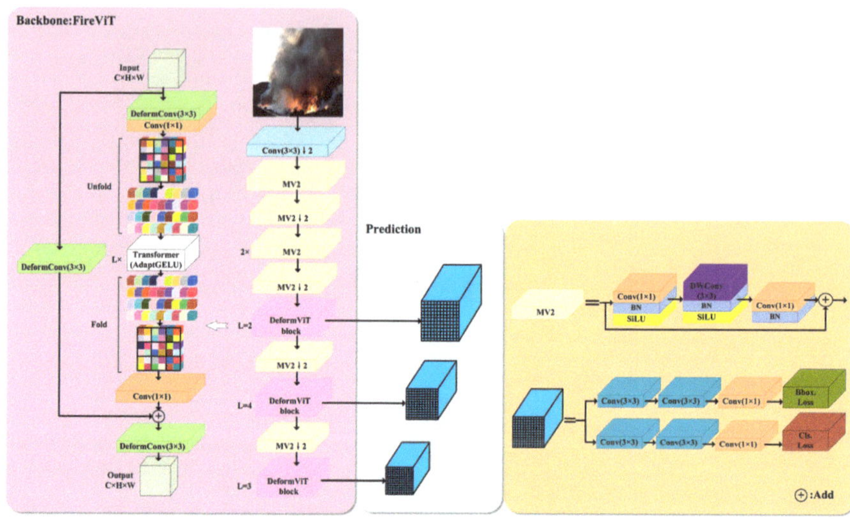

Figure 3. Detailed structure of the fire detection network using FireViT as the backbone network; *C* represents the number of channels of the feature map, *H* represents the height of the feature map, *W* represents the width of the feature map, DWConv stands for depthwise separable convolution, DeformConv stands for Deformable Convolution, (1×1) and (3×3) are the convolutional kernel sizes, BN stands for batch normalization, "↓2" indicates the downsampling operation, and "L" stands for the number of layers in the transformer block.

3.1. Adaptive Lightweight Backbone Network Module: DeformViT Block

The DeformViT block aims to combine the advantages of deformable convolutions for local feature extraction with those of transformers for global feature extraction and to perform feature extraction on the input tensor with fewer parameters. In [27], the authors proposed a Deformable Attention Transformer; however, the number of parameters in its minimal model reached 29 M. In addition to MobileViT, there are a number of network models [14,15] that attempt to combine the benefits of convolution and transformers; however, these use fixed-shaped convolutional modules, limiting feature extraction to the irregularly varying smoke and flames of a fire. Our proposed DeformViT block uses deformable convolution in combination with a transformer to enable better and more fine-grained extraction of the ever-changing features of flame and smoke in fire scenes.

Deformable convolution for capturing better local features. It is obvious that the standard fixed-size convolution kernel is not well suited for this task of adapting to irregular smoke and flames at a fire scene; thus, we use deformable convolution, which adapts the structure of the convolution kernel by learning the offsets, as shown in Figure 4.

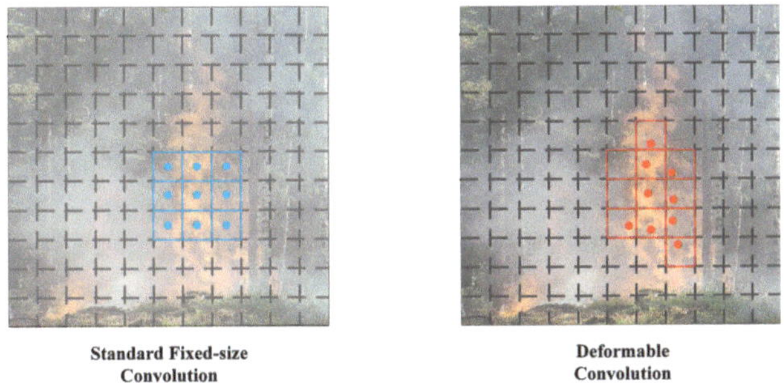

Figure 4. Illustration of sampling for the 3 × 3 standard fixed-size convolution kernel and the deformable convolution kernel. The (**left**) side of Figure 4 shows the sampling points and grid for the 3 × 3 fixed convolution kernel (blue dots and boxes), while the (**right**) side of Figure 4 shows the sampling points and grid for the deformable convolution with positional offset (red dots and boxes).

A standard fixed-size convolution can be divided into the following two-step operation.

Step 1: Sample the pixel point gridding from the input feature map $X \in \mathbb{R}^{C \times H \times W}$ (where C is the number of feature map channels, H is the feature map height, and W is the feature map width). Assuming that a 2D $K \times K$ convolution is used for sampling, the position of the pixel point in the sensory field can be denoted as P_n, as shown in Equation (2):

$$P_n = \{P_{ij}\} \quad (0 \leqslant i, j \leqslant K-1; 0 \leqslant n \leqslant K^2-1; i, j, n \in N). \tag{2}$$

Step 2: Output each position P_{out} on the feature map Y after the convolution operation, as shown in Equation (3):

$$Y(P_{out}) = \sum_{n=0}^{P_n} \omega(P_n) \cdot x(P_{out} + P_n), \tag{3}$$

where $\omega(\cdot)$ denotes the value learned by the network in the convolution and $x(\cdot)$ denotes the value after grid sampling on the input feature map.

The essence of deformable convolution lies in the modification of the sampling results to achieve a variation in the convolutional effect, as illustrated in Figure 5. The right side of Figure 5 shows the value of the convolution kernel offset predicted by the convolution. In deformable convolution, ΔP_n is used to offset the position of the point P_n on the feature map, the weight coefficients $\Delta\gamma$ are added at each sampling point to reduce the interference of irrelevant information on the feature map; at this point, deformable convolution is computed as in Equation (4):

$$Y(P_{out}) = \sum_{n=0}^{P_n} \omega(P_n) \cdot x(P_{out} + P_n + \Delta P_n) \cdot \Delta\gamma_n. \tag{4}$$

The offset value predicted by the convolution operation is often a small number that cannot be sampled directly from the feature map X. Here, bilinear interpolation (Equation (5)) is used to ensure that the feature map can be sampled after the offset:

$$x(P) = \sum_{P_s} B(P_s, P) \cdot x(P_s), \tag{5}$$

where P is the position after offset; $P = P_{out} + P_n + \Delta P_n$; P_s is the site on the space of all integrals of the feature map X; and $B(\cdot,\cdot)$ is a 2D bilinear interpolation kernel, as shown in Equation (6):

$$B(P_s, P) = b(P_{sx}, P_x) \cdot b(P_{sy}, P_y), \quad (6)$$

where $b(e, f) = \max(0, 1 - |e - f|)$.

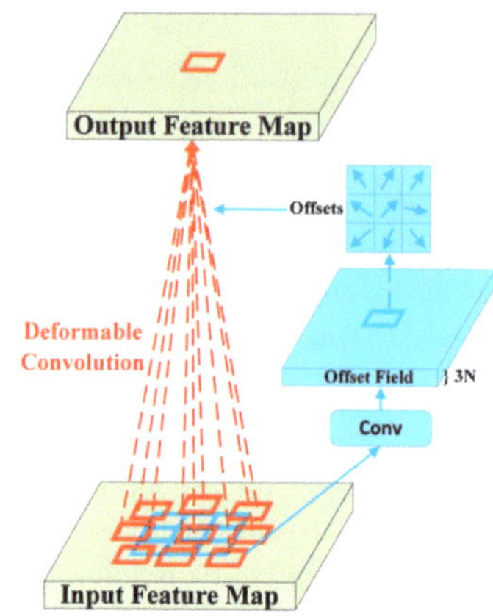

Figure 5. Diagram of the 3 × 3 deformable convolution structure; N is the number of pixels in the convolution kernel, and the figure shows a 3 × 3 convolution with N = 9.

Unfold–Transformer–Fold operation focusing on global features. The emergence of DETR [28] has opened the door to the use of transformers for computer vision target detection. After the emergence of ViT, transformers were identified as a neural network architecture that performs very well in the field of computer vision. While ViT requires attention for each token (high computational cost), our network uses deformable convolution prior to the transformer block to better grasp the local features; in this way, the computational cost can be reduced by dividing the feature map into multiple patches during global modeling of the feature map $X \in \mathbb{R}^{C \times H \times W}$ (where C is the number of feature map channels, H is the feature map height, and W is the feature map width), then self-attention can be performed for the pixels in the same position in each patch. We call this the Unfold–Transformer–Fold operation. The patch has dimensions of (h, w) (ignoring the number of channels C), where h is the height of the patch and w is its width. The unfold operation spreads the pixels at the same position in each patch in a sequence, then the attention of each sequence is calculated in parallel by the transformer and finally collapsed back to the size of the original feature map by the fold operation, as shown in Figure 6.

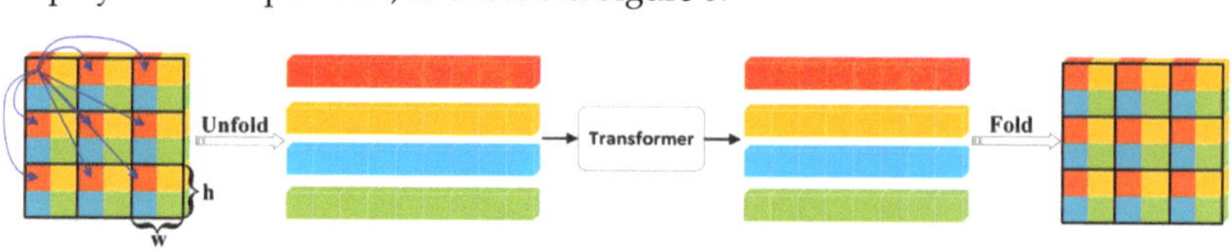

Figure 6. Schematic diagram of the Unfold–Transformer–Fold operation. The dimension (h, w) of each patch in the figure is $(2, 2)$, i.e., each patch consists of four pixels (shown in the figure in red, yellow, blue, and green). The token (pixel) in each patch only calculates attention with its own token of the same position (the color block of the same color in the figure), as indicated by the dark blue arrow. The feature map X dimensions are (C, H, W) and the cost according to self-attention alone is $O(CHW)$. According to the Unfold–Transformer–Fold calculation, at this time, $Patch = 2 \times 2 = 4$ and the calculation cost is $O(\frac{CHW}{4})$, amounting to $\frac{1}{4}$ of the original calculation cost. The unfold and fold operations reshape the data to satisfy the self-attention computation.

In the fire occurrence scenario, we use deformable convolution to better capture the local features and the Unfold–Transformer–Fold operation to capture the global features; we propose an adaptive lightweighting module called DeformViT that combines them to improve the accuracy of fire detection. We conducted a comparative experiment on the eight forms of DeformViT modules designed in Section 4.4, finally choosing scheme VIII, which is shown in Figure 7.

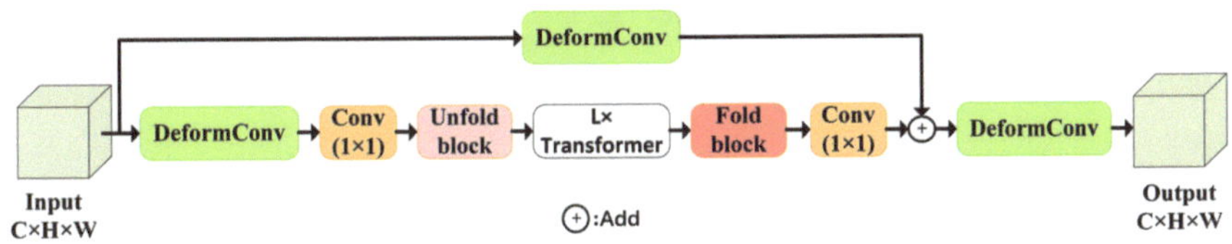

Figure 7. Structure of the DeformViT module, where *L* denotes the number of layers in the transformer block, *C* denotes the number of channels in the feature map, *H* denotes the height of the feature map, and *W* denotes its width.

3.2. Adaptive Activation Function: Adaptive GELU (AdaptGELU)

Activation functions are of paramount importance in deep learning, as they enhance the neural network's capacity for nonlinear expression. Gaussian Error Linear Units (GELUs) [29] are beginning to attract attention in applications such as Google's BERT [30] and OpenAI's GPT-2 [31]. A graphical representation of the GELU function is provided in Figure 8, while its mathematical expression is provided in Equation (7):

$$GELU(G) = x \cdot PR(G \leqslant x) = x \cdot \Phi(x), x(0,1), \quad (7)$$

where x is the input, G is a Gaussian random variable with zero mean and unit variance, $PR(G \leqslant x)$ is the probability of G being less than or equal to a given value of x, and $\Phi(x) = \frac{1}{2}\left[1 + erf\left(\frac{x}{\sqrt{2}}\right)\right]$ is the cumulative distribution function of the standard normal distribution.

The approximate solution of Equation (7) is calculated as shown in Equation (8):

$$GELU(G) = 0.5x\left(1 + \tanh\left[\sqrt{\frac{\pi}{2}}\left(x + 0.044715x^3\right)\right]\right). \quad (8)$$

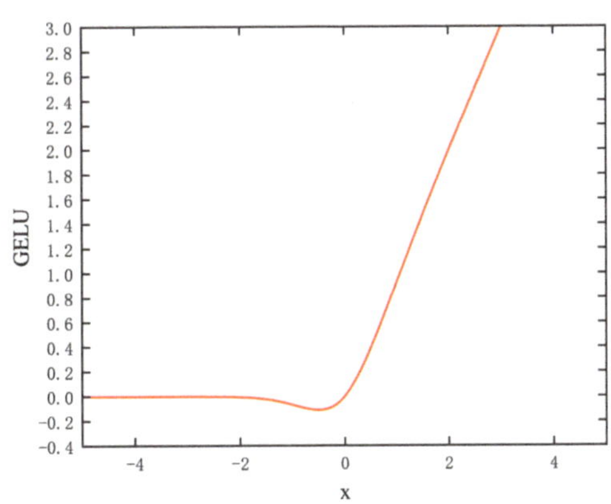

Figure 8. GELU activation function.

GELU is smoother compared to ReLU [32], and is better at mitigating vanishing gradients and supporting network training and optimization in deep neural network models. However, as GELU is not optimal in certain cases, we propose Adaptive GELU

(AdaptGELU) by introducing a trainable parameter *a* (with the initial value of *a* set to 1.0) to further improve the performance of the network. AdaptGELU replaces the original activation function in the feedforward network in the transformer. We propose two versions of AdaptGELU, which we respectively name AdaptGELUv1 and AdaptGELUv2. The morphology of the two versions of AdaptGELU with different values of *a* is shown in Figure 9.

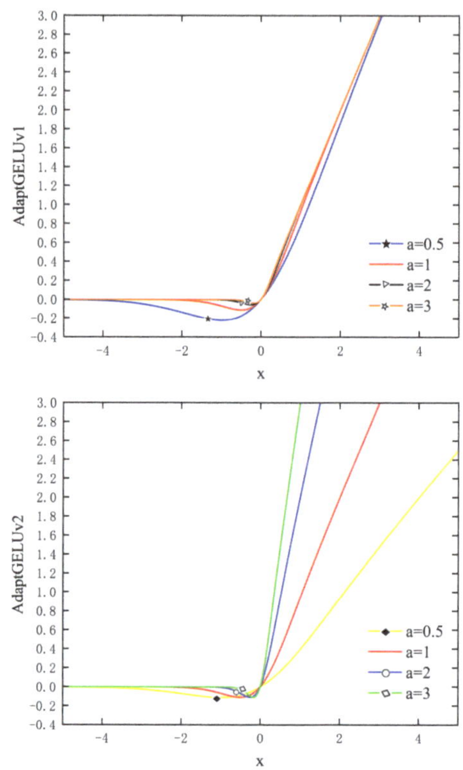

Figure 9. Morphological maps of the two versions of AdaptGELU with different values for the *a* parameter.

AdaptGELUv1 is represented by Equation (9):

$$AdaptGELUv1(G) = \frac{1}{2}x\left[1 + erf\left(\frac{ax}{\sqrt{2}}\right)\right]. \tag{9}$$

AdaptGELUv2 is represented by Equation (10):

$$AdapeGELUv2(G) = \frac{1}{2}ax\left[1 + erf\left(\frac{ax}{\sqrt{2}}\right)\right]. \tag{10}$$

Section 4.4 describes our comparative fire detection experiments, where we replaced the original SiLU [33] activation function in the transformer's feedforward network with Sigmoid, ReLU, GELU, AdaptGELUv1, and AdaptGELUv2 functions. Our comparative analysis of reveals that the network with AdaptGELUv2 achieves better accuracy; thus, we ultimately selected AdaptGELUv2 as the activation function for our model in the feedforward network of the transformer. Detailed experimental results and analysis are presented in Section 4.4.

4. Experiments and Results

This section first describes the labeled fire dataset we collected and produced, along with details of the implementation and evaluation metrics. Then, we describe the validation of our proposed model components through a series of comparative experiments. Finally, the effectiveness of our model is further demonstrated by comparing the experimental results and visualizations of different lightweight backbone networks used in fire detection experiments on the fire natural light and the fire infrared datasets.

4.1. Dataset

The FireViT backbone network we designed is used in this paper to study supervised fire detection. The overall fire detection model using FireViT as the backbone network is able to identify and localize smoke and flame in a fire. Therefore, preparing labeled public datasets of fires or constructing a self-made labeled fire dataset is an essential step. Due to the severe shortage of publicly available labeled fire datasets at present, we constructed a self-made labeled fire dataset by extensively reviewing the relevant literature and collecting fire-related data from various sources. The labeled fire dataset we collected and constructed is currently the richest labeled fire dataset containing fire occurrence scenarios, as well as the dataset containing the largest number of fire images. The fire dataset we collected and constructed is divided into a fire natural light dataset and fire infrared dataset.

The data in the fire natural light dataset come from: (A) the fire image and PNG still image dataset from the 2018 study of Dunnings and Breckton [34]; (B) VisiFire [35]; (C) the KMU fire and smoke database [36]; (D) video smoke detection [37]; (E) the flame dataset: aerial imagery pile burn detection using drones dataset [38]; and (F) the fire public welfare web platform. All of these data were unlabeled for fires, as shown in the top half of Figure 10, which depicts a partial demonstration of the collected fire natural light data. These data consist of both image format and video format data; as our focusing here is on fire detection from images, the collected video data were processed into images by exporting frames at 0.05 s intervals. In order to improve the robustness of the fire detection models and further enrich the fire data, we randomly selected three fire data sources from A, B, C, D, E, and F. Next, we randomly decided whether or not to carry out the corresponding operation with 50% probability through Rotate–Flip–Affine transformation (as shown in Figure 11) one by one for each of the three fire images in order to expand the dataset.

The data in the fire infrared dataset mainly consist of data captured by the infrared thermal imager in E and simulated fire data captured by an infrared structured light depth camera. The first three rows of the fire infrared data section shown in Figure 10, from top to bottom, show portions of the Fusion, GreenHot, and WhiteHot data captured by an infrared thermal imager, while the last row shows portions of the simulated fire data captured by an infrared structured light depth camera. We similarly first cut the frames of the captured video data at 0.05 s intervals in order to process them into images. Next, we randomly selected two fire data sources from Fusion, GreenHot, WhiteHot, and infrared structured light data, and finally selected each fire image from these two fire data sources one by one in all the modes of operation of the Rotate–Flip–Affine transformation with 50% probability to perform the corresponding operation to expand the dataset.

All of the above data were labeled using the LabelImage data labeling tool. We constructed datasets containing both fire natural light data (121,339 images) and fire infrared data (96,112 images) to fulfill different application requirements. Detailed information is presented in Table 1. The dataset was divided into training, validation, and test sets in a ratio of 8:1:1 to constructed the final labeled fire dataset in VOC data format. The fire natural light dataset contains the labels "fire" and "smoke", while the fire infrared dataset uses only the "fire" label, as smoke is much harder to capture at night. All of the fire natural light data were used in the experiments, while only the Fusion data were used in the fire infrared data (all of the following are expressed as fire infrared data).

Figure 10. Partial presentation of the collected fire data.

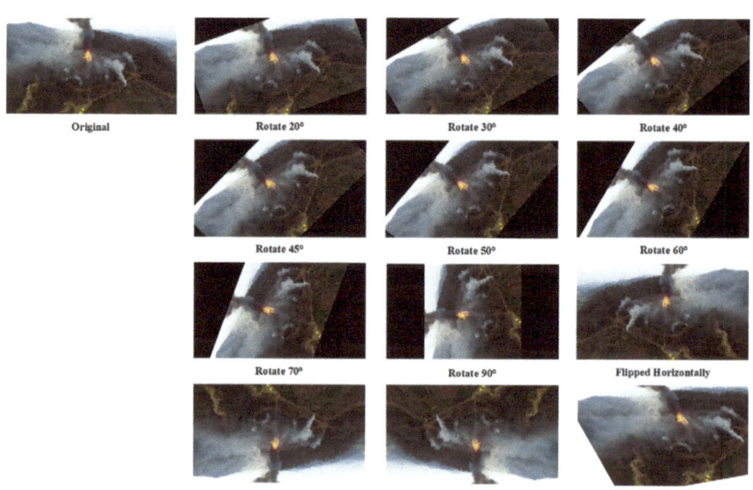

Figure 11. Rotate–Flip–Affine transformation.

Table 1. Fire dataset details: (a) fire natural light dataset information and (b) fire infrared dataset information. Here, "√" indicates that the Rotate-Flip-Affine transformation was used, while "×" indicates that it was not used. Data sources: (A) PNG still image fire image dataset from Dunnings and Breckton, 2018; (B) VisiFire; (C) KMU fire and smoke database; (D) video smoke detection; (E) the flame dataset: aerial imagery pile burn detection using drones; and (F) the fire public welfare web platform.

(a) Fire natural light dataset				
Data Sources	Rotate-Flip-Affine Transformation	Total Number	Number of "smoke" labels	Number of "fire" labels
A	×	10,048	27,221	19,103
B	√	54,968	61,324	20,297
C	√	29,222	25,345	17,711
D	√	6488	12,124	5453
E	×	12,201	23,615	72,660
F	×	8412	6064	34,159

(b) Fire infrared dataset			
Infrared data	Rotate-Flip-Affine Transformation	Total Number	Number of "fire" labels
Fusion	√	75,219	214,005
GreenHot	×	5011	26,386
WhiteHot	×	5891	22,590
Structured Infrared Light	√	9991	10,198

4.2. Implementation Details

All models were trained and experimented with on a system running Ubuntu 20.04, Python 3.8, CUDA 11.3, PyTorch1.11.0, and an NVIDIA RTX4090 GPU. The models were trained with fixed random number seeds and without pretraining weights. The parameter settings used for model training are shown in Table 2.

Table 2. Parameter settings during model training.

Training Parameter Settings	Particulars
Initialization	MSRA initialization [39]
Input image dimensions	(640, 640, 3)
Optimizer	SGD
Momentum	0.937
Initial learning rate	0.01
Weight Decay	0.0005
Number of images per batch	8
Epochs	50

4.3. Evaluation Metrics

To evaluate the effectiveness of FireViT as a backbone network for fire detection, we used the following metrics: Precision, Recall, Parameters, Average Precision (AP), mAP, and Floating Point Operations (FLOPs). AP is the area under the Precision–Recall (PR) curve, where we used a value of 0.5 for the Intersection over Union (IoU = 0.5). mAP is the mean of AP calculated for each individual class. Higher AP and mAP values indicate better performance.

Precision is shown in Equation (11):

$$Precision = \frac{TP}{TP + FP}. \tag{11}$$

Recall is shown in Equation (12):

$$Recall = \frac{TP}{TP + FN}. \tag{12}$$

Above, TP (True Positive) represents the number of correctly identified positive samples, FP (False Positive) represents the number of samples that are actually negative and predicted as positive, and FN (False Negative) represents the number of samples that are actually positive and predicted as negative.

FLOPs are used to measure the complexity of the algorithm, as shown in Equation (13):

$$FLOPs = (2 \times C_{in} \times K^2 - 1) \times H \times W \times C_{out}, \tag{13}$$

where C_{in} represents the number of input channels, C_{out} represents the number of output channels, K represents the size of the convolutional kernel, and H and W respectively denote the height and width of the feature map.

4.4. Ablation Experiments on the Fire Natural Light Dataset

We designed eight build scenarios for the DeformViT module, as shown in Figure 12. We replaced the DeformViT block with the MobileViT block in the MobileViT network and used the SiLU activation function in the network. In addition, we replaced the MobileViT block in the MobileViT network with the DeformViT block and used the SiLU activation function in the network, at which point, forming a backbone network that we name FireViT-SiLU. Fire detection was performed on the bottom layer of the FireViT-SiLU network using the three decoupled detection heads mentioned in Section 2.2 The results of the ablation experiments are shown in Table 3.

Table 3. Comparative experimental results for fire detection using the FireViT-SiLU backbone feature extraction network composed of DeformViT blocks with different architectural configurations.

Options	mAP	Params	GFLOPs
MobileViT block	90.9%	1.9M	13.8
I	91.7%	2.1M	12.5
II	91.6%	1.8M	12.2
III	91.5%	1.8M	12.2
IV	91.1%	1.6M	11.9
V	91.2%	1.6M	11.9
VI	91.5%	1.8M	12.2
VII	91.5%	1.8M	12.2
VIII	91.8%	2.1M	12.5

From Figure 12 and Table 3, it can be seen that the fire detection network formed by our proposed building scheme on the Type I DeformViT block has an overall advantage over the fire detection network with MobileViT as the backbone network; the number of parameters improves by only 0.2 M, GFLOPs decrease by 1.3, and mAP improves by 0.8%. The comparison of the fire detection backbone network scheme consisting of the Type II DeformViT block to the construction scheme of the Type VII DeformViT block provided us with further ideas. Therefore, we replaced the copy operation in the construction scheme of the type I DeformViT block with deformable convolution to form the type VIII construction scheme of the DeformViT block. The mAP of the fire detection network formed using the type VIII building scheme for the DeformViT block was further improved with respect to the fire detection network based on the type I building scheme of the DeformViT block. Based on these results, we selected the construction scheme using the type VIII DeformViT block as the final build scheme in our network.

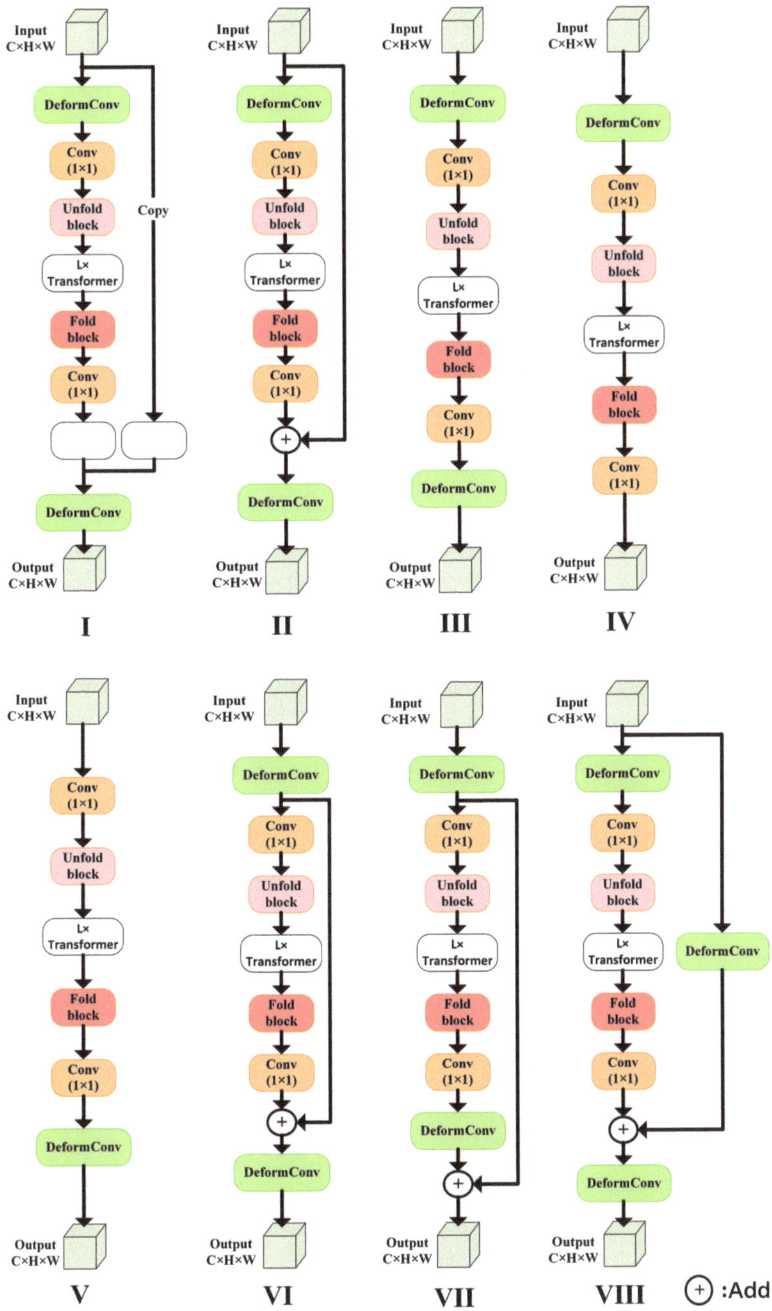

Figure 12. Different architectural configurations used for the DeformViT block.

FireViT-SiLU uses the SiLU global activation function. In this section, we used the Sigmoid, ReLU, GELU, and our improved adaptive AdaptGELUv1 and AdaptGELUv2 activation functions to replace the SiLU activation function in the feedforward network of FireViT-SiLU's transformer block. The three decoupled detection heads mentioned in Section 2.2 were used for fire detection at the bottom layer of the overall network. The comparative results of the detection experiments are shown in Table 4. Because only the activation function in the transformer's feedforward network was replaced, the amount of variation in the number of parameters and GFLOPs in the overall network is almost negligible.

Table 4. Comparative experimental results of different activation functions replacing the SiLU activation function of the transformer's feedforward network in FireViT-SiLU for fire detection. Because only the activation function in the transformer's feedforward network was replaced, the amount of variation in the number of parameters and GFLOPs in the overall network is almost negligible.

Model	mAP	Params	GFLOPs
FireViT-SiLU	91.82%	2.1 M	12.5
FireViT-Sigmoid	91.38%	2.1 M	12.5
FireViT-ReLU	91.89%	2.1 M	12.5
FireViT-GELU	91.88%	2.1 M	12.5
FireViT-AdaptGELUv1	92.09%	2.1 M	12.5
FireViT-AdaptGELUv2	92.14%	2.1 M	12.5

From Table 4, it can be seen that our proposed improved GELU activation functions, AdaptGELUv1 and AdaptGELUv2, have advantages over other activation functions in the fire detection context of this paper. When FireViT-SiLU, FireViT-Sigmoid, FireViT-ReLU, and FireViT-GELU were applied to fire detection, FireViT-ReLU achieved the highest mAP at 91.89%, FireViT-AdaptGELUv1 a 0.2% higher mAP than FireViT-ReLU, and FireViT-AdaptGELUv2 a 0.25% higher mAP than FireViT-ReLU. After comparing the experimental results, we chose AdaptGELUv2 as the activation function for use in the transformer's feedforward networkm, and used FireViT-AdaptGELUv2 as the backbone network for the final FireViT fire detection model.

We conducted fire detection comparison experiments by comparing our proposed approach with several mainstream lightweight convolutional backbone network algorithms: GhostNetV2 [40], PP-LCNet [41], ShuffleNetV2 [42], MobileNetV3 [43], and EfficientNet [44]. The results of our experiments are shown in Table 5. The best fire detection among GhostNetV2, PP-LCNet, ShuffleNetV2, MobileNetV3, and EfficientNet as backbone networks was PP-LCNet, with an mAP of 90.25%. Although the GFLOPs of our proposed FireViT backbone network for fire target detection were 1.7 higher than PP-LCNet, FireViT had a 1.85% higher mAP and 0.9 M fewer parameters than PP-LCNet. Our proposed FireViT network backbone achieves a good balance between model complexity and detection accuracy for fire target detection.

Table 5. Experimental results comparing FireViT to mainstream lightweight convolutional backbone networks for fire target detection on our fire natural light dataset.

Model	AP_{fire}	AP_{smoke}	mAP	Params	GFLOPs
GhostNetV2	89.2%	91.0%	90.1%	3.8 M	6.3
PP-LCNet	89.6%	90.9%	90.25%	3.0 M	10.8
ShuffleNetV2	89.4%	90.1%	89.75%	3.0 M	10.7
MobileNetV3	89.0%	90.2%	89.6%	3.1 M	5.5
EfficientNet	89.0%	91.0%	90.0%	3.4 M	8.4
FireViT(ours)	91.3%	92.9%	92.1%	2.1 M	12.5

We next analyzed and compared FireViT with mainstream lightweight backbone network algorithms based on the transformer architecture: EfficientViT-M0 [14], SwinTransformer [45] (the model underwent depth compression of 0.33 and width compression of 0.25), EfficientFormerV2-S0 [15], and MobileViT-XS [16]. The results of these experiments are shown in Table 6. The results show that our proposed FireViT network backbone for fire target detection can improve detection accuracy while reducing the computational complexity of the model. In the case where the models have roughly the same number of parameters, MobileViT-XS has the highest mAP among EfficientViT-M0, EfficientFormerV2-S0, SwinTransformer and MobileViT-XS when used as the backbone network for fire target detection. Our proposed FireViT network backbone has a 1.2% higher mAP and 1.3 fewer GFLOPs relative to MobileViT-XS.

In order to obtain a more intuitive understanding of the features learned by various networks for detecting smoke and flames in fire incidents, we conducted heatmap visualization for the following backbone networks: GhostNetV2, PP-LCNet, ShuffleNetV2, MobileNetV3, EfficientNet, EfficientViT-M0, SwinTransformer (the model underwent depth compression of 0.33 and width compression of 0.25), EfficientFormerV2-S0, MobileViT-XS, and our proposed FireViT. The visualization results are depicted in Figure 13. From the heatmap visualization results of each network model, it is clear that our proposed FireViT is better able to capture the characteristics of smoke and flame in fire detection scenarios.

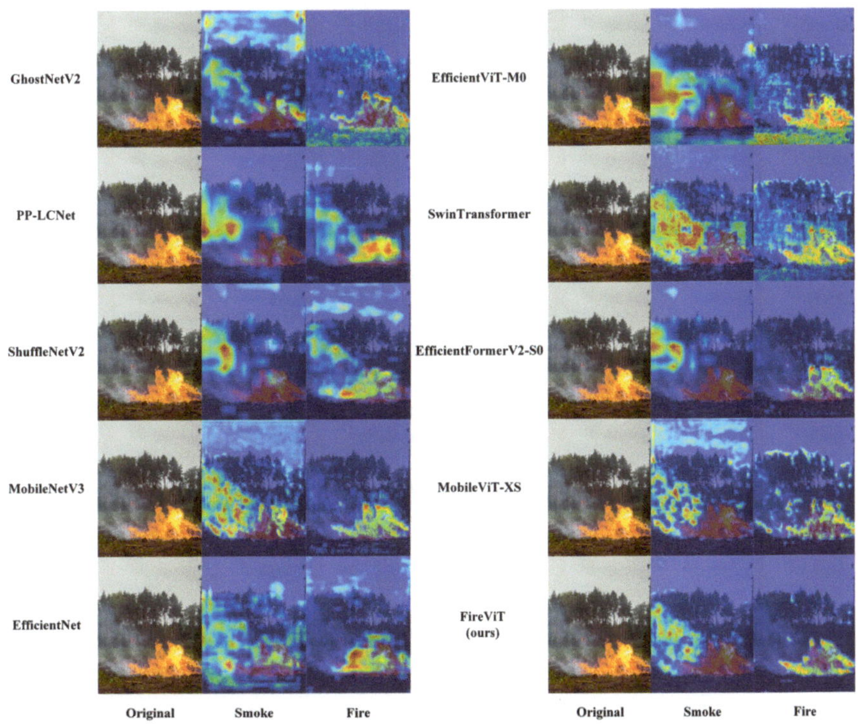

Figure 13. Heat map visualization results of each lightweight backbone network model for fire target detection.

The fire detection performance of FireViT as the network backbone is illustrated in Figure 14. From the first detection picture from top to bottom in the second column and the third picture from top to bottom in the fourth column of Figure 14, it can be seen that FireViT is able to achieve good detection and recognition performance even for smaller fire targets. In the third column of Figure 14, from top to bottom, the first fire picture contains an obvious "smoke" target, an obvious "fire" target, and another fuzzy "fire" target; FireViT can easily detect the two obvious targets, and is able to detect the fuzzy "fire" target as well. Overall, it can be seen from the figure that FireViT used as a network backbone for fire target detection is well suited for detection tasks in fire scenarios.

Table 6. Experimental results of comparison between FireViT and mainstream lightweight backbone networks based on the transformer architecture for fire target detection on our fire natural light dataset; the SwinTransformer model underwent depth and width compression, with 0.33 for depth and 0.25 for width.

Model	AP_{fire}	AP_{smoke}	mAP	Params	GFLOPs
EfficientViT-M0	89.0%	90.7%	89.85%	2.8 M	7.4
SwinTransformer	88.9%	90.1%	89.5%	2.2 M	6.9
EfficientFormerV2-S0	89.4%	91.2%	90.3%	3.8 M	9.0
MobileViT-XS	91.0%	90.8%	90.9%	1.9 M	13.8
FireViT(ours)	91.3%	92.9%	92.1%	2.1 M	12.5

Figure 14. Detection results of FireViT used as the backbone network for fire detection.

4.5. Comparison Experiments on Fire Infrared Datasets

Our previous fire detection comparison experiments used the fire natural light dataset. In order to further validate the generalization performance of FireViT, we used GhostNetV2, PP-LCNet, ShuffleNetV2, MobileNetV3, EfficientNet, EfficientViT-M0, SwinTransformer (the model underwent depth compression of 0.33 and width compression of 0.25), EfficientFormerV2-S0, MobileViT-XS, and FireViT as network backbones for fire detection on our fire infrared dataset. The results of these comparison experiments are shown in Table 7.

On the fire infrared dataset, other than our proposed FireViT lightweight network backbone model, MobileViT-XS was the most effective for fire detection, with its fire detection accuracy reaching 94.3%; PP-LCNet was second to MobileViT-XS, with a detection accuracy of 94.1%. Our proposed FireViT model for fire detection achieved 0.8% better detection accuracy on the infrared dataset than MobileViT-XS and 1% better detection accuracy than PP-LCNet. Overall, our proposed FireViT showed good generalization performance on the infrared fire dataset when used as the network backbone for fire detection.

Table 7. Comparison results of FireViT and mainstream lightweight backbone networks for fire detection on our fire infrared dataset.

Model	AP_{fire}	Params	GFLOPs
GhostNetV2-Infrared	93.7%	3.8 M	6.3
PP-LCNet-Infrared	94.1%	3.0 M	10.8
ShuffleNetV2-Infrared	94.0%	3.0 M	10.7
MobileNetV3-Infrared	93.5%	3.1 M	5.5
EfficientNet-Infrared	93.9%	3.4 M	8.4
EfficientViT-M0-Infrared	93.8%	2.8 M	7.4
SwinTransformer-Infrared	93.3%	2.2 M	6.9
EfficientFormerV2-S0-Infrared	93.9%	3.8 M	9.0
MobileViT-XS-Infrared	94.3%	1.9 M	13.8
FireViT-Infrared (ours)	95.1%	2.1 M	12.5

5. Conclusions

In this study, we have proposed an adaptive lightweight network backbone that can be used for fire detection, along with presentation of an improved adaptive activation function and a collection of labeled fire datasets containing the richest fire scenarios and the largest number of fire images built to date. First, in order to address the relatively small number of publicly available labeled fire datasets, as part of this research we collected and established a fire dataset that contains the richest fire scenes and the largest number of fire images currently available, for which we used the Rotate–Flip–Affine transformation operation. Our full fire dataset consists of a fire natural light dataset and fire infrared dataset, thereby meeting different application requirements. Second, in order to solve the problem of insufficient extraction of smoke and flame features that change irregularly in the fire scene, we propose the DeformViT block, a lightweight module that combines deformable convolution and a transformer to better grasp the features of smoke and flame in fire scenes both locally and holistically. Finally, we propose an improved adaptive activation function to further enhance the detection accuracy and nonlinear representation of the network. Our experimental results indicate that the FireViT adaptive lightweight network backbone proposed in this paper has high accuracy in fire detection scenarios. When used as the network backbone for fire detection, FireViT achieved an mAP of 92.1% on the fire natural light dataset and 95.1% on the fire infrared dataset with a model computational complexity of 12.5 GFLOPs. Based on these results, FireViT has important application value for fire warning, and can provide an effective solution for early warning in intelligent firefighting.

Author Contributions: Conceptualization, P.S., N.S. and K.H.; methodology, N.S., K.H. and X.Y.; software, P.S., N.S. and K.H.; validation, P.S. and N.S.; formal analysis, P.S., N.S. and P.W.; investigation, P.S., N.S. and Q.X.; resources, N.S., X.Y. and P.S.; data curation, P.S., K.H. and C.W.; writing—original draft preparation, P.S., Q.X. and C.W.; writing—review and editing, P.S., X.Y. and P.W.; supervision, N.S. and P.W. All authors have read and agreed to the published version of the manuscript.

Funding: The research in this article was supported by the National Natural Science Foundation of China (No. 42275156 and 42205150), Jiangsu Natural Science Foundation (No. BK20210661), Jiangsu Postgraduate Innovation Project (No. SJCX23_0386), and Qing Lan Project of Jiangsu Province.

Data Availability Statement: The data and code used to support the findings of this study are available from the corresponding author upon request (001764@cwxu.edu.cn).

Conflicts of Interest: The authors declare no conflict of interest.

References

1. Rachman, F.Z.; Yanti, N.; Hadiyanto, H.; Suhaedi, S.; Hidayati, Q.; Widagda, M.E.P.; Saputra, B.A. Design of the early fire detection based fuzzy logic using multisensor. *Conf. Ser. Mater. Sci. Eng.* **2020**, *732*, 012039. [CrossRef]
2. Ye, H.; Xiaogang, W.; Shuchuan, G. Design and Evaluation Method of Wireless Fire Detection Node Based on Multi-Source Sensor Data Fusion. *Int. J. Sens. Sens. Netw.* **2021**, *9*, 19. [CrossRef]
3. Solórzano, A.; Eichmann, J.; Fernández, L.; Ziems, B.; Jiménez-Soto, J. M.; Marco, S.; Fonollosa, J. Early fire detection based on gas sensor arrays: Multivariate calibration and validation. *Sens. Actuators B Chem.* **2022**, *352*, 130961. [CrossRef]
4. Li, Y.; Yu, L.; Zheng, C.; Ma, Z.; Yang, S.; Song, F.; Tittel, F.K. Development and field deployment of a mid-infrared CO and CO_2 dual-gas sensor system for early fire detection and location. *Spectrochim. Acta Part A Mol. Biomol. Spectrosc.* **2022**, *270*, 120834. [CrossRef]
5. Liu, X.; Sun, B.; Xu, Z. D.; Liu, X.; Xu, D. An intelligent fire detection algorithm and sensor optimization strategy for utility tunnel fires. *J. Pipeline Syst. Eng. Pract.* **2022**, *13*, 04022009. [CrossRef]
6. Qiu, T.; Yan, Y.; Lu, G. An autoadaptive edge-detection algorithm for flame and fire image processing. *IEEE Trans. Instrum. Meas.* **2011**, *61*, 1486–1493. [CrossRef]
7. Ji-neng, O.; Le-ping, B.; Zhi-kai, Y.; Teng, W. An early flame identification method based on edge gradient feature. In Proceedings of the 2018 2nd IEEE Advanced Information Management, Communicates, Electronic and Automation Control Conference (IMCEC), Xi'an, China, 25–27 May 2018; pp. 642–646. [CrossRef]
8. Khalil, A.; Rahman, S. U.; Alam, F.; Ahmad, I.; Khalil, I. Fire detection using multi color space and background modeling. *Fire Technol.* **2021**, *57*, 1221–1239. [CrossRef]
9. Majid, S.; Alenezi, F.; Masood, S.; Ahmad, M.; Gündüz, E.S.; Polat, K. Attention based CNN model for fire detection and localization in real-world images. *Expert Syst. Appl.* **2022**, *189*, 116114. [CrossRef]
10. Chen, G.; Cheng, R.; Lin, X.; Jiao, W.; Bai, D.; Lin, H. LMDFS: A Lightweight Model for Detecting Forest Fire Smoke in UAV Images Based on YOLOv7. *Remote Sens.* **2023**, *15*, 3790. [CrossRef]
11. Dogan, S.; Barua, P.D.; Kutlu, H.; Baygin, M.; Fujita, H.; Tuncer, T.; Acharya, U.R. Automated accurate fire detection system using ensemble pretrained residual network. *Expert Syst. Appl.* **2022**, *203*, 117407. [CrossRef]
12. Li, A.; Zhao, Y.; Zheng, Z. Novel Recursive BiFPN Combining with Swin Transformer for Wildland Fire Smoke Detection. *Forests* **2022**, *13*, 2032. [CrossRef]
13. Huang, J.; Zhou, J.; Yang, H.; Liu, Y.; Liu, H. A Small-Target Forest Fire Smoke Detection Model Based on Deformable Transformer for End-to-End Object Detection. *Forests* **2023**, *14*, 162. [CrossRef]
14. Liu, X.; Peng, H.; Zheng, N.; Yang, Y.; Hu, H.; Yuan, Y. EfficientViT: Memory Efficient Vision Transformer with Cascaded Group Attention. In Proceedings of the IEEE/CVF Conference on Computer Vision and Pattern Recognition (CVPR), Vancouver, BC, Canada, 18–22 June 2023; pp. 14420–14430. [CrossRef]
15. Li, Y.; Hu, J.; Wen, Y.; Evangelidis, G.; Salahi, K.; Wang, Y.; Ren, J. Rethinking vision transformers for mobilenet size and speed. *arXiv* **2022**, arXiv:2212.08059. [CrossRef]
16. Mehta, S.;Rastegari, M. Mobilevit: Light-weight, general-purpose, and mobile-friendly vision transformer. *arXiv* **2021**, arXiv:2110.02178. [CrossRef]
17. Wang, R.; Shivanna, R.; Cheng, D.; Jain, S.; Lin, D.; Hong, L.; Chi, E. Dcn v2: Improved deep and cross network and practical lessons for web-scale learning to rank systems. *arXiv* **2020**, arXiv:2008.13535.
18. Dosovitskiy, A.; Beyer, L.; Kolesnikov, A.; Weissenborn, D.; Zhai, X.; Unterthiner, T.; Houlsby, N. An image is worth 16x16 words: Transformers for image recognition at scale. *arXiv* **2020**, arXiv:2010.11929. [CrossRef]
19. Zhuang, J.; Qin, Z.; Yu, H.; Chen, X. Task-Specific Context Decoupling for Object Detection. *arXiv* **2023**, arXiv:2303.01047. [CrossRef]
20. Ge, Z.; Liu, S.; Wang, F.; Li, Z.; Sun, J. Yolox: Exceeding yolo series in 2021. *arXiv* **2021**, arXiv:2107.08430.[CrossRef]
21. Redmon, J.; Farhadi, A. Yolov3: An incremental improvement. *arXiv* **2018**, arXiv:1804.02767. [CrossRef]
22. Xu, S.; Wang, X.; Lv, W.; Chang, Q.; Cui, C.; Deng, K.; Lai, B. PP-YOLOE: An evolved version of YOLO. *arXiv* **2022**, arXiv:2203.16250. [CrossRef]
23. Li, C.; Li, L.; Jiang, H.; Weng, K.; Geng, Y.; Li, L.; Wei, X. YOLOv6: A single-stage object detection framework for industrial applications. *arXiv* **2022**, arXiv:2209.02976. [CrossRef]
24. Ultralytics-YOLOv8. Available online: https://github.com/ultralytics/ultralytics (accessed on 26 June 2023).
25. Zheng, Z.; Wang, P.; Liu, W.; Li, J.; Ye, R.; Ren, D. Distance-IoU loss: Faster and better learning for bounding box regression. In Proceedings of the AAAI conference on artificial intelligence, New York, NY, USA, 7–12 February 2020; Volume 34, pp. 12993–13000. [CrossRef]
26. Li, X.; Wang, W.; Wu, L.; Chen, S.; Hu, X.; Li, J.; Yang, J. Generalized focal loss: Learning qualified and distributed bounding boxes for dense object detection. *arXiv* **2020**, arXiv:2006.04388. [CrossRef]
27. Xia, Z.; Pan, X.; Song, S.; Li, L.E.; Huang, G. Vision transformer with deformable attention. In Proceedings of the IEEE/CVF Conference on Computer Vision and Pattern Recognition, New Orleans, LA, USA, 18–24 June 2022; pp. 4794–4803. [CrossRef]
28. Carion, N.; Massa, F.; Synnaeve, G.; Usunier, N.; Kirillov, A.; Zagoruyko, S. End-to-end object detection with transformers. In Proceedings of the European Conference on Computer Vision, Glasgow, UK, 23–28 August 2020; pp. 213–229. [CrossRef]
29. Hendrycks, D.; Gimpel, K. Gaussian error linear units (gelus). *arXiv* **2016**, arXiv:1606.08415. [CrossRef]

30. Devlin, J.; Chang, M.W.; Lee, K.; Toutanova, K. Bert: Pre-training of deep bidirectional transformers for language understanding. *arXiv* **2018**, arXiv:1810.04805. [CrossRef]
31. Radford, A.; Wu, J.; Child, R.; Luan, D.; Amodei, D.; Sutskever, I. Language models are unsupervised multitask learners. *OpenAI blog* **2019**, *1*, 9.
32. Glorot, X.; Bordes, A.; Bengio, Y. Deep sparse rectifier neural networks. In Proceedings of the Fourteenth International Conference on Artificial Intelligence and Statistics, Fort Lauderdale, FL, USA, 11–13 April 2011; pp. 315–323.
33. Elfwing, S.; Uchibe, E.; Doya, K. Sigmoid-weighted linear units for neural network function approximation in reinforcement learning. *Neural Netw.* **2018**, *107*, 3–11. [CrossRef]
34. Dunnings, A.; Breckon, T.P. *Fire Image Data Set for Dunnings 2018 Study-PNG Still Image Set*; Durham University: Durham, UK, 2018. [CrossRef]
35. Dedeoglu, N.; Toreyin, B.U.; Gudukbay, U.; Cetin, A.E. Real-time fire and flame detection in video. In Proceedings of the IEEE International Conference on Acoustics, Speech, and Signal Processing (ICASSP'05), Philadelphia, PA, USA, 18–23 March 2005; Volume 2, pp. 669–672. [CrossRef]
36. Ko, B.; Kwak, J.Y.; Nam, J.Y. Wildfire smoke detection using temporospatial features and random forest classifiers. *Opt. Eng.* **2012**, *51*, 017208. [CrossRef]
37. Zhang, Q.X.; Lin, G.H.; Zhang, Y.M.; Xu, G.; Wang, J.J. Wildland forest fire smoke detection based on faster R-CNN using synthetic smoke images. *Procedia Eng.* **2018**, *211*, 441–446. [CrossRef]
38. Shamsoshoara, A.; Afghah, F.; Razi, A.; Zheng, L.; Fulé, P.Z.; Blasch, E. Aerial imagery pile burn detection using deep learning: The FLAME dataset. *Comput. Netw.* **2021**, *193*, 108001. [CrossRef]
39. He, K.; Zhang, X.; Ren, S.; Sun, J. Delving deep into rectifiers: Surpassing human-level performance on imagenet classification. In Proceedings of the IEEE International Conference on Computer Vision, Santiago, Chile, 7–13 December 2015; pp. 1026–1034. [CrossRef]
40. Tang, Y.; Han, K.; Guo, J.; Xu, C.; Xu, C.; Wang, Y. GhostNetv2: Enhance cheap operation with long-range attention. *Adv. Neural Inf. Process. Syst.* **2022**, *35*, 9969–9982. [CrossRef]
41. Cui, C.; Gao, T.; Wei, S.; Du, Y.; Guo, R.; Dong, S.; Ma, Y. PP-LCNet: A lightweight CPU convolutional neural network. *arXiv* **2021**, arXiv:2109.15099. [CrossRef]
42. Ma, N.; Zhang, X.; Zheng, H.T.; Sun, J. Shufflenet v2: Practical guidelines for efficient cnn architecture design. In Proceedings of the European Conference on Computer Vision (ECCV), Munich, Germany, 8–14 September 2018; pp. 116–131. [CrossRef]
43. Howard, A.; Sandler, M.; Chu, G.; Chen, L.C.; Chen, B.; Tan, M.; Adam, H. Searching for mobilenetv3. In Proceedings of the IEEE/CVF International Conference on Computer Vision, Seoul, Republic of Korea, 27 October–2 November 2019; pp. 1314–1324. [CrossRef]
44. An, M.; Le, Q. Efficientnet: Rethinking model scaling for convolutional neural networks. In Proceedings of the International Conference on Machine Learning, Long Beach, CA, USA, 9–15 June 2019; pp. 6105–6114. [CrossRef]
45. Liu, Z.; Lin, Y.; Cao, Y.; Hu, H.; Wei, Y.; Zhang, Z.; Guo, B. Swin transformer: Hierarchical vision transformer using shifted windows. In Proceedings of the IEEE/CVF International Conference on Computer Vision, Virtual, 11–17 October 2021; pp. 10012–10022. [CrossRef]

forests

MDPI

Article

Enhancing Tree Species Identification in Forestry and Urban Forests through Light Detection and Ranging Point Cloud Structural Features and Machine Learning

Steffen Rust [1,*] and Bernhard Stoinski [2]

1 Faculty of Resource Management, University of Applied Sciences and Arts, Büsgenweg 1a, 37077 Göttingen, Germany
2 Private Institute for Dynamic Logic, Herforder Straße 15, 50737 Köln, Germany; bernhard.stoinski@pifdl.eu
* Correspondence: steffen.rust@hawk.de

Abstract: As remote sensing transforms forest and urban tree management, automating tree species classification is now a major challenge to harness these advances for forestry and urban management. This study investigated the use of structural bark features from terrestrial laser scanner point cloud data for tree species identification. It presents a novel mathematical approach for describing bark characteristics, which have traditionally been used by experts for the visual identification of tree species. These features were used to train four machine learning algorithms (decision trees, random forests, XGBoost, and support vector machines). These methods achieved high classification accuracies between 83% (decision tree) and 96% (XGBoost) with a data set of 85 trees of four species collected near Krakow, Poland. The results suggest that bark features from point cloud data could significantly aid species identification, potentially reducing the amount of training data required by leveraging centuries of botanical knowledge. This computationally efficient approach might allow for real-time species classification.

Keywords: urban forest inventory; tree species classification; bark features; machine learning; point cloud data; LiDAR scanning

Citation: Rust, S.; Stoinski, B. Enhancing Tree Species Identification in Forestry and Urban Forests through Light Detection and Ranging Point Cloud Structural Features and Machine Learning. *Forests* **2024**, *15*, 188. https://doi.org/10.3390/f15010188

Academic Editors: Guojie Wang and Zengxin Zhang

Received: 13 December 2023
Revised: 9 January 2024
Accepted: 15 January 2024
Published: 17 January 2024

1. Introduction

The identification of tree species is a fundamental aspect of measuring and monitoring forests and urban trees, providing critical insights into ecosystem dynamics, species-specific ecosystem services, and economic considerations such as timber values. Accurately identifying tree species is crucial for making informed decisions, ecological conservation, and sustainable resource management in urban and rural environments. With the growing adoption of remote sensing technologies in the fields of forestry and urban forestry, the task of automatic species classification has gained paramount importance.

Traditionally, species identification for inventories has relied on labor-intensive field surveys and manual observations by trained foresters or botanists. However, the emergence of remote sensing technologies has revolutionized this process by enabling large-scale non-invasive data collection. In this context, the automatic classification of tree species has become a critical challenge and a focus of extensive research efforts.

To date, several methods have been successfully employed for tree species classification, with many of them primarily relying on RGB images as the primary data source. While these methods have yielded promising results, they often overlook valuable information contained within the point cloud structure of a tree's stem. Point clouds, generated through LiDAR (Light Detection and Ranging) or photogrammetry, offer a three-dimensional representation of the forest environment, including the stems of individual trees, when using a terrestrial platform [1].

Laser scanning is used in forestry and urban forestry for various applications, including forest inventory, tree mapping, and monitoring of individual tree characteristics. In forestry, laser scanning techniques, such as Airborne Laser Scanning (ALS), Terrestrial Laser Scanning (TLS), including mobile LiDAR systems such as handheld systems, and Mobile Laser Scanning (MLS), have been widely investigated for applications in forest inventory [2,3]. These techniques provide efficient means for acquiring detailed three-dimensional (3D) data from vegetation, enabling the extraction of tree and forest parameters such as tree height, crown dimensions, and biomass [4–16].

In urban forestry, laser scanning is used for mapping and monitoring single tree characteristics, providing a convenient tool for measuring tree attributes in cities and urban forests [12,17–20]. ALS can be used to generate high-resolution spatially explicit maps of urban forest structure, including the detection, mapping, and characterization of individual trees [21]. TLS has also been applied in urban forestry for capturing detailed 3D tree structures and monitoring tree growth and health [22,23].

Tree species classification using LiDAR data is typically based on extracting specific features from the point cloud data, such as geometric, radiometric, and full-waveform features [24]. These features can be used to differentiate between tree species based on their unique structural and reflectance properties. Researchers have developed various methods for tree species classification using LiDAR data, including deep learning models [25–29], individual tree segmentation and shape fitting [30], and also combined LiDAR with data from other sensor types like hyper- or multispectral data [31].

For example, a study using a 3D deep learning approach achieved an overall accuracy of 92.5% in tree species classification directly using ALS point clouds to derive the structural features of trees [27]. Another study proposed a method based on the crown shape of segmented individual trees extracted from ALS point clouds to identify tree species [30]. These and other studies, e.g., [24–26,28,31–58], demonstrate the potential of LiDAR technology in providing accurate and efficient species classification in forestry and ecological applications.

However, individual tree segmentation and shape fitting methods using LiDAR data for species classification can face challenges in dense forests [32], be sensitive to data quality [59], have limitations in capturing species-specific features [47], and require significant computational resources [60]. Integrating additional data sources and developing algorithms that are more advanced can help address these drawbacks and improve the accuracy and efficiency of tree species classification using LiDAR data.

While textural features of tree organs like leaves or flowers vary notably because of seasonal change, the morphology of tree bark remains constant across seasons. Most previous work has been based on RGB images of tree stems. While this approach has been very successful, reaching accuracies well above 90% [61,62], the quality of such images can vary with contrasting or insufficient lighting conditions of the trunk, and such images can only be acquired during daylight hours. Although LiDAR point clouds should not be affected by lighting as much as RGB images are and can be collected even at night, the potential of utilizing structural bark characteristics derived from LiDAR point cloud data for tree species identification remains largely untapped.

While [44,63] have already used structural bark features, this paper tries to address this critical gap by exploring the utility of structural features traditionally used by experts to identify tree species based on their bark and stem characteristics. Apart from bark color, botanists use structural features like ridges, fissures, peeling, or scales, which have been described, amongst others, in [64]. To our knowledge, this is the first attempt to describe these structures mathematically and to derive features for machine learning from this description.

We test whether machine learning models can differentiate between various common Central European tree species (*Acer platanoides* L., *Fraxinus excelsior* L., *Robinia pseudoacacia* L., *Larix decidua* Mill., *Fagus sylvatica* L.). Our study goes beyond the conventional reliance on RGB images and demonstrates the potential of adding LiDAR point cloud information to enhance the automatic classification of tree species.

If successful, this approach has the potential to make a significant contribution to the field of remote sensing applications in forestry and urban forestry. Providing an additional tool for accurate tree species identification will have implications for improved forest management, conservation efforts, and sustainable urban planning.

2. Materials and Methods

2.1. Trees and Scanning

LiDAR point clouds of 85 trees (Table 1) were acquired from mature trees in forests around Krakow by the company ProGea 4D, Poland, using a Faro laser scanner.

Table 1. Number of individuals sampled per species.

Species	*n*	Mean Diameter, m
Acer platanoides	18	0.29 ± 0.06
Fraxinus excelsior	14	0.22 ± 0.05
Robinia pseudoacacia	17	0.59 ± 0.15
Larix decidua	17	0.30 ± 0.05
Fagus sylvatica	19	0.72 ± 0.29

TLS point clouds were not very noisy. There were no moving objects in the scans (e.g., people or animals), and the scanning was performed on a windless day. The only noise that was created was probably in the tree canopy, at the edges of the leaves. Filter tools available in the Faro Scene 2023 software were completely sufficient to avoid processing the TLS cloud in external software (e.g., SOR filter in Cloud Compare 2.12.4).

All trees had fully developed mature bark features.

2.2. Feature Creation

The bark analysis method for tree species identification that is presented here is based on set theory and algebraic mathematical methods. The bark is characterized by several parameters. These parameters form a model of the bark. More precisely, it is a vector space that is provided with a distance measure, a metric space $(\mathbf{X}, d)$ [65–67].

Each tree and its associated bark structure are represented as a vector in a vector space. The elements of a vector are the parameters listed in Table 2.

Further information exists for each point, which is determined using LiDAR scanning. This includes, for example, time stamps and color values. However, these are of secondary importance for the bark analysis. The tree species therefore form a real subset of X. Each tree species has a number of comparable elements per vector. The metric d of the vector space is required to enable the tree species sets to be separated by a distance value.

To analyze tree bark surfaces using digital point clouds, they must be described and quantified through standardized methods. This project aims to propose a method for standardization that evaluates the parameters of rib structure, spacing, spatial orientation, and appearance.

Definition 1. *Ribs are elevations that differ from a "smooth" bark surface by an additive positive value ε as the difference value.*

The clusters represent open subsets. This is the case because a cluster does not contain all accumulation points. The general concept of a vector space can therefore be made more precise. It is a topological space M in which the separation axiom T_2 [68], the Hausdorff separation axiom, applies. This means that no cluster exists that is identical to either its predecessor or its successor. Mathematically speaking, a point cloud in which only disjoint clusters exist is a Hausdorff space.

Table 2. Features derived from point clouds.

Parameter	Acronym	Explanation
Geometric description		
Follow cluster left/right	CL-LR	The subsequent cluster of the rib is oriented to the right or left.
Follow clusters form diagonal left/right	CL-D	The sum of the subsequent clusters of a rib form diagonals that run to the right or left.
Follow clusters vertical	CL-V	The subsequent cluster of a rib is located vertically below its predecessor.
Follow clusters horizontal	CL-H	The subsequent cluster of a rib is located horizontally next to its predecessor.
Follow clusters branched	CL-B	The subsequent clusters of a rib form a branch.
Follow clusters in right angle left	CL-AL	The angle formed between a subsequent cluster (to the left of the perpendicular) and a perpendicular that intersects the predecessor.
Follow clusters in right angle right	CL-AR	The angle formed between a subsequent cluster (to the right of the perpendicular) and a perpendicular that intersects the predecessor.
Rib characteristics		
Horizontal behavior		The horizontal proportion of gradients, slopes, and horizontal components in a bark grid.
Rib spacing		The distance between two rib clusters.
Roughness		The roughness is defined by the sum of all cluster points with a higher or lower ε value of a rib.
Proportion of smooth surface		Proportion of smooth to rough (ribbed) surface.

These clusters must be adjacent, i.e., they must be spatially close to each other and parallel to the trunk diameter, the abscissa axis. A point, as defined in this model, is a 15-dimensional vector. An important element is the cluster size C_{Size}. This specification makes it possible to form clusters with similar properties that satisfy the following conditions:

$$C_{Size} := \left(\sum_{i=1}^{n} x_i | z(x_i) > z(x_{i-1}) \wedge y(x_i) > y(y_{min}) \right) \tag{1}$$

with x ∈ trunk diameter, y ∈ bark depth, z ∈ trunk length, and

$$y_{min} := \frac{\sum_{i=1}^{n} (y_i - y_{i-1}) > 0}{n} \tag{2}$$

The direction of x, y, and z coordinates is illustrated in Figure 1. By using clusters with similar sizes and properties, it is possible to trace structures along z. All clusters with similar structures represent elements of a meta-cluster, which in turn represents the bark rib. The clustering of the meta-cluster and the analysis of the bark structure is carried out using the AI software Dylogos 2.0. With the help of the software, it is possible to analyze the individual data of a point cloud.

For the study of the similarity of ribs, the following properties are considered:

- Rib width;
- Depth between two adjacent ribs;
- Distance between the ribs;
- Shape of the ribs, fissures, roughness, etc.

In the model described here, 11 evaluation criteria are defined which serve to classify the species (see Table 2).

These 11 evaluation criteria plus the three spatial axes and the cluster size form the 15 elements of the vector and thus of the cluster.

The following cluster axioms constitute the model:

- Axiom 1: Each model has at least one cluster.
- Axiom 2: Every model has no zero cluster.
- Axiom 3: More than one successor cluster can exist.
- Axiom 4: Each cluster contains information about its predecessors.
- Axiom 5: If no subsequent cluster exists, the number of predecessors defines the length of the ridge. This forms the 15th element of the cluster vector.
- Axiom 6: All clusters with similar properties and a spatial proximity form elements of a bark rib.

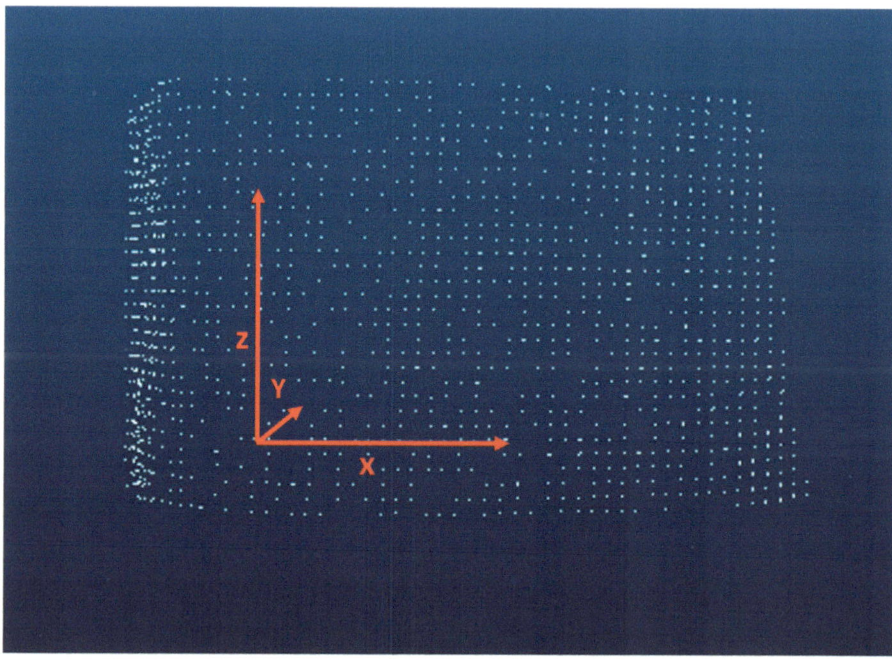

Figure 1. Illustration of the orientation of x, y, and z coordinates.

Figure 2 (Figures 2 and 3 are visualizations of the clusters identified by Dylogos) of *Robinia* bark will explain this in more detail. It shows the bark structure as a cluster cloud. Based on this, axioms 1–6 will be shown in the following. Axioms 1 and 2 are fulfilled. The first cluster in the red circle has several successors (axiom 3 and 4).

The cluster A is the last cluster of the row (axiom 5) and is the end element of the rib (axiom 6). The elements of the green circle form a meta-cluster structure. By this structure, we can also recognize the spatial orientation. In this case, it is diagonal. The clusters in the yellow circle are examples of horizontal as well as vertical structures. The middle cluster in the second row of the red circle is an example of branching. Structures that are more complex may have a combination of features. Each bark is individual, like a fingerprint, but shows characteristic relationships per species. These are shown in the figure below as an example for a *Robinia pseudoacacia*. The data shown therein are an excerpt from the training data.

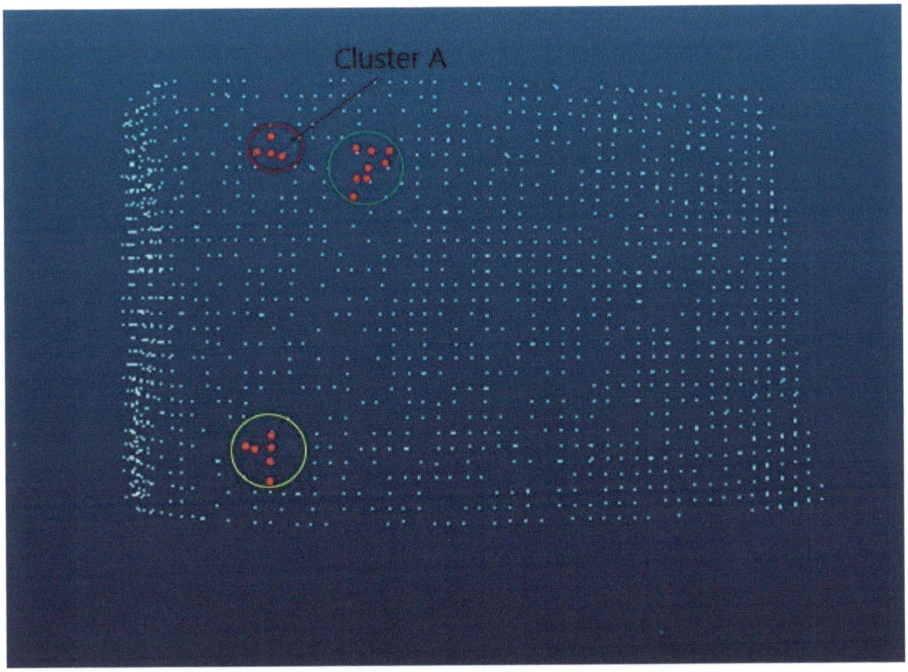

Figure 2. *Robinia pseudoacacia* bark as cluster representation.

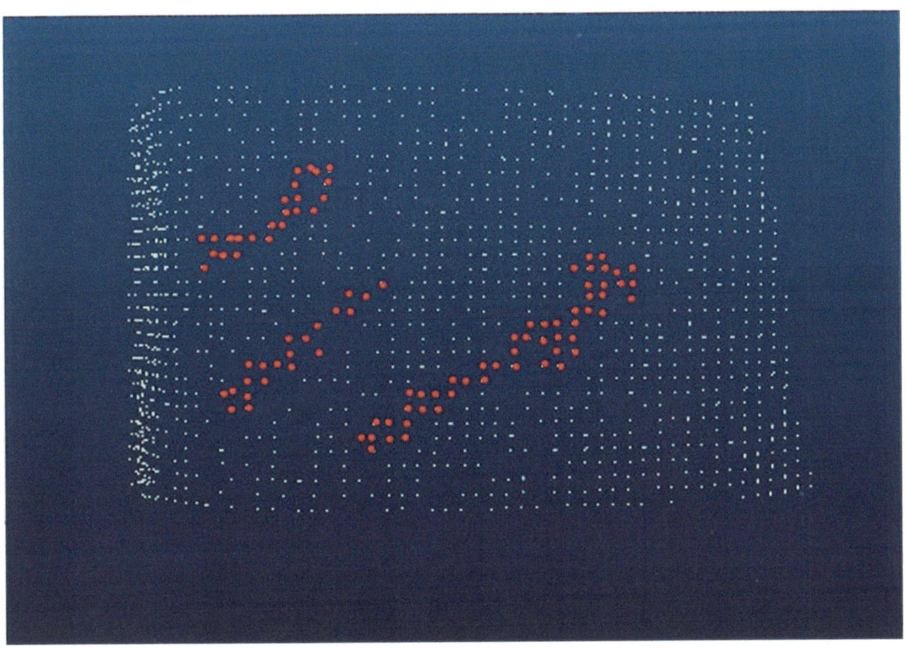

Figure 3. *Robinia pseudoacacia* bark with diagonal, vertical, and branching alignments.

In Figure 3, which is based on Figure 2, the diagonal and vertical alignments as well as the branching have been highlighted.

Areas without cluster points are areas that lie below the average bark surface. These points are suppressed in the display but are taken into account for the evaluation of roughness. The average bark surface is determined using average values from the decision grid and forms a meta-level for the entire trunk.

Figure 3 shows all clusters that have a positive distance from the average bark surface, the ϵ value. Figure 4 shows a section of Figure 2, in which the meta-plane, the average bark surface, and an example of ε are plotted.

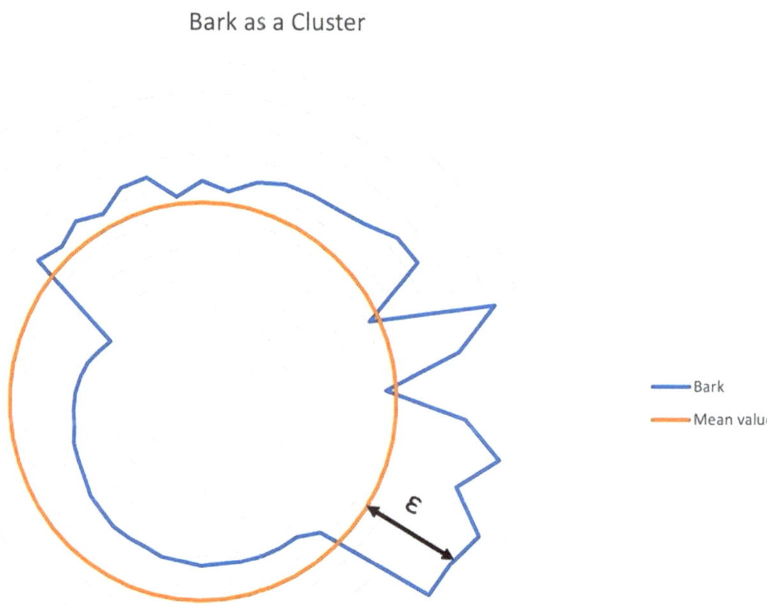

Figure 4. *Robinia pseudoacacia* bark with the mean value (ochre color).

Figure 4 can be generated from Figure 3, using the mapping rule $\mathcal{M}_{Cluster} := \mathcal{P}_{oint} - \alpha > 0$ with α = mean value and $\mathcal{M}_{Cluster} \subseteq \mathcal{P}_{oint}$ The mean value represents the average roughness of the meta-cluster, i.e., the reference bark. The circle is the graphical representation of the mean value of the meta-cluster. The blue bark curve, which is shown as a deviation from the mean value of the meta-cluster, is not formed by individual points but by clusters. The clusters represent a large number of points. All cluster elements that lie above the meta level are considered for the analysis; these are the elements of the set $\mathcal{M}_{Cluster}$. All cluster elements are determining elements of the evaluation and recognition of the bark using this method. The only value that has no reference to the meta-level is that of the gradient. This value is an evaluation of the surface of each individual ridge. The evaluation consists of looking at each rib cluster and its orientation in space in comparison to its predecessor and successor. The number of successor clusters that show positive, negative, or no change along the Y-axis is decisive for the evaluation.

The LiDAR data were available as LAS files. For the bark analysis, a trunk section of 2 m was used, which was divided into 40 × 40 cm grids. These grids were used to determine an average bark grid and thus an average bark pattern. All parameter values were determined from this grid. If the bark is evenly comparable over large parts or the entire trunk, all grids are combined into one grid.

The evaluation was carried out using the AI software Dylogos 2.0. The Dylogos software transforms the LAS data into XYZ data. These are then clustered into the two groups:

1. Geometric description of the bark;
2. Rib characteristics.

The features created in this way were further analyzed with R.

2.3. Machine Learning

For the quality of the decision making, the clustering of the training data is of particular importance. In the following, the four methods applied in this study and their structural differences will be applied to the bark model. These are as follows:

- Decision trees;
- Random forests;
- XGBoost;
- Support vector machines.

Their results can be characterized by two parameters. These are the accuracy and the predictive power of the trained system. The way of calculating the accuracy by means of the confusion matrix is identical for all four methods. A confusion matrix is a 2×2 matrix scheme. The elements of the matrix are as follows:

- True Positive (TP);
- False Positive (FP);
- True Negative (RN);
- False Negative (FN).

All four elements are taken into account in the machine learning process.

Here, the evaluation of a condition and its future development is judged. The assessment can be true or false, and the respective expression can turn out to be positive or negative for the assessment model in the future. The rows are filled with the actual condition and the columns with the predicted condition.

Each data set is now classified into one of the four elements (classes) of the matrix according to the model generated using the training data. The accuracy value is determined as follows:

$$Accuracy = \frac{TP - TN}{TP + TN + FP + FN} \tag{3}$$

The accuracy value is between 0 and 1. A value > 0.9 is a good value. A value > 0.7 is good, and 0.7 is a fair result.

The four methods differ in the prediction condition. The decision tree and random forests methods use the roc auc value (Compute Area Under the Receiver Operating Characteristic Curve) to parameterize the prediction. The roc auc value is between 0 and 1. A value of 0.5 represents a random estimate.

The XGBoost and the support vector machine methods use κ (kappa) for the quality of the prediction. κ or Cohen's kappa is a measure of interrater reliability and thus a parameter that reflects the agreement or disagreement between two observers on a decision. The authors of [69] suggest that $\kappa < 0$ = "poor agreement", $0 < \kappa < 0.2$ = "slight agreement", $0.21 < \kappa < 0.40$ = "fair agreement", $0.41 < \kappa < 0.60$ = "moderate agreement", $0.61 < \kappa < 0.80$ = "substantial agreement", and $\kappa > 0.81$ = "almost perfect agreement".

Table 2 shows the parameters used. This structure corresponds to the generated evaluation database. They are divided into two blocks:

- Geometric description of the bark;
- Rib characteristics.

These give the geometry of the bark, a description of the bark's appearance, and the description of the individual ribs.

The machine learning methods we investigated differ not only in their functionalities, which will be discussed in more detail, but also in the different weighting and therefore relevance of the various parameters. The type and number of parameters taken into account vary from method to method.

2.3.1. Decision Trees

The decision tree method is based on the assumption that all information important for a decision is available in the training data sets. The trees have a uniform structure in the form that the leaves of the tree describe classes and the branches form conjunctions of features that then lead to a class.

Decisions are thus better structured. The path of a decision is not a linear path but has nodes with branches. The choice of which branch to select is made by means of a decision function that is derived from the training data. In most cases, the decision function separates within a cluster whether values are larger or smaller than a target value.

2.3.2. Random Forests

In the random forest method, the samples used to determine the tree structure are randomly selected from the training data. After a new node is created, the samples are added back to the training set. At each node, a randomly selected subset of criteria from the entire set of criteria is used to make a decision (branch). The selection of features is performed to minimize the impurity of the overall model. Due to the randomness, multiple decision trees are created per training set, forming a decision forest. The predictions of the individual trees are then aggregated to produce an overall prediction.

2.3.3. XGBoost

In the XGBoost model, the fitting of a tree structure is performed using a loss function. The tree structure is generated, starting from a starting point, by means of the Newton method. Each new node is considered as a new model and optimized by a loss function.

2.3.4. Support Vector Machines

A support vector machine (SVM) is a discriminative machine learning model that uses a hyperplane to separate training data into two classes. Unlike the DBSCAN clustering method, which searches for the elements with the smallest distance to the hyperplane, SVMs search for the elements with the widest distance to the hyperplane. This results in data clusters with sharp boundaries.

The analyses were implemented in R [70]. Models were fitted with 10-fold cross-validation on a training data set of 75% of the samples and tested on the remaining 25%. Features with near-zero variation and closely correlated features were removed prior to analysis. When necessary, features were Yeo-Johnson transformed. All features were normalized.

Finally, Figure 5 illustrates the entire process as a workflow for greater clarity.

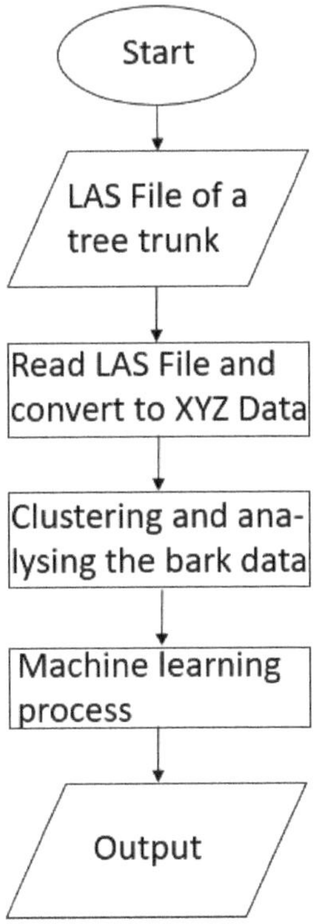

Figure 5. Scheme of the workflow used in this study.

3. Results

3.1. Features

The calculation of the features was computationally very efficient and was conducted on a recent desktop PC (DELL Latitude, CPU Intel Core i5-6300, 8,00 GB Ram).

Some of the proposed features were correlated, e.g., roughness and smoothness (r = 0.81), behavior and smoothness (r = 0.91), or behavior and CL-LR (r = 0.85). Only one of each pair was used for further analyses. Before transformation, the distributions of most features were highly skewed.

Fagus sylvatica had the most distinctive set of bark features. It was the species with the smoothest bark (Figure 6a) and the lowest spacing between ribs (Figure 6b).

3.2. Machine Learning

The decision tree (Figure 7) performed less well than the other approaches and reached an accuracy of 83% and a roc auc score of 94%. *R. pseudoacacia* with its very distinctive bark was the first species that was split from the others in this model.

The remaining three methods performed equally well, with accuracies between 92% and 96% even for this small data set of 85 trees (Table 3 and Figure 8).

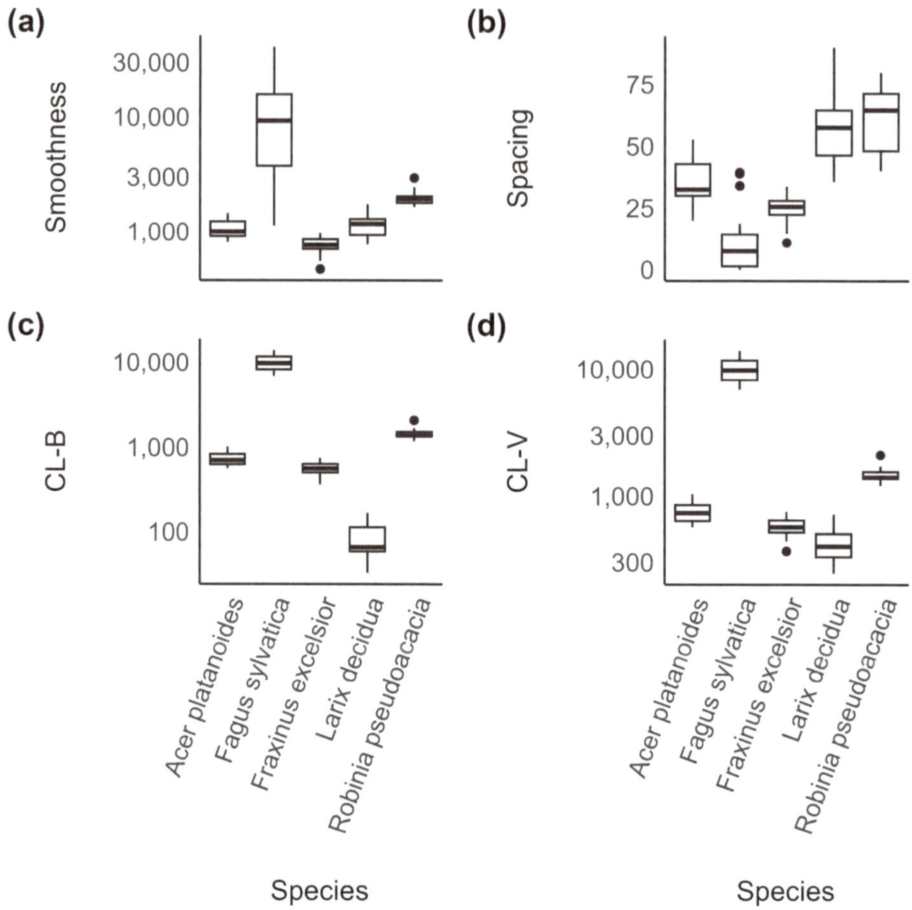

Figure 6. Distribution of four selected features ((**a**): smoothness, (**b**): spacing, (**c**): branching, and (**d**): vertical alignment of clusters) of all species in the sample. They illustrate how the features differentiate between species.

Table 3. Performance of the machine learning methods used in this study.

	Accuracy	**Roc Auc**	κ
Decision tree	0.83	0.943	✗
Random forests	0.916	0.985	✗
XGBoost	0.96	✗	0.95
Support vector machines	0.92	✗	0.9

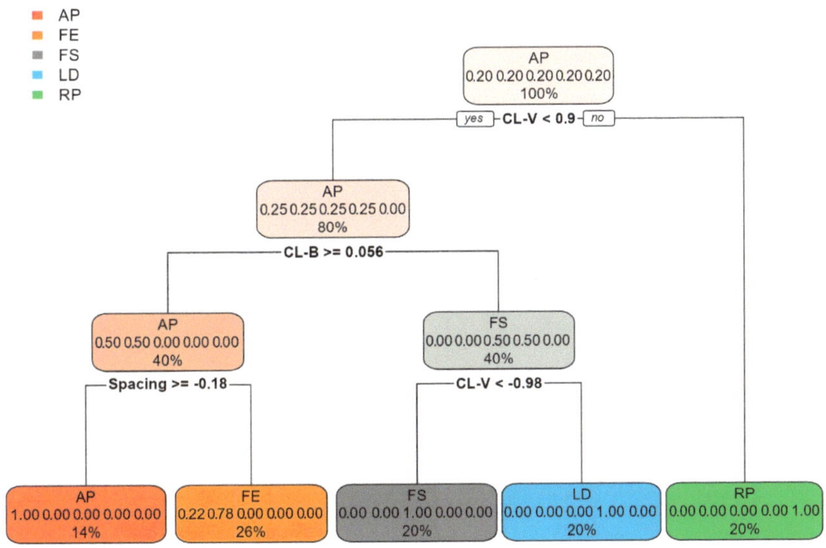

Figure 7. Result of the classification based on a decision tree, illustrating the subset of features used by this model. AP: *Acer pseudoplatanus*, FE: *Fraxinus excelsior*, FS: *Fagus sylvatica*, LD: *Larix decidua*, RP: *Robinia pseudoacacia*.

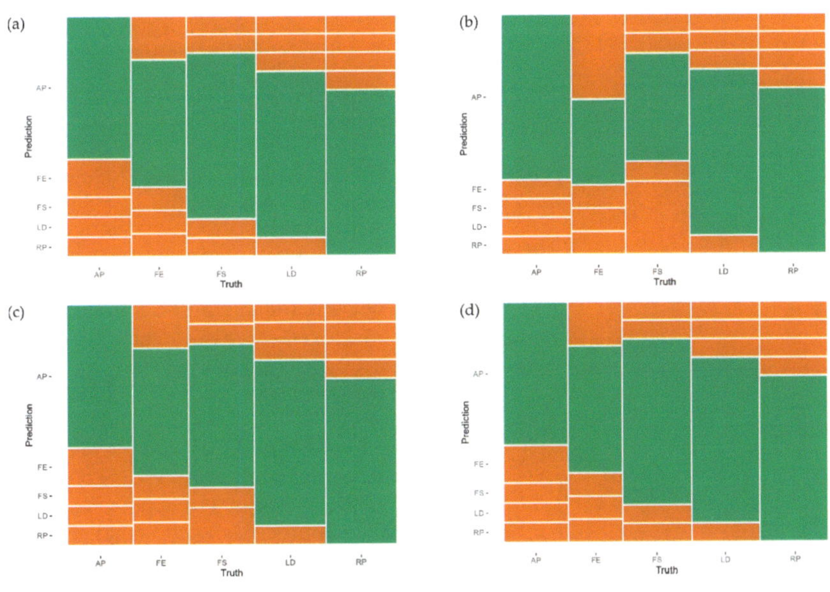

Figure 8. Confusion matrices of (**a**) decision tree, (**b**) random forest, (**c**) SVM, and (**d**) XGBoost. AP: *Acer pseudoplatanus*, FE: *Fraxinus excelsior*, FS: *Fagus sylvatica*, LD: *Larix decidua*, RP: *Robinia pseudoacacia*. The green area represents the proportion of members of the test sample set (not used in the training of the models) classified correctly, while red represents false classifications.

The other more complex methods achieved 96% accuracy in predicting the species of the test population. However, the predictions of the random forest model were comparatively poor for *F. excelsior* (Figure 8b). Overall, the XGBoost model had the best results

(Table 3 and Figure 8d). The difference between the accuracy value and κ, the value of the prediction, is the smallest of all the presented models.

The ranking of feature importance was not similar for all models. The most important feature for the XGBoost model was "Cluster cluster left/right" (CL-RL), followed by "clusters vertical" (CL-V) and bark roughness (Figure 9). The decision tree, on the other hand, used only the features "CL-V", "CL-B", and "Spacing" to classify species (Figure 7).

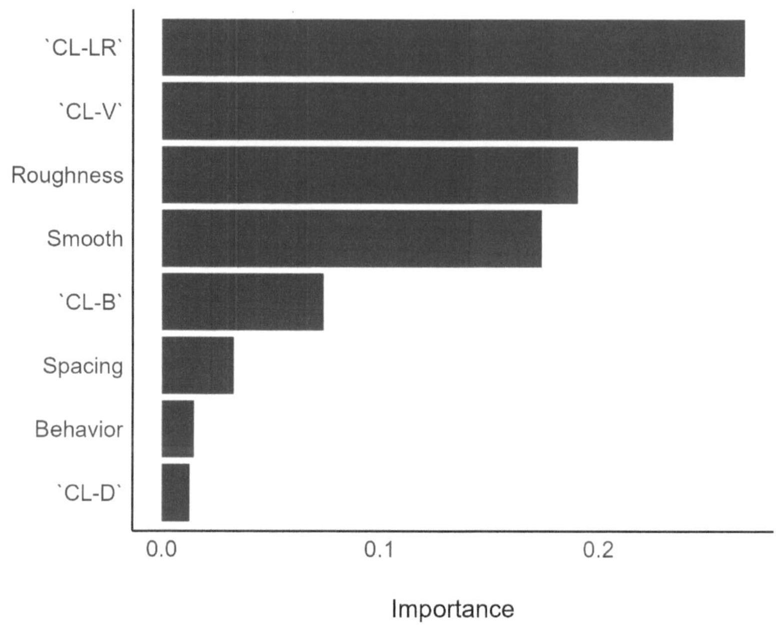

Figure 9. Feature importance of the XGBoost model. Note that compared to the decision tree (Figure 7) a different subset of features was selected by this algorithm.

4. Discussion

Trees provide a wide range of ecosystem services in urban areas, including air and water purification, noise reduction, and temperature regulation [71]. Up-to-date tree inventories are essential for effective tree management and monitoring of their ecosystem services [72]. Terrestrial Laser scanning (TLS) can make tree inventory data collection much more efficient than traditional methods [73], but species identification from TLS point clouds is still challenging.

In this study, we hypothesized that bark features traditionally used by botanists, like ridges, crevices, and smoothness, could be described mathematically and applied to species identification from TLS point clouds. The approach we presented differs from the very few other recently published approaches, such as that of [44], in the kind and number of bark features considered.

Since different machine learning methods are structurally different, we tested the performance of several approaches. The accuracy of all methods was high, despite the rather small data set used. Our results suggest that the mathematical description of bark features used by botanists could be used to complement, or even provide advantages over, the black-box approaches used so far.

Correct species identification is essential for tree inventories, as it is the basis for, amongst other things, ecosystem services calculations, tree maintenance, and tree risk management. Our method will complement other approaches to identify tree species based on remote sensing data, will help to increase overall accuracy, and thus, will support the more efficient creation of tree inventories. This assumption is based on a timesaving potential as well as the future possibility of the partial automation of this process.

However, our study has some limitations. First, the number of trees and tree species in our data set was relatively small. Secondly, we only used trees with fully developed mature bark characteristics. This means that the trees examined had diameters corresponding to the age class. Future research should focus on increasing the size and diversity of the training data set to improve the accuracy and robustness of the machine learning models. Additionally, the models should be tested on younger trees because the bark structure can change significantly during the lifetime of a tree [74]. Although present on some trees, we did not study the effects of epiphytes growing on the bark on the accuracy of species identification. Furthermore, the models should be integrated into existing tree inventory workflows to assess their feasibility for practical use.

5. Conclusions

This study provides promising evidence that explainable bark features can be used to identify species from TLS point clouds. A model of bark features based on expert knowledge could potentially reduce the number of required samples in comparison to black-box approaches. This could result in more efficient and accurate collection of tree inventory data, which is crucial for the effective management and monitoring of their ecosystem services in urban areas. This computationally efficient approach might allow for real-time species classification.

Author Contributions: Conceptualization, S.R.; methodology, S.R. and B.S.; software, B.S.; validation, S.R.; formal analysis, S.R.; data curation, B.S.; writing—original draft preparation, S.R. and B.S.; writing—review and editing, S.R. and B.S.; visualization, S.R. and B.S. All authors have read and agreed to the published version of the manuscript.

Funding: This research received no external funding.

Data Availability Statement: Data are unavailable due to privacy restrictions.

Acknowledgments: We thank ProGea 4D, Krakow, Poland, for providing the point cloud data.

Conflicts of Interest: The authors declare no conflicts of interest.

References

1. Kushwaha, S.K.P.; Singh, A.; Jain, K.; Cabo, C.; Mokros, M. Integrating Airborne and Terrestrial Laser Scanning for Complete 3D Model Generation in Dense Forest. In Proceedings of the IGARSS 2023—2023 IEEE International Geoscience and Remote Sensing Symposium, Pasadena, CA, USA, 16–21 July 2023; pp. 3137–3140.
2. Chen, S.; Liu, H.; Feng, Z.; Shen, C.; Chen, P. Applicability of Personal Laser Scanning in Forestry Inventory. *PLoS ONE* **2019**, *14*, e0211392. [CrossRef]
3. Hyyppä, J.; Yu, X.; Kaartinen, H.; Kukko, A.; Jaakkola, A.; Liang, X.; Wang, Y.; Holopainen, M.; Vastaranta, M.; Hyyppa, H. *Forest Inventory Using Laser Scanning*; Shan, J., Toth, C.K., Eds.; CRC Press-Taylor & Francis Group: Boca Raton, FL, USA, 2018; pp. 379–412, ISBN 978-1-4987-7228-0.
4. Brede, B.; Terryn, L.; Barbier, N.; Bartholomeus, H.M.; Bartolo, R.; Calders, K.; Derroire, G.; Krishna Moorthy, S.M.; Lau, A.; Levick, S.R.; et al. Non-Destructive Estimation of Individual Tree Biomass: Allometric Models, Terrestrial and UAV Laser Scanning. *Remote Sens. Environ.* **2022**, *280*, 113180. [CrossRef]
5. Arseniou, G.; MacFarlane, D.W.; Calders, K.; Baker, M. Accuracy Differences in Aboveground Woody Biomass Estimation with Terrestrial Laser Scanning for Trees in Urban and Rural Forests and Different Leaf Conditions. *Trees* **2023**, *37*, 761–779. [CrossRef]
6. Wang, F.; Sun, Y.; Jia, W.; Li, D.; Zhang, X.; Tang, Y.; Guo, H. A Novel Approach to Characterizing Crown Vertical Profile Shapes Using Terrestrial Laser Scanning (TLS). *Remote Sens.* **2023**, *15*, 3272. [CrossRef]
7. Demol, M.; Verbeeck, H.; Gielen, B.; Armston, J.; Burt, A.; Disney, M.; Duncanson, L.; Hackenberg, J.; Kükenbrink, D.; Lau, A.; et al. Estimating Forest Above-ground Biomass with Terrestrial Laser Scanning: Current Status and Future Directions. *Methods Ecol. Evol.* **2022**, *13*, 1628–1639. [CrossRef]
8. Vazirabad, Y.F.; Karslioglu, M.O. *Lidar for Biomass Estimation*; Matovic, D., Ed.; Intech Europe: Rijeka, Croatia, 2011; pp. 3–26. ISBN 978-953-307-492-4.
9. Calders, K.; Verbeeck, H.; Burt, A.; Origo, N.; Nightingale, J.; Malhi, Y.; Wilkes, P.; Raumonen, P.; Bunce, R.G.H.; Disney, M. Laser Scanning Reveals Potential Underestimation of Biomass Carbon in Temperate Forest. *Ecol. Solut. Evid.* **2022**, *3*, e12197. [CrossRef]

10. Dassot, M.; Barbacci, A.; Colin, A.; Fournier, M.; Constant, T. Tree Architecture and Biomass Assessment from Terrestrial LiDAR Measurements: A Case Study for Some Beech Trees (*Fagus sylvatica*). *Silvilaser Full Proc.* **2010**, 206–215.
11. Sun, C.; Huang, C.; Zhang, H.; Chen, B.; An, F.; Wang, L.; Yun, T. Individual Tree Crown Segmentation and Crown Width Extraction from a Heightmap Derived From Aerial Laser Scanning Data Using a Deep Learning Framework. *Front. Plant Sci.* **2022**, *13*, 914974. [CrossRef]
12. Wu, L.; Shi, Y.; Zhang, F.; Zhou, Y.; Ding, Z.; Lv, S.; Xu, L. Estimating Carbon Stocks and Biomass Expansion Factors of Urban Greening Trees Using Terrestrial Laser Scanning. *Forests* **2022**, *13*, 1389. [CrossRef]
13. Imai, Y.; Setojima, M.; Yamagishi, Y.; Fujiwara, N. Tree-Height Measuring Characteristics of Urban Forests by Lidar Data Different in Resolution. In Proceedings of the Geo-Imagery Bridging Continents, Istanbul, Turkey, 12–23 July 2004; pp. 2–5.
14. Guo, Y.; Luka, A.; Wei, Y. Modeling Urban Tree Growth for Digital Twins: Transformation of Point Clouds into Parametric Crown Models. *J. Digit. Landsc. Archit.* **2022**, 213–223. [CrossRef]
15. Uzquiano, S.; Barbeito, I.; San Martín, R.; Ehbrecht, M.; Seidel, D.; Bravo, F. Quantifying Crown Morphology of Mixed Pine-Oak Forests Using Terrestrial Laser Scanning. *Remote Sens.* **2021**, *13*, 4955. [CrossRef]
16. Ma, Q.; Lin, J.; Ju, Y.; Li, W.; Liang, L.; Guo, Q. Individual Structure Mapping over Six Million Trees for New York City USA. *Sci. Data* **2023**, *10*, 102. [CrossRef] [PubMed]
17. Münzinger, M.; Prechtel, N.; Behnisch, M. Mapping the Urban Forest in Detail: From LiDAR Point Clouds to 3D Tree Models. *Urban For. Urban Green.* **2022**, *74*, 127637. [CrossRef]
18. Zieba-Kulawik, K.; Skoczylas, K.; Wezyk, P.; Teller, J.; Mustafa, A.; Omrani, H. Monitoring of Urban Forests Using 3D Spatial Indices Based on LiDAR Point Clouds and Voxel Approach. *Urban For. Urban Green.* **2021**, *65*, 127324. [CrossRef]
19. Zieba-Kulawik, K.; Wezyk, P. Monitoring 3D Changes in Urban Forests Using Landscape Metrics Analyses Based on Multi-Temporal Remote Sensing Data. *Land* **2022**, *11*, 883. [CrossRef]
20. César de Lima Araújo, H.; Silva Martins, F.; Tucunduva Philippi Cortese, T.; Locosselli, G.M. Artificial Intelligence in Urban Forestry—A Systematic Review. *Urban For. Urban Green.* **2021**, *66*, 127410. [CrossRef]
21. Matasci, G.; Coops, N.C.; Williams, D.A.R.; Page, N. Mapping Tree Canopies in Urban Environments Using Airborne Laser Scanning (ALS): A Vancouver Case Study. *For. Ecosyst.* **2018**, *5*, 31. [CrossRef]
22. Plowright, A.A.; Coops, N.C.; Eskelson, B.N.I.; Sheppard, S.R.J.; Aven, N.W. Assessing Urban Tree Condition Using Airborne Light Detection and Ranging. *Urban For. Urban Green.* **2016**, *19*, 140–150. [CrossRef]
23. Degerickx, J.; Roberts, D.A.; McFadden, J.P.; Hermy, M.; Somers, B. Urban Tree Health Assessment Using Airborne Hyperspectral and LiDAR Imagery. *Int. J. Appl. Earth Obs. Geoinf.* **2018**, *73*, 26–38. [CrossRef]
24. Michałowska, M.; Rapiński, J. A Review of Tree Species Classification Based on Airborne LiDAR Data and Applied Classifiers. *Remote Sens.* **2021**, *13*, 353. [CrossRef]
25. Sakharova, E.K.; Nurlyeva, D.D.; Fedorova, A.A.; Yakubov, A.R.; Kanev, A.I. Issues of Tree Species Classification from LiDAR Data Using Deep Learning Model. In *Advances in Neural Computation, Machine Learning, and Cognitive Research V*; Kryzhanovsky, B., Dunin-Barkowski, W., Redko, V., Tiumentsev, Y., Klimov, V.V., Eds.; Springer International Publishing: Cham, Switzerland, 2022; pp. 319–324.
26. Hartling, S.; Sagan, V.; Sidike, P.; Maimaitijiang, M.; Carron, J. Urban Tree Species Classification Using a WorldView-2/3 and LiDAR Data Fusion Approach and Deep Learning. *Sensors* **2019**, *19*, 1284. [CrossRef] [PubMed]
27. Liu, M.; Han, Z.; Chen, Y.; Liu, Z.; Han, Y. Tree Species Classification of LiDAR Data Based on 3D Deep Learning. *Measurement* **2021**, *177*, 109301. [CrossRef]
28. Xi, Z.; Hopkinson, C.; Rood, S.B.; Peddle, D.R. See the Forest and the Trees: Effective Machine and Deep Learning Algorithms for Wood Filtering and Tree Species Classification from Terrestrial Laser Scanning. *ISPRS J. Photogramm. Remote Sens.* **2020**, *168*, 1–16. [CrossRef]
29. Seidel, D.; Annighöfer, P.; Thielman, A.; Seifert, Q.E.; Thauer, J.-H.; Glatthorn, J.; Ehbrecht, M.; Kneib, T.; Ammer, C. Predicting Tree Species From 3D Laser Scanning Point Clouds Using Deep Learning. *Front. Plant Sci.* **2021**, *12*, 635440. [CrossRef]
30. Qian, C.; Yao, C.; Ma, H.; Xu, J.; Wang, J. Tree Species Classification Using Airborne LiDAR Data Based on Individual Tree Segmentation and Shape Fitting. *Remote Sens.* **2023**, *15*, 406. [CrossRef]
31. Fassnacht, F.E.; Latifi, H.; Stereńczak, K.; Modzelewska, A.; Lefsky, M.; Waser, L.T.; Straub, C.; Ghosh, A. Review of Studies on Tree Species Classification from Remotely Sensed Data. *Remote Sens. Environ.* **2016**, *186*, 64–87. [CrossRef]
32. Qiao, Y.; Zheng, G.; Du, Z.; Ma, X.; Li, J.; Moskal, L. Tree-Species Classification and Individual-Tree-Biomass Model Construction Based on Hyperspectral and LiDAR Data. *Remote Sens.* **2023**, *15*, 1341. [CrossRef]
33. Hakula, A.; Ruoppa, L.; Lehtomäki, M.; Yu, X.; Kukko, A.; Kaartinen, H.; Taher, J.; Matikainen, L.; Hyyppä, E.; Luoma, V.; et al. Individual Tree Segmentation and Species Classification Using High-Density Close-Range Multispectral Laser Scanning Data. *ISPRS Open J. Photogramm. Remote Sens.* **2023**, *9*, 100039. [CrossRef]
34. Ahlswede, S.; Schulz, C.; Gava, C.; Helber, P.; Bischke, B.; Förster, M.; Arias, F.; Hees, J.; Demir, B.; Kleinschmit, B. TreeSatAI Benchmark Archive: A Multi-Sensor, Multi-Label Dataset for Tree Species Classification in Remote Sensing. *Earth Syst. Sci. Data* **2023**, *15*, 681–695. [CrossRef]
35. Liu, B.; Huang, H.; Su, Y.; Chen, S.; Li, Z.; Chen, E.; Tian, X. Tree Species Classification Using Ground-Based LiDAR Data by Various Point Cloud Deep Learning Methods. *Remote Sens.* **2022**, *14*, 5733. [CrossRef]

36. Liu, B.; Chen, S.; Huang, H.; Tian, X. Tree Species Classification of Backpack Laser Scanning Data Using the PointNet++ Point Cloud Deep Learning Method. *Remote Sens.* **2022**, *14*, 3809. [CrossRef]
37. Hell, M.; Brandmeier, M.; Briechle, S.; Krzystek, P. Classification of Tree Species and Standing Dead Trees with Lidar Point Clouds Using Two Deep Neural Networks: PointCNN and 3DmFV-Net. *PFG* **2022**, *90*, 103–121. [CrossRef]
38. Cetin, Z.; Yastikli, N. The Use of Machine Learning Algorithms in Urban Tree Species Classification. *ISPRS Int. J. Geo-Inf.* **2022**, *11*, 226. [CrossRef]
39. Faizal, S. Automated Identification of Tree Species by Bark Texture Classification Using Convolutional Neural Networks. *Int. J. Res. Appl. Sci. Eng. Technol.* **2022**, *10*, 1384–1392. [CrossRef]
40. Misra, D.; Crispim-Junior, C.; Tougne, L. Patch-Based CNN Evaluation for Bark Classification. In *Computer Vision—ECCV 2020 Workshops*; Bartoli, A., Fusiello, A., Eds.; Springer: Cham, Switzerland, 2020; pp. 197–212. ISBN 978-3-030-65413-9.
41. Abdollahnejad, A.; Panagiotidis, D. Tree Species Classification and Health Status Assessment for a Mixed Broadleaf-Conifer Forest with UAS Multispectral Imaging. *Remote Sens.* **2020**, *12*, 3722. [CrossRef]
42. Terryn, L.; Calders, K.; Disney, M.; Origo, N.; Malhi, Y.; Newnham, G.; Raumonen, P.; Åkerblom, M.; Verbeeck, H. Tree Species Classification Using Structural Features Derived from Terrestrial Laser Scanning. *ISPRS J. Photogramm. Remote Sens.* **2020**, *168*, 170–181. [CrossRef]
43. Xu, Z.; Shen, X.; Cao, L.; Coops, N.C.; Goodbody, T.R.H.; Zhong, T.; Zhao, W.; Sun, Q.; Ba, S.; Zhang, Z.; et al. Tree Species Classification Using UAS-Based Digital Aerial Photogrammetry Point Clouds and Multispectral Imageries in Subtropical Natural Forests. *Int. J. Appl. Earth Obs. Geoinf.* **2020**, *92*, 102173. [CrossRef]
44. Mizoguchi, T.; Ishii, A.; Nakamura, H. Individual Tree Species Classification Based on Terrestrial Laser Scanning Using Curvature Estimation and Convolutional Neural Network. *Int. Arch. Photogramm. Remote Sens. Spat. Inf. Sci.* **2019**, *XLII-2-W13*, 1077–1082. [CrossRef]
45. Nguyen, H.M.; Demir, B.; Dalponte, M. A Weighted SVM-Based Approach to Tree Species Classification at Individual Tree Crown Level Using LiDAR Data. *Remote Sens.* **2019**, *11*, 2948. [CrossRef]
46. Marrs, J.; Ni-Meister, W. Machine Learning Techniques for Tree Species Classification Using Co-Registered LiDAR and Hyperspectral Data. *Remote Sens.* **2019**, *11*, 819. [CrossRef]
47. Dalponte, M.; Frizzera, L.; Gianelle, D. Individual Tree Crown Delineation and Tree Species Classification with Hyperspectral and LiDAR Data. *PeerJ* **2019**, *6*, e6227. [CrossRef] [PubMed]
48. Wang, K.; Wang, T.; Liu, X. A Review: Individual Tree Species Classification Using Integrated Airborne LiDAR and Optical Imagery with a Focus on the Urban Environment. *Forests* **2018**, *10*, 1. [CrossRef]
49. Mizoguchi, T.; Ishii, A.; Nakamura, H.; Inoue, T.; Takamatsu, H. Lidar-Based Individual Tree Species Classification Using Convolutional Neural Network. In *Proceedings of the Videometrics, Range Imaging, and Applications XIV*; Remondino, F., Shortis, M.R., Eds.; SPIE: Bellingham, WA, USA, 2017; Volume 10332, p. UNSP103320O.
50. Shen, X.; Cao, L. Tree-Species Classification in Subtropical Forests Using Airborne Hyperspectral and LiDAR Data. *Remote Sens.* **2017**, *9*, 1180. [CrossRef]
51. Matsuki, T.; Yokoya, N.; Iwasaki, A. Hyperspectral Tree Species Classification of Japanese Complex Mixed Forest with the Aid of LiDAR Data. *IEEE J. Sel. Top. Appl. Earth Obs. Remote Sens.* **2015**, *8*, 2177–2187. [CrossRef]
52. Othmani, A.; Voon, L.F.C.L.Y.; Stolz, C.; Piboule, A. Single Tree Species Classification from Terrestrial Laser Scanning Data for Forest Inventory. *Pattern Recognit. Lett.* **2013**, *34*, 2144–2150. [CrossRef]
53. Yao, W.; Krzystek, P.; Heurich, M. Tree Species Classification and Estimation of Stem Volume and DBH Based on Single Tree Extraction by Exploiting Airborne Full-Waveform LiDAR Data. *Remote Sens. Environ.* **2012**, *123*, 368–380. [CrossRef]
54. Dalponte, M.; Bruzzone, L.; Gianelle, D. Tree Species Classification in the Southern Alps Based on the Fusion of Very High Geometrical Resolution Multispectral/Hyperspectral Images and LiDAR Data. *Remote Sens. Environ.* **2012**, *123*, 258–270. [CrossRef]
55. Heinzel, J.; Koch, B. Exploring Full-Waveform LiDAR Parameters for Tree Species Classification. *Int. J. Appl. Earth Obs. Geoinf.* **2011**, *13*, 152–160. [CrossRef]
56. Korpela, I.; Ørka, H.; Maltamo, M.; Tokola, T.; Hyyppä, J. Tree Species Classification Using Airborne LiDAR—Effects of Stand and Tree Parameters, Downsizing of Training Set, Intensity Normalization, and Sensor Type. *Silva Fenn.* **2010**, *44*, 156. [CrossRef]
57. Weiss, U.; Biber, P.; Laible, S.; Bohlmann, K.; Zell, A. Plant Species Classification Using a 3D LIDAR Sensor and Machine Learning. In Proceedings of the 2010 Ninth International Conference on Machine Learning and Applications, Washington, DC, USA, 12–14 December 2010; pp. 339–345.
58. Voss, M.; Sugumaran, R. Seasonal Effect on Tree Species Classification in an Urban Environment Using Hyperspectral Data, LiDAR, and an Object-Oriented Approach. *Sensors* **2008**, *8*, 3020–3036. [CrossRef]
59. Luo, H.; Khoshelham, K.; Chen, C.; He, H. Individual Tree Extraction from Urban Mobile Laser Scanning Point Clouds Using Deep Pointwise Direction Embedding. *ISPRS J. Photogramm. Remote Sens.* **2021**, *175*, 326–339. [CrossRef]
60. Hamraz, H.; Contreras, M.A.; Zhang, J. Forest Understory Trees Can Be Segmented Accurately within Sufficiently Dense Airborne Laser Scanning Point Clouds. *Sci. Rep.* **2017**, *7*, 6770. [CrossRef] [PubMed]
61. Cui, Z.; Li, X.; Li, T.; Li, M. Improvement and Assessment of Convolutional Neural Network for Tree Species Identification Based on Bark Characteristics. *Forests* **2023**, *14*, 1292. [CrossRef]

62. Kim, T.K.; Hong, J.; Ryu, D.; Kim, S.; Byeon, S.Y.; Huh, W.; Kim, K.; Baek, G.H.; Kim, H.S. Identifying and Extracting Bark Key Features of 42 Tree Species Using Convolutional Neural Networks and Class Activation Mapping. *Sci. Rep.* **2022**, *12*, 4772. [CrossRef]
63. Othmani, A.A.; Jiang, C.; Lomenie, N.; Favreau, J.-M.; Piboule, A.; Voon, L.F.C.L.Y. A Novel Computer-Aided Tree Species Identification Method Based on Burst Wind Segmentation of 3D Bark Textures. *Mach. Vis. Appl.* **2016**, *27*, 751–766. [CrossRef]
64. Whitmore, T.C. Studies in Systematic Bark Morphology. I. Bark Morphology in Dipterocarpaceae. *New Phytol.* **1962**, *61*, 191–207. [CrossRef]
65. Zeidler, E. *Springer-Handbuch Der Mathematik I–IV*; Springer Fachmedien: Wiesbaden, Germany, 2013.
66. Brieskorn, E. *Lineare Algebra Und Analytische Geometrie III*; Springer Spektrum: Wiesbaden, Germany, 2019; Volume III.
67. Bronstein, I.N.; Semendjajew, K.A.; Zeideler, E.; Herausgeber, W. *Teubner-Taschenbuch Der Mathematik*; B. G. Teubner Stuttgart: Leibzig, Germany, 1996.
68. Barner, M.; Flor, F. *Analysis II*; de Gruyter: Berlin, Germany; New York, NY, USA, 1989.
69. Landis, J.R.; Koch, G.G. The Measurement of Observer Agreement for Categorical Data. *Biometrics* **1977**, *33*, 159–174. [CrossRef]
70. *R Core Team R: A Language and Environment for Statistical Computing*; R Foundation for Statistical Computing: Vienna, Austria, 2021.
71. Salbitano, F.; Borelli, S.; Conigliaro, M.; Chen, Y. *Guidelines on Urban and Peri-Urban Forestry*; FAO Forestry Paper; Food and Agriculture Organization of the United Nations: Rome, Italy, 2016; ISBN 978-92-5-109442-6.
72. Edgar, C.B.; Nowak, D.J.; Majewsky, M.A.; Lister, T.W.; Westfall, J.A.; Sonti, N.F. Strategic National Urban Forest Inventory for the United States. *J. For.* **2020**, *119*, 86–95. [CrossRef]
73. Morgenroth, J.; Östberg, J. Measuring and Monitoring Urban Trees and Urban Forests. In *Routledge Handbook of Urban Forestry*; Routledge: London, UK, 2017; pp. 33–48.
74. Whitmore, T.C. Studies in Systematic Bark Morphology: IV. The Bark of Beech, Oak and Sweet Chestnut. *New Phytol.* **1963**, *62*, 161–169. [CrossRef]

forests

MDPI

Article

A Lightweight Pine Wilt Disease Detection Method Based on Vision Transformer-Enhanced YOLO

Quanbo Yuan [1,2], Suhua Zou [2], Huijuan Wang [2,*], Wei Luo [2], Xiuling Zheng [2], Lantao Liu [2] and Zhaopeng Meng [1]

[1] College of Intelligence and Computing, Tianjin University, Tianjin 300072, China; yqb_cs2018@tju.edu.cn (Q.Y.)

[2] School of Computer, North China Institute of Aerospace Engineering, Langfang 065000, China

* Correspondence: wanghj@nciae.edu.cn

Abstract: Pine wilt disease (PWD) is a forest disease characterized by rapid spread and extremely high lethality, posing a serious threat to the ecological security of China's forests and causing significant economic losses in forestry. Given the extensive forestry area, limited personnel for inspection and monitoring, and high costs, utilizing UAV-based remote sensing monitoring for diseased trees represents an effective approach for controlling the spread of PWD. However, due to the small target size and uneven scale of pine wilt disease, as well as the limitations of real-time detection by drones, traditional disease tree detection algorithms based on RGB remote sensing images do not achieve an optimal balance among accuracy, detection speed, and model complexity due to real-time detection limitations. Consequently, this paper proposes Light-ViTeYOLO, a lightweight pine wilt disease detection method based on Vision Transformer-enhanced YOLO (You Only Look Once). A novel lightweight multi-scale attention module is introduced to construct an EfficientViT feature extraction network for global receptive field and multi-scale learning. A novel neck network, CACSNet(Content-Aware Cross-Scale bidirectional fusion neck network), is designed to enhance the detection of diseased trees at single granularity, and the loss function is optimized to improve localization accuracy. The algorithm effectively reduces the number of parameters and giga floating-point operations per second (GFLOPs) of the detection model while enhancing overall detection performance. Experimental results demonstrate that compared with other baseline algorithms, Light-ViTeYOLO proposed in this paper has the least parameter and computational complexity among related algorithms, with 3.89 MFLOPs and 7.4 GFLOPs, respectively. The FPS rate is 57.9 (frames/s), which is better than the original YOLOv5. Meanwhile, its mAP@0.5:0.95 is the best among the baseline algorithms, and the recall and mAP@0.5 slightly decrease. Our Light-ViTeYOLO is the first lightweight method specifically designed for detecting pine wilt disease. It not only meets the requirements for real-time detection of pine wilt disease outbreaks but also provides strong technical support for automated forestry work.

Keywords: pine wilt disease; YOLOv5; Light-ViTeYOLO; EfficientViT; real-time detection

Citation: Yuan, Q.; Zou, S.; Wang, H.; Luo, W.; Zheng, X.; Liu, L.; Meng, Z. A Lightweight Pine Wilt Disease Detection Method Based on Vision Transformer-Enhanced YOLO. *Forests* **2024**, *15*, 1050. https://doi.org/10.3390/f15061050

Academic Editors: Guojie Wang and Zengxin Zhang

Received: 13 May 2024
Revised: 13 June 2024
Accepted: 14 June 2024
Published: 18 June 2024

1. Introduction

Pine wilt disease (PWD), caused by the pine wood nematode, is a forest disease characterized by high pathogenicity, rapid spread, and a wide transmission pathway, resulting in severe damage to pine forest resources in China [1]. This disease has been classified as a quarantine pest in more than 40 countries, with China experiencing substantial direct economic losses and ecological service value depletion [2]. Owing to the extensive forested area and the high costs and limited scope of manual inspection and monitoring, there is a need for efficient, cost-effective, and accurate monitoring techniques. In recent years, the advancement of UAV (Unmanned Aerial Vehicle) remote sensing technology has demonstrated significant potential for application in the monitoring of pine wood nematode disease, leveraging its operational ease, adaptability, extensive coverage, and real-time capabilities [3].

The use of UAV remote sensing for monitoring pine blight outbreaks has undergone significant evolution over the past few decades. Traditional machine learning algorithms, such as SVM (Support Vector Machine), RF (Random Forest), and ANNs (Artificial Neural Networks), have been developed and optimized by integrating spectral and spatial features. These methods have been successfully employed for identifying pine blight tree damage in Multi-Spectral Imagery (MSI) and Hyper-Spectral Imagery (HSI) datasets. However, classical machine learning necessitates intricate feature selection and combination work, posing challenges for leveraging in-depth image information [4].

In recent years, with the development of deep learning object classification and detection technology, researchers have gradually applied it to PWD detection [5,6]. For instance, Qin et al. [7] utilized a proposed SCANet (spatial-context-attention network) to diagnose pine nematode disease in UAV-based MSI datasets, achieving an average overall accuracy of 79.33%. Wu et al. [8] used Faster R-CNN (Region-CNN) and YOLOv3 for early diagnosis of infected trees, demonstrating that YOLOv3 is more suitable for PWD detection. Gong et al. [9] identified pine blight spots affected by pine wilt using YOLOv5, achieving a mean Average Precision (mAP) of 84.5%. Similarly, Sun et al. [10] utilized the improved MobileNetv2-YOLOv4 algorithm to identify abnormal discoloration blight caused by pine wilt nematode disease, and the improved model achieved higher detection accuracy of 86.85%.

Although current deep learning methods have achieved some results in disease detection, realizing real-time detection on UAV platforms still faces great challenges. Changes in UAV flight altitude and speed lead to too small and different scales of disease and pest targets in trees, making detection difficult. In addition, limited by the computational resources, storage, and communication capabilities of the UAV platform, it is difficult for the existing deep learning-based methods to achieve a balance between detection accuracy and speed due to the complexity of their models.

Aiming at the above problems and difficulties, this paper takes the YOLOv5 model as the baseline network, redesigns and optimizes the feature extraction network, neck network, and loss function, and proposes ViTeYOLO, a lightweight pine wilt detection method based on Vision Transformer-enhanced YOLO, to improve its detection accuracy for PWD and achieve light weight. The main contributions of this paper are as follows:

1. A lightweight Multi-Scale Attention module (MSA) is introduced to construct an EfficientViT feature extraction network, which achieves efficient global information extraction and multi-scale learning through efficient hardware operations, reducing network computational complexity;
2. A Content-Aware Cross-Scale bidirectional fusion neck network (CACSNet) is proposed, which uses the Content-Aware Reassembly Feature Enhancement (CARAFE) operator to replace the bilinear difference in PANET (Path Aggregation Network) for upsampling, and uses cross-scale weighting for feature fusion to improve the expression ability of fine-grained features of diseased trees, prevent small target feature loss, and improve detection accuracy;
3. Optimization of the loss function and introduction of EIOU (Efficient Intersection over Union) loss to help the model better balance the size and shape information of the target, improving the accuracy and robustness of PWD detection.

2. Related Works

2.1. Visual Transformer in Remote Sensing

Transformer [11] employs an attention-based architecture that first demonstrated its great impact on sequence modeling and machine translation tasks, and has evolved to become the primary deep learning model for many natural language processing (NLP) tasks. Inspired by these significant achievements, Transformer has been applied to the field of computer vision (CV) and has led to some groundbreaking work, giving rise to Visual Transformer (ViT) [12].

ViT has shown exceptional performance in various computer vision tasks. ViT relies on a self-attention mechanism to skillfully capture global interactions by utilizing the connections between elements in the input sequence. According to studies [13,14], the results demonstrate the ability to model content-dependent remote interactions. It possesses the property of being able to flexibly adjust the sensory field so as to be able to adapt to multiple complexities in the data and learn effective feature representations. As a result, ViT and its variants have been successfully applied to several computer vision tasks such as classification, detection, and segmentation. With the success of ViT in the field of computer vision, the remote sensing community has also observed a significant growth in the application of transformer-based frameworks for multiple tasks. This has triggered a promising wave of research in remote sensing, where researchers have adopted a variety of approaches [15,16] to train and analyze remote sensing data using visual transformers. Hong et al. [17] developed SpectralFormer based on ViT to obtain advanced classification results for hyperspectral images. Liu et al. [18] introduced a Deep Spatial Spectral Transformer (DSS-TRM) for end-to-end hyperspectral image classification. In addition, a hybrid approach based on a combination of transformers and cellular neural networks was used to detect changes in dual-time images [17] as well as to detect small objects in remotely sensed images in complex backgrounds [19]. The above ViT-based methods, while achieving advanced performance in remote sensing, also bring a greater number of parameters and computational effort.

2.2. Lightweight Multi-Scale Attention

Multiscale learning and global receptive fields during feature extraction can effectively improve the performance of tasks such as semantic segmentation and target detection, but their computation is quadratic in the resolution of the input image, and sometimes requires special support for hardware to achieve good efficiency [18]. Cai et al. [20] proposed a lightweight MSA module for semantic segmentation that requires only hardware operation for global sense field and multi-scale learning, and shows significant speedup on edge devices. As shown in Figure 1, the lightweight MSA uses ReLU-based lightweight attention for global receptive fields, and the feature map, after obtaining Q/K/V tokens through a linear projection layer, is aggregated with nearby tokens through lightweight convolution of a small kernel to generate multiscale tokens.

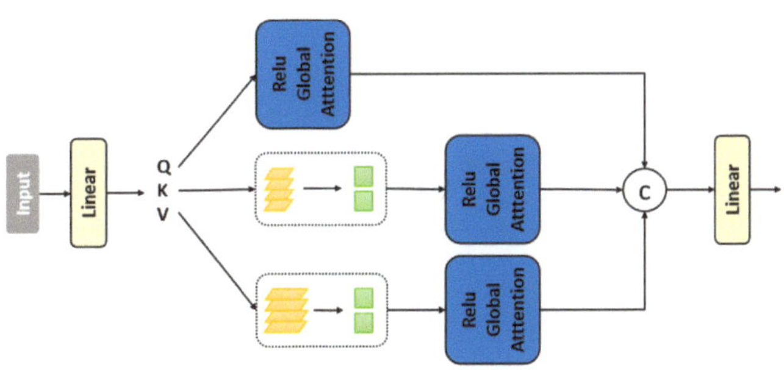

Figure 1. Lightweight multi-scale attention.

ReLU-based global attention (Equation (1)) is applied to the multiscale token and connects and inputs the output to the final linear projection layer for feature fusion.

$$W_i = \sum_{j=1}^{N} \frac{\mathrm{Relu}(Q_i)\mathrm{Relu}(K_j)^T}{\sum_{j=1}^{N} \mathrm{Relu}(Q_i)\mathrm{Relu}(K_j)^T} V_j \tag{1}$$

Here, $Q = xW^Q$, $K = xW^K$, and $V = xW^V$, where WQ, WK, and WV are learnable linear projection matrices. W_i represents the i-th row of the matrix W. By leveraging the associativity property of matrix multiplication, it is possible to reduce the computational complexity and memory usage from quadratic to linear without altering the functionality. As a result, Equation (1) can be expressed as:

$$W_i = \frac{\mathrm{Relu}(Q_i)\left(\sum_{j=1}^{N}\mathrm{Relu}(K_j)^T V_j\right)}{\mathrm{Relu}(Q_i)\left(\sum_{j=1}^{N}\mathrm{Relu}(K_j)^T\right)} \quad (2)$$

This module leverages lightweight ReLU-based attention [21] as an alternative to the more complex self-attention [11], facilitating the establishment of a global receptive field with linear computational complexity. This approach notably enhances the speed of model inference on mobile devices compared to softmax attention, leading to a considerable reduction in model latency and consequently faster detection. Incorporating this module into the existing ViT model proves to be an effective strategy for accelerating model inference without sacrificing accuracy.

3. Materials and Methods

In this paper, we redesigned the feature extraction network based on the baseline network YOLOv5 and proposed a lightweight pine wilt detection method based on ViT-enhanced YOLO. Firstly, we constructed a lightweight EfficientViT feature extraction network with lightweight MSA as the core to replace YOLOv5's CSPDarkNet53(DarkNet53 with Cross-stage Partial Connections). Secondly, a cross-scale feature fusion neck network (CACSNET) was designed, which uses the CARAFE operator to replace the bilinear difference in the original model for upsampling, and then performs cross scale feature fusion. Finally, EIOU was introduced to optimize the loss function. The comprehensive architecture of the proposed Light-ViTeYOLO is shown in Figure 2.

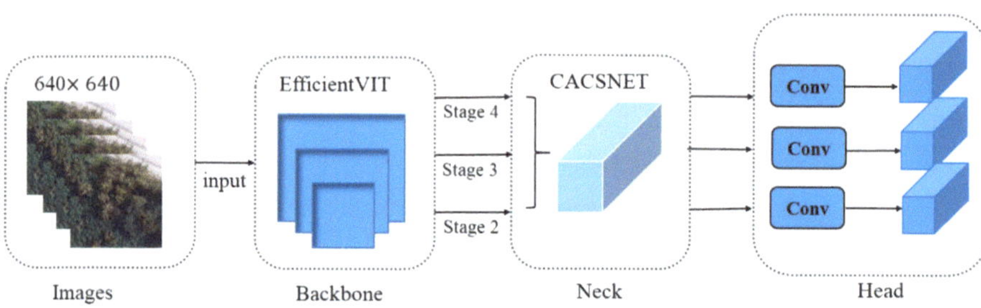

Figure 2. Overall architecture of the lightweight PWD model based on YOLOv5.

Below, we will analyze YOLOv5 and provide a detailed explanation of the proposed Light-ViTeYOLO.

3.1. Baseline Network YOLOv5

The network structure of YOLOv5 can be divided into the following three parts: the backbone, the neck segment, and the head output segment, with the specific structure depicted in Figure 3.

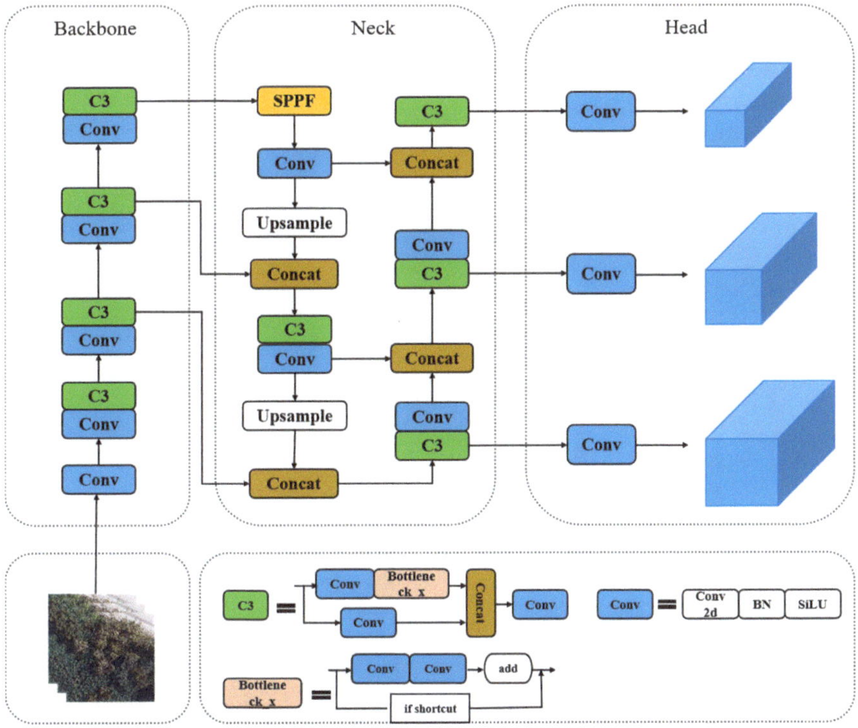

Figure 3. Network architecture of YOLOv5.

After the input image undergoes preprocessing, it is fed into the backbone feature extraction network, CSPDarkNet53, associated with multiple convolutional operations. This process transforms the image into a feature map and facilitates the extraction of semantic and structural information from the input image. Subsequently, at the neck layer, a feature pyramid PANet is established at varying scales, with each feature map comprising different resolutions corresponding to receptive fields of different scales. Finally, YOLOv5 utilizes the output of detection frame processed by the NMS (Non-maximum Suppression)- as the ultimate target detection result.

However, YOLOv5's use of a series of convolutional modules for feature extraction results in a complex network and the inability to effectively capture global information. Consequently, the bilinear interpolation used in the neck network cannot utilize the semantic information of the feature maps, and the perception domain is limited to the sub-pixel domain. The above design cannot be suitable for the real-time detection requirements of multi-scale and small targets in PWD detection tasks. In light of this, the following improvements will be carried out.

3.2. Redesign of Backbone Feature Extraction Network

In this paper, an examination of the structure of the visual transformer (ViT) has revealed that the main computational bottleneck is the softmax attention module, which exhibits quadratic computational complexity with respect to the input resolution. To address this issue, the lightweight Multi-Scale Attention (MSA) module introduced in Section 2.2 is specifically designed to enhance execution speed, delivering a substantial inference speedup while maintaining accuracy.

Based on this, we construct the EfficientViT module with lightweight MSA as the core, which is used for the design of the feature extraction network in this paper. The redesigned

EfficientViT feature extraction model is shown in Figure 4 (left), with the EfficientViT module shown in Figure 4 (right).

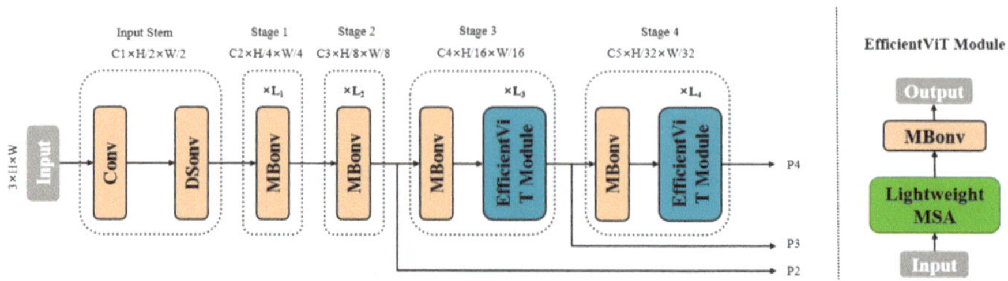

Figure 4. Macro architecture of EfficientViT (**left**) and illustration of EfficientViT's building blocks (**right**).

The EfficientViT module comprises the lightweight MSA module and the MBConv module [22]. The lightweight MSA module is employed for contextual information extraction, while the MBConv module facilitates local information extraction. Notably, the linear attention utilized by the lightweight MSA module has limitations in capturing localized details, potentially leading to a notable loss in accuracy. To mitigate this shortcoming, a deep convolutional MBConv module is integrated behind the MSA to enhance linear attention. This strategy incurs low computational overhead while significantly augmenting the capability of linear attention in local feature extraction.

The EfficientViT model adheres to the standard backbone head/decoder architecture, reflecting the following design features:

(1) The backbone network incorporates an input backbone and four stages, characterized by diminishing feature map size and escalating channel numbers;
(2) Lightweight MSAs are integrated into stages 3 and 4;
(3) For downsampling, the model employs MBConv with a step size of 2.

The outputs of Stage 2, Stage 3, and Stage 4 collectively generate a feature map pyramid, serving as the input for feature fusion in the neck network. The detailed architecture configurations of EfficientViT variants is shown in the following Table 1.

Table 1. Detailed architecture configurations of different EfficientViT variants.

Variants	Input Stem	Stage 1	Stage 2	Stage 3	Stage 4
C	C0 = 8	C1 = 16	C2 = 32	C3 = 64	C4 = 128
L	L0 = 1	L1 = 2	L2 = 2	L3 = 2	L4 = 2
H	640	640	640	640	640
W	640	640	640	640	640

Here, C denotes the number of channels, L denotes the number of blocks, H denotes the feature map height, and W denotes the feature map width.

In this paper, the above-designed EfficientViT model replaces the feature extraction network CSPDarkNet53 of YOLOv5, aiming to achieve efficient hardware operation through a lightweight MSA design to improve the accelerated inference performance of the model, to achieve global awareness and multi-scale learning to ensure that performance is not sacrificed, and ultimately, to enable the proposed model to realize the real-time PWD detecting task.

3.3. Design of CACSNet Neck Networks

YOLOv5 uses PANET as a neck network for feature extraction and fusion, and as a key operation of the feature pyramid, the feature upsampling method uses bilinear interpolation. This method is unable to utilize the semantic information of the feature map and the perceptual domain is limited to the sub-pixel domain. In order to further optimize the performance, in this paper, PANET is improved and a content-aware cross-scale bi-directional fusion network (CACSNet) is designed as a new neck network. The specific improvements are described as follows.

Firstly, we use the CARAFE [23] operator as the new up-sampling kernel to complete the up-sampling operation of the neck network (P7_u,P6_u,P5_u,P4_u in Figure 5b) to realize the up-sampling based on the input content. The implementation specifically consists of two steps: the first step is to predict the reorganization kernel of each target location based on its content, and the second step is to restructure the features with the predicted kernel.

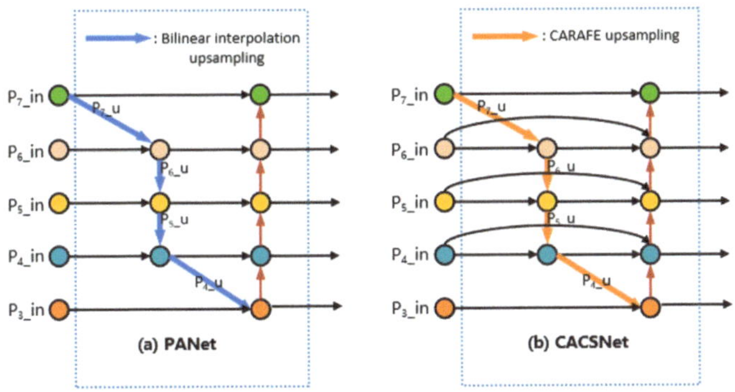

Figure 5. Illustrating the design of the feature network. (**a**) PANet adds an extra bottom-up path atop FPN. (**b**) CACSNet utilizes feature graph semantic information to implement a top-down process based on input content.

Given a feature map X of size $C \times H \times W$ and an up-sampling rate α (α is an integer), CARAFE will generate a new feature map X' of size $C \times \alpha H \times \alpha W$ and, for any target location $l' = (i', j')$ of X', its corresponding original location is $l = (i, j)$, where $i = \left\lfloor \frac{i'}{\sigma} \right\rfloor$ and $j = \left\lfloor \frac{j'}{\sigma} \right\rfloor$. Here, we denote $N(X_l, k)$ as the $k \times k$ subregion of X centered at location l, i.e., the neighbors of X_l.

In the first step, the kernel prediction module ψ predicts the spatially variant kernels $W_{l'}$ for each position l' based on the neighborhoods of X_l, as shown in Equation (3). The second step is the restructuring step shown in Equation (4), where ϕ is the content-aware reassembly module, which reassembles the neighborhoods of X_l with the kernel $W_{l'}$.

$$W_{l'} = \psi(N(X_l, k_{encoder})) \tag{3}$$

$$X'_{l'} = \phi(N(X_l, k_{up}), W_{l'}) \tag{4}$$

where weights are generated in a content-aware manner. In addition, for each location there exists multiple sets of such up-sampling weights, and then feature up-sampling is accomplished by generating features rearranged into spatial blocks. CARAFE up-sampling can aggregate and reorganize the contextual information around the target within a large perceptual field, which improves the ability to express feature details and introduces little computational overhead.

Furthermore, to prevent the loss of feature information related to small targets during the feature extraction process, the paper incorporates cross-scale weighting for feature fusion in the neck layer (see Figure 5). This is achieved by introducing additional connections (depicted as curved edges in Figure 5) between the feature input nodes from the backbone network and the output nodes of the neck network at the same level. This approach facilitates the fusion of more original image features to maximize the retention of features related to individual diseased trees.

3.4. Optimization of Loss Function

In target detection, the loss function is critical in quantifying the disparity between the model's predicted output and the actual target, driving continual learning during training to enhance the performance of the detection task. Typically, loss functions in object detection encompass bounding box regression loss, classification loss, and object presence loss. While YOLOv5 employs the CIOU (complete concatenated intersection) loss function for bounding box regression, this approach has limitations in handling variations in object location and size. As the CIOU loss function does not directly consider target location information, the model may prioritize the wrong bounding box location during optimization, leading to mismatches between the detected and actual disease areas and affecting detection accuracy. Moreover, the CIOU loss function exhibits reduced sensitivity to the degree of deformation in small targets, resulting in suboptimal model performance for detecting small targets.

To address these limitations, this paper adopts the EIOU loss function as an alternative. EIOU loss better balances detection accuracy by integrating position and size information of the target frame. By combining width and height information of the target frame and considering the intersection region-to-minimum closure region ratio, the EIOU loss function effectively addresses target size changes and deformation issues, enhancing detection accuracy and robustness. The EIOU loss function is calculated as follows:

$$L_{EIOU} = L_{IOU} + L_{dis} + L_{asp} = 1 - IOU + \frac{\rho^2(b, b^{gt})}{(w^c)^2 + (h^c)^2} + \frac{\rho^2(w, w^{gt})}{(w^c)^2} + \frac{\rho^2(h, h^{gt})}{(h^c)^2} \quad (5)$$

The loss function comprises three components: the overlap loss (L_{IOU}), the center distance loss (L_{dis}), and the width-height loss (L_{asp}). The first two components follow the approach used in CIOU. However, the width-height loss directly minimizes the disparity between the widths and heights of the target box and the predicted box, thereby accelerating convergence. Here, w^c and h^c are the width and height of the minimum enclosing box covering both boxes. $\rho^2(b, b^{gt})$ represents the Euclidean distance between the center points of the anchor box and the ground truth box, $\rho^2(w, w^{gt})$ represents the Euclidean distance between the width of the anchor box and the ground truth box, and $\rho^2(h, h^{gt})$ represents the Euclidean distance between the height of the anchor box and the ground truth box.

4. Experiment and Performance Analysis

4.1. Research Area and Data Acquisition

The image data for this study were obtained from the forest field of Zhuanshanzi, Tai'an City, Shandong Province, China (latitude 31°14′ N, longitude 117°01′ E, altitude 40 m). Figure 6 illustrates a schematic of the data acquisition site. In order to mitigate the effects of wind, shadows, strong light, weak light, and reflections on image quality, we selected the time frame from 2 PM to 5 PM on 3 May and 4 May 2022 for image data collection. This specific time period was chosen due to its favorable meteorological conditions.

We utilized a DJI Mavic Air 2 drone, which is outfitted with a 48-megapixel visible light camera, to capture the image data. The camera of this drone boasts a maximum flight time of 34 min, a maximum flight range of 18.5 km, and a maximum flight speed of 19 m per second. The resulting images were stored in JPEG format with a resolution of 6000 × 4000 pixels. Throughout the flights, the drone's speed and direction were manually controlled, while the camera remained fixed perpendicular to the ground at a 90-degree angle. The drone was equipped with precise GPS and GLONASS positioning capabilities, enabling accurate recording of the location and altitude of each image. The flight altitude of the drone was maintained at approximately 300 m. Figure 7 shows examples of forestry images captured by the drone, showcasing a resolution of 6000 × 4000 pixels.

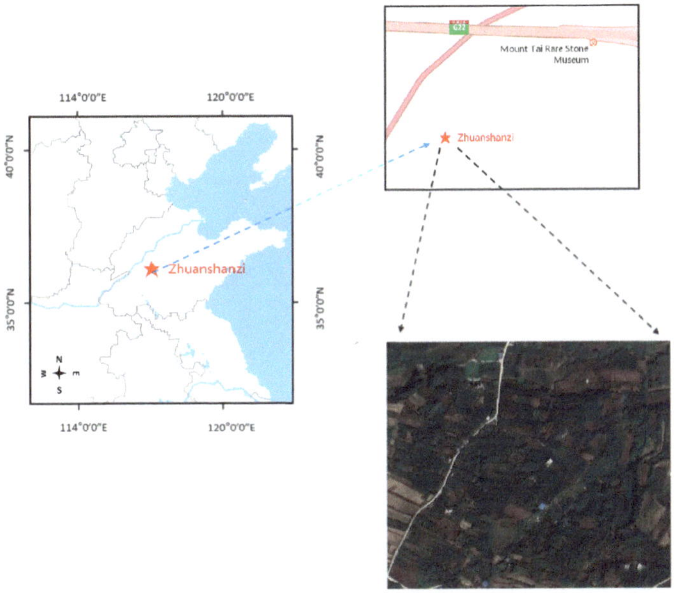

Figure 6. Data acquisition area.

Figure 7. Examples of UAV tree images.

To enhance the usability of the collected drone images, we adhered to the steps outlined below. Firstly, due to the images' high resolution and extensive spatial coverage, it would necessitate significant computational resources to directly use all images for training a network model, given the limited number of sample images. Consequently, we opted to extract image patches of 640 × 640 pixels from 300 drone forestry images gathered in the study area. Subsequently, 10,000 image patches were randomly selected for the training set, while 1200 image patches were utilized for the validation set. In order to perform image analysis and applications, we use the image annotation tool LabelImg and annotate the diseased trees in the images under the guidance of forestry experts. A segmented tree image is shown in Figure 8.

Figure 8. Illustrations of cropping tree images.

4.2. Experimental Configuration

We implemented and trained the neural network model using the PyTorch deep learning framework on the Linux operating system. Table 2 presents the exhaustive hardware and software environment arrangement for the experiments.

Table 2. Hardware and software environment arrangement.

Platform	Configuration
Operating system	Linux 3.10.0
CPU	Intel(R) Xeon(R) Gold 6138 CPU @ 2.00 GHz
GPU	Tesla V100-PCIE-32GB
GPU accelerator	CUDA 10.2
Deep learning frame	PyTorch 1.10.1
Compilers	PyCharm and Anaconda
Scripting language	Python 3.7

4.3. Experimental Indicators

To ensure a precise assessment of the new model's performance, we utilized several performance evaluation metrics: Average Precision (AP), recall, model parameters, Giga Floating-point Operations Per second (GFLOPs), and Frames Per Second (FPS).

AP represents the average precision of a single target class, providing an overall measure of the model's detection performance. The AP is calculated using the following formula:

$$AP = \int_0^1 Precision(Recall)d(Recall) \tag{6}$$

Here, Precision denotes the proportion of correctly predicted boxes to the total predicted boxes, while Recall represents the proportion of predicted boxes to all actual boxes. To further assess the accuracy of the detector, we employed two metrics: Average Precision at an IOU threshold of 0.5 (AP@0.5) and Average Precision with IOU thresholds ranging from 0.5 to 0.95 (AP@0.5:0.95).

We use the model parameters and GFLOPs to measure the model complexity and size, while FPS is used to measure the running speed of algorithms, representing the number of images that can be processed per second. The smaller the number of model parameters and GFLOPs, the lower the model complexity and size. The larger the FPS, the faster the algorithm processing speed, which is more conducive to the deployment of the model on edge devices.

Assuming a convolutional layer with a size of $h \times w \times c_i \times c_o$ (c_i is the number of input channels, c_o is the number of output channels) and an output feature map size of $H' \times W'$, the formula for calculating the parameters of the convolutional layer is as follows:

$$Params = c_o \times (h \times w \times c_i + 1) \quad (7)$$

The formula for calculating the FLOPs of the convolutional layer is as follows:

$$FLOPs = H' \times W' \times c_o \times (h \times w \times c_i + 1) \quad (8)$$

While GFLOPs = FLOPs $\times$ 10^9.

4.4. *Performance Comparison of Different Methods*

4.4.1. Performance Comparison of Different Methods

In order to evaluate the effectiveness of our proposed model, this paper lists several representative methods and compares them with the algorithm proposed in this paper in terms of model detection performance and model complexity. The specific results can be seen in Tables 3 and 4. Tables 3 and 4 show the comparison results with YOLOv5 [9], Faster R-CNN [24], RetinaNet [25], YOLOv5, YOLOv6 [26], YOLOv7 [27], and YOLOx [28] on the test set in terms of recall, mAP, Parameter, GFLOPs, and FPS.

Table 3. The comparison of the detection accuracy between Light-ViTeYOLO and existing methods.

Method	Recall (%)	mAP@0.5(%)	mAP@0.5:0.95(%)
Faster-RCNN	75.4	75.2	66.2
RetinaNET	96.6	95.9	92.5
YOLOV6	96.7	95.9	80.8
YOLOv7	93.9	82.5	55.9
YOLOX	96.9	96.0	84.3
YOLOv5	96.1	97.6	90.8
Light-ViTeYOLO	95.7	97.2	94.3

Table 4. Comparison of calculation quantity and parameter quantity between Light-ViTeYOLO and existing models.

Method	Parameters (M)	GFLOPs	FPS (Frames/s)
Faster-RCNN	41.1	78.1	15.5
RetinaNET	36.1	81.6	12.3
YOLOV6	17.1	21.8	26.3
YOLOv7	6.5	13.9	39.5
YOLOX	8.9	13.3	46.5
YOLOv5	7.1	15.8	67.0
Light-ViTeYOLO	3.89	7.4	57.9

From Table 3, it can be seen that the proposed algorithm exhibits a significant improvement in mAP@0.5:0.95 compared to other algorithms, with slightly lower recall rates than RetinaNET, YOLOv6, YOLOX, and only slightly lower mAP@0.5 compared to YOLOv5. In the task of pine wilt disease (PWD) detection, accurate detection of diseased areas is a prerequisite for subsequent disease control, and a higher mAP@0.5:0.95 demonstrates that the proposed algorithm achieves good detection results in the PWD task.

From Table 4, it is evident that the proposed algorithm significantly reduces parameter number and computational complexity compared to other algorithms. The YOLOv7 and YOLOX, which are the most lightweight, have reduced by more than 40%. At the same time, the algorithm's inference speed is superior to all other models except YOLOv5, meeting the real-time requirements of drone scenarios.

Based on Tables 3 and 4, compared with other algorithms, Light-ViTeYOLO proposed in this paper achieves the minimum number of parameters and computational complexity, with suboptimal inference speed. While achieving a lightweight model, mAP@0.5 0.95 (%) reaches its maximum, mAP@0.5 (%) reached the second highest. Although the recall and mAP@0.5 (%) of Light-ViTeYOLO did not reach the optimal level, it is only slightly lower than the optimal algorithm. Light-ViTeYOLO has obvious advantages in lightweight level. The FPS of Light-ViTeYOLO is lower than that of YOLOv5. However, the number of parameters and computation is nearly 50% less than that of YOLOv5, and mAP@0.5:0.95(%) is almost 4% higher than that of YOLOv5.

The impact of lightweight networks on the detection performance of powdery mildew was also compared. The experimental results are shown in Section 4.4.3.

Based on the above analysis, the proposed algorithm ensures detection accuracy, has high detection precision, and simultaneously significantly reduces model complexity and inference speed. Light-ViTeYOLO is more suitable for PWD detection tasks than other algorithms.

4.4.2. Ablation Experiment

Light-ViTeYOLO proposed in this paper redesigned the feature extraction network of YOLOv5, proposed a neck network, and optimized the loss function. In order to evaluate the effectiveness of each module of our method, comparative experiments were conducted on the PWD dataset, and the improvement scheme was incrementally added. The specific experimental results are shown in Table 5.

Table 5. Detection effects of different modules on the model.

Model	Parameters (M)	GFLOPs	mAP@0.5(%)	mAP@0.5:0.95(%)
baseline	7.02	15.8	97.6	90.8
+EfficientViT	3.74	6.8	97.22	93.6
+CACSNet	3.89	7.4	97.20	94.0
+EIOU	3.89	7.4	97.27	94.3

From Table 5, it can be observed that after using EfficientViT for global feature extraction, the parameter count and GFLOPs were effectively reduced. While mAP@0.5 showed a slight decrease, there was a significant improvement in mAP@0.5:0.95, indicating that EfficientViT significantly improved efficiency in feature extraction without sacrificing performance. After optimizing the neck network and loss function, there was no change in the parameter count and GFLOPs, while mAP@0.5 and mAP@0.5:0.95 were further improved. This indicates that the new neck network and optimization of the loss function have improved the performance of object detection without increasing the number of parameters and calculations, and does not affect the operational efficiency of the model. Compared to the original YOLOv5 model, mAP@0.5:0.95 of proposed Light-ViTeYOLO increased by more than 3 percentage points, while the model's parameter count reduced by 44.6%, and computational complexity reduced by 53.2%.

4.4.3. Feature Extraction Performance Analysis of EfficientViT

We integrate multiple typical visual Transformer models and lightweight networks on the core architecture of the YOLOv5 network for experimental comparison in order to analyze object detection performance and model complexity of the EfficientViT in pine wilt disease detection. As shown in Table 6, ViT [12], BoTNet [29], and CoNet [30] are typical ViT models, while Shufflenetv2 [31], Mobilenetv3 [32], RepVGG [33], and GhostNet [34] are classic lightweight feature extraction networks. It can be seen from the experimental results that the AP@0.5 0.95(%) of three ViT models is about 7.5 percent higher than the four classic lightweight networks, but the ViT parameter count and computational complexity are significantly higher than those of lightweight networks. After integrating EfficientViT, the model has almost the same number of parameters and computational complexity as the lightweight network, which is about 50% lower than the ViT model. However, its detection performance is improved by nearly 6 percentage points compared to the lightweight network, slightly lower than ViT. Overall, the improved EfficientViT method, combined with the optimization of the neck network and loss function, achieves the best performance in terms of model accuracy and complexity.

Table 6. Comparison of pine wilt detection performance between EfficientViT and other feature extraction networks.

Model	Parameters (M)	GFLOPs	AP@0.5(%)	AP@0.5:0.95(%)
YOLOv5 +ViT	7.02	15.6	97.2	95.8
YOLOv5 +BoTNet	6.69	15.5	97.2	95.9
YOLOv5 +CoNet	8.19	16.8	97.2	95.4
YOLOv5 +Shufflenetv2	3.79	7.9	96.07	87.87
YOLOv5 +Mobilenetv3	3.19	5.9	94.19	86.83
YOLOv5 +RepVGG	7.19	16.3	97.13	85.93
YOLOv5 +GhostNet	3.68	8.1	96.49	86.06
YOLOv5 +EfficientViT	3.74	6.8	97.22	93.6

4.4.4. Performance Analysis of the Training Process

We compared the changes in mAP@0.5 and mAP@0.5:0.95 during the training process of the original YOLOv5 model and its iterations with the inclusion of EfficientViT, optimization of the CACSNET neck network, and EIOU loss function, as shown in Figure 9; the left graph reflects that the mAP@0.5 of the four models sharply increases at the beginning of training, then tends to plateau around 10 iterations. The mAP@0.5 of the improved models remains similar to that of the original model in the final iterations, indicating that the improved models start to change faster than the original model.

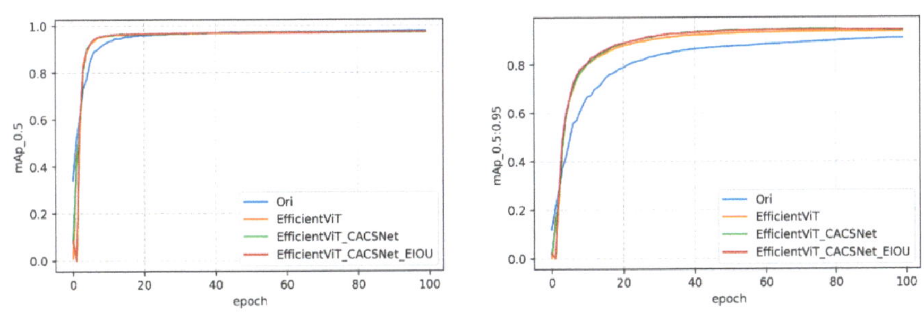

Figure 9. The detection performance change of the algorithm during training after the improved scheme superposition.

In the right graph, the mAP@0.5:0.95 of the improved models quickly surpasses the original model, after which the values of all four models gradually increase and plateau around 60 iterations. Ultimately, the mAP@0.5:0.95 of the improved models is significantly better than that of the original model, with the proposed model achieving the highest mAP@0.5:0.95. From the above analysis, it is evident that the models exhibit faster training accuracy after the improvements compared to the original model. This demonstrates that the various optimizations proposed in this article have a promoting effect on the model's performance, and the reduction in model parameters and computational complexity has not had a major impact on the model's performance. This highlights the strong generalization ability of the lightweight model.

We compared the loss values, precision, recall, and AP@0.5:0.95 of the proposed model with YOLOv5 by plotting curves during the training process. As depicted in Figure 10, it is observed that during the initial training phase, the values of each model metric undergo rapid changes, and after approximately 100 iterations, the loss function values exhibit less fluctuation, signifying a relatively stable state. Concurrently, the model's precision and recall also reach a relatively balanced state. Although the mAP continues to increase gradually, the rate of change is minimal, suggesting that the model has essentially converged at this juncture. The trends of the various metrics in the graph reveal that the proposed model matches or surpasses YOLOv5 in all metrics, underscoring the robustness of the proposed model's object representation.

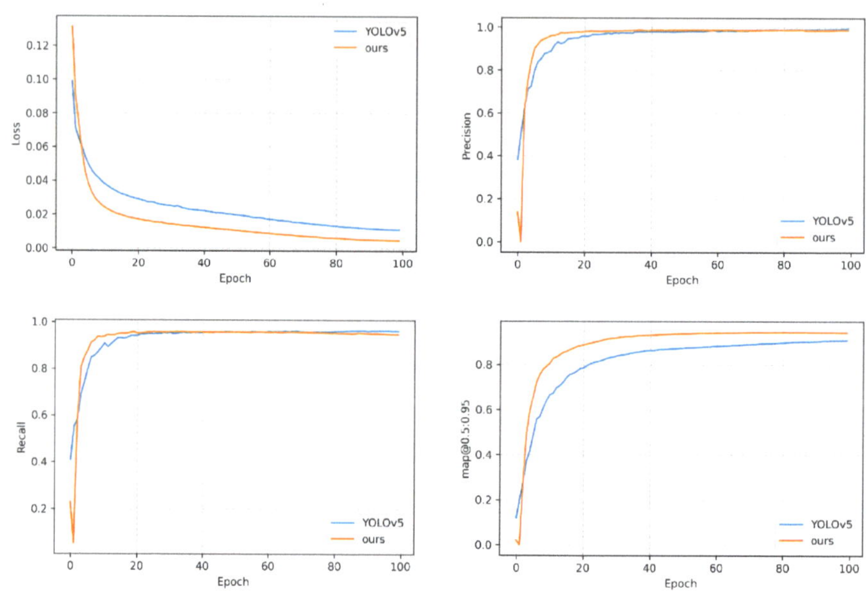

Figure 10. Curves of the loss values, precision, recall, and AP@0.5:0.95 for Light-ViTeYOLO during the training process.

5. Conclusions

In current methods for pine wilt disease detection, convolutional neural networks (CNNs) are commonly utilized for network architecture, leveraging their strong performance in feature extraction. However, the sensory field of CNNs is constrained by kernel size and network depth, limiting their capacity to effectively model long-term dependencies. On the other hand, Transformers are adept at capturing global and rich contextual information, but their high computational demands hinder their practicality for real-time monitoring scenarios, such as UAV-based applications. To address these challenges, this paper introduces Light-ViTeYOLO, a lightweight PWD detection method based on Vision Transformer-enhanced YOLOv5. By incorporating a lightweight Multi-Scale Attention (MSA) to redesign the backbone network process and enhancing the neck and head, the proposed method achieves impressive performance in terms of detection accuracy, model complexity, and inference speed. Notably, this approach manages to exceed the detection accuracy of many target detectors even with significantly reduced parameters. This achievement marks a successful balance between model accuracy and efficiency, underscoring its strong robustness. The use of drones carrying our detection method for real-time detection of pine wilt disease-discolored wood may lead to higher economic results, including benefits in terms of improved detection efficiency, reduced costs, reduced risk of disease transmission, and optimized decision support. However, the specific economic effects still need to be professionally assessed based on actual applications and relevant cost data. Therefore, we have the following outlook for future work:

1. The method proposed in this paper has been experimentally verified on a standard platform. The next step is to deploy the application on a drone hardware platform through algorithms to further verify its feasibility and potential economic benefits;
2. Combining the method proposed in this paper with satellite-based forest monitoring to further strengthen the monitoring of pine tree discoloration caused by pine wilt disease. Integrating drone images with satellite images for multi-scale analysis from both macroscopic and local perspectives, comprehensively monitoring diseases through data fusion and analysis;
3. Applying the method proposed in this paper to the detection of other forest diseases, such as bark beetle damage.

Author Contributions: Conceptualization, Q.Y. and S.Z.; methodology, Q.Y.; software, S.Z.; validation, Q.Y., Z.M. and H.W.; formal analysis, X.Z. and Q.Y.; investigation, L.L.; resources, H.W. and W.L.; data curation, S.Z. and X.Z; writing—original draft preparation, Q.Y. and S.Z.; writing—review and editing, H.W. and W.L.; visualization, S.Z.; supervision, Q.Y. and H.W.; project administration, Q.Y. and H.W.; funding acquisition, Q.Y. and H.W. All authors have read and agreed to the published version of the manuscript.

Funding: This work was supported by the Fund Project of Central Government Guided Local Science and Technology Development under Grant No. 226Z0302G and the Special Project of Langfang Key Research and Development under Grant No. 2023011005B.

Data Availability Statement: The data presented in this study are available on request from the corresponding author due to project privacy.

Conflicts of Interest: The authors declare no conflicts of interest.

References

1. Pan, C. Development of studies on pinewood nematodes diseases. *J. Xiamen Univ.* **2011**, *50*, 476–483.
2. Liu, F.; Su, H.; Ding, T.; Huang, J.; Liu, T.; Ding, N.; Fang, G. Refined Assessment of Economic Loss from Pine Wilt Disease at the Subcompartment Scale. *Forests* **2023**, *14*, 139. [CrossRef]
3. Duarte, A.; Borralho, N.; Cabral, P.; Caetano, M. Recent advances in forest insect pests and diseases monitoring using UAV-based data: A systematic review. *Forests* **2022**, *13*, 911. [CrossRef]
4. Zhang, X.; Yang, H.; Cai, P.; Chen, G.; Li, X.; Zhu, K. Research progress on remote sensing monitoring of pine wilt disease. *Trans. Chin. Soc. Agric. Eng* **2022**, *38*, 184–194.
5. Cai, P.; Chen, G.; Yang, H.; Li, X.; Zhu, K.; Wang, T.; Liao, P.; Han, M.; Gong, Y.; Wang, Q.; et al. Detecting Individual Plants Infected with Pine Wilt Disease Using Drones and Satellite Imagery: A Case Study in Xianning, China. *Remote Sens.* **2023**, *15*, 2671. [CrossRef]
6. You, J.; Zhang, R.; Lee, J. A deep learning-based generalized system for detecting pine wilt disease using RGB-based UAV images. *Remote Sens.* **2021**, *14*, 150. [CrossRef]
7. Qin, J.; Wang, B.; Wu, Y.; Lu, Q.; Zhu, H. Identifying pine wood nematode disease using UAV images and deep learning algorithms. *Remote Sens.* **2021**, *13*, 162. [CrossRef]
8. Wu, B.; Liang, A.; Zhang, H.; Zhu, T.; Zou, Z.; Yang, D.; Tang, W.; Li, J.; Su, J. Application of conventional UAV-based high-throughput object detection to the early diagnosis of pine wilt disease by deep learning. *For. Ecol. Manag.* **2021**, *486*, 118986. [CrossRef]
9. Gong, H.; Ding, Y.; Li, D.; Wang, W.; Li, Z. Recognition of Pine Wood Affected by Pine Wilt Disease Based on YOLOv5. In Proceedings of the 2022 China Automation Congress (CAC), Xiamen, China, 25–27 November 2022; IEEE: Piscataway, NJ, USA, 2022; pp. 4753–4757.
10. Sun, Z.; Ibrayim, M.; Hamdulla, A. Detection of pine wilt nematode from drone images using UAV. *Sensors* **2022**, *22*, 4704. [CrossRef] [PubMed]
11. Vaswani, A.; Shazeer, N.; Parmar, N.; Uszkoreit, J.; Jones, L.; Gomez, A.N.; Kaiser, Ł.; Polosukhin, I. Attention is all you need. In *Advances in Neural Information Processing Systems*; MIT Press: Cambridge, MA, USA, 2017; Volume 30.
12. Dosovitskiy, A.; Beyer, L.; Kolesnikov, A.; Weissenborn, D.; Zhai, X.; Unterthiner, T.; Dehghani, M.; Minderer, M.; Heigold, G.; Gelly, S.; et al. An image is worth 16x16 words: Transformers for image recognition at scale. *arXiv* **2020**, arXiv:2010.11929.
13. Naseer, M.M.; Ranasinghe, K.; Khan, S.H.; Hayat, M.; Shahbaz Khan, F.; Yang, M.H. Intriguing properties of vision transformers. In *Advances in Neural Information Processing Systems*; MIT Press: Cambridge, MA, USA, 2021; Volume 34, pp. 23296–23308.
14. Park, N.; Kim, S. How do vision transformers work? *arXiv* **2022**, arXiv:2202.06709.
15. Hao, S.; Wu, B.; Zhao, K.; Ye, Y.; Wang, W. Two-stream swin transformer with differentiable sobel operator for remote sensing image classification. *Remote Sens.* **2022**, *14*, 1507. [CrossRef]

16. Ma, J.; Li, M.; Tang, X.; Zhang, X.; Liu, F.; Jiao, L. Homo–heterogenous transformer learning framework for RS scene classification. *IEEE J. Sel. Top. Appl. Earth Obs. Remote Sens.* **2022**, *15*, 2223–2239. [CrossRef]
17. Xie, E.; Wang, W.; Yu, Z.; Anandkumar, A.; Alvarez, J.M.; Luo, P. SegFormer: Simple and efficient design for semantic segmentation with transformers. In *Advances in Neural Information Processing Systems;* MIT Press: Cambridge, MA, USA, 2021; Volume 34, pp. 12077–12090.
18. Guo, M.H.; Lu, C.Z.; Hou, Q.; Liu, Z.; Cheng, M.M.; Hu, S.M. Segnext: Rethinking convolutional attention design for semantic segmentation. In *Advances in Neural Information Processing Systems;* MIT Press: Cambridge, MA, USA, 2022; Volume 35, pp. 1140–1156.
19. Wang, G.; Li, B.; Zhang, T.; Zhang, S. A network combining a transformer and a convolutional neural network for remote sensing image change detection. *Remote Sens.* **2022**, *14*, 2228. [CrossRef]
20. Cai, H.; Li, J.; Hu, M.; Gan, C.; Han, S. EfficientViT: Lightweight Multi-Scale Attention for On-Device Semantic Segmentation. *arXiv* **2023**, arXiv:2205.14756.
21. Katharopoulos, A.; Vyas, A.; Pappas, N.; Fleuret, F. Transformers are rnns: Fast autoregressive transformers with linear attention. In Proceedings of the International Conference on Machine Learning, Virtual Event, 13–18 July 2020; PMLR; pp. 5156–5165.
22. Sandler, M.; Howard, A.; Zhu, M.; Zhmoginov, A.; Chen, L.C. Mobilenetv2: Inverted residuals and linear bottlenecks. In Proceedings of the IEEE Conference on Computer Vision and Pattern Recognition, Salt Lake City, UT, USA, 18–23 June 2018; pp. 4510–4520.
23. Wang, J.; Chen, K.; Xu, R.; Liu, Z.; Loy, C.C.; Lin, D. Carafe: Content-aware reassembly of features. In Proceedings of the IEEE/CVF International Conference on Computer Vision, Seoul, Republic of Korea, 27 October–2 November 2019; pp. 3007–3016.
24. Ren, S.; He, K.; Girshick, R.; Sun, J. Faster r-cnn: Towards real-time object detection with region proposal networks. In *Advances in Neural Information Processing Systems;* MIT Press: Cambridge, MA, USA, 2015; Volume 28.
25. Lin, T.Y.; Goyal, P.; Girshick, R.; He, K.; Dollár, P. Focal loss for dense object detection. In Proceedings of the IEEE International Conference on Computer Vision, Venice, Italy, 22–29 October 2017; pp. 2980–2988.
26. Li, C.; Li, L.; Jiang, H.; Weng, K.; Geng, Y.; Li, L.; Ke, Z.; Li, Q.; Cheng, M.; Nie, W.; et al. YOLOv6: A single-stage object detection framework for industrial applications. *arXiv* **2022**, arXiv:2209.02976.
27. Wang, C.Y.; Bochkovskiy, A.; Liao, H.Y.M. YOLOv7: Trainable bag-of-freebies sets new state-of-the-art for real-time object detectors. In Proceedings of the IEEE/CVF Conference on Computer Vision and Pattern Recognition, Vancouver, BC, Canada, 17–24 June 2023; pp. 7464–7475.
28. Ge, Z.; Liu, S.; Wang, F.; Li, Z.; Sun, J. Yolox: Exceeding yolo series in 2021. *arXiv* **2021**, arXiv:2107.08430.
29. Srinivas, A.; Lin, T.Y.; Parmar, N.; Shlens, J.; Abbeel, P.; Vaswani, A. Bottleneck transformers for visual recognition. In Proceedings of the IEEE/CVF Conference on Computer Vision and Pattern Recognition, Nashville, TN, USA, 20–25 June 2021; pp. 16519–16529.
30. Li, Y.; Yao, T.; Pan, Y.; Mei, T. Contextual transformer networks for visual recognition. *IEEE Trans. Pattern Anal. Mach. Intell.* **2022**, *45*, 1489–1500. [CrossRef] [PubMed]
31. Ma, N.; Zhang, X.; Zheng, H.T.; Sun, J. Shufflenet v2: Practical guidelines for efficient cnn architecture design. In Proceedings of the European Conference on Computer Vision (ECCV), Munich, Germany, 8–14 September 2018; pp. 116–131.
32. Howard, A.; Sandler, M.; Chu, G.; Chen, L.C.; Chen, B.; Tan, M.; Wang, W.; Zhu, Y.; Pang, R.; Vasudevan, V.; et al. Searching for mobilenetv3. In Proceedings of the IEEE/CVF International Conference on Computer Vision, Seoul, Republic of Korea, 27 October–2 November 2019; pp. 1314–1324.
33. Ding, X.; Zhang, X.; Ma, N.; Han, J.; Ding, G.; Sun, J. Repvgg: Making vgg-style convnets great again. In Proceedings of the IEEE/CVF Conference on Computer Vision and Pattern Recognition, Nashville, TN, USA, 20–25 June 2021; pp. 13733–13742.
34. Tang, Y.; Han, K.; Guo, J.; Xu, C.; Xu, C.; Wang, Y. GhostNetv2: Enhance cheap operation with long-range attention. In *Advances in Neural Information Processing Systems;* MIT Press: Cambridge, MA, USA, 2022; Volume 35, pp. 9969–9982.

forests

MDPI

Article

Multi-Agent Reinforcement Learning for Stand Structure Collaborative Optimization of *Pinus yunnanensis* Secondary Forests

Shuai Xuan [1], Jianming Wang [1,*], Jiting Yin [2], Yuling Chen [3] and Baoguo Wu [4]

1 School of Mathematics and Computer Science, Dali University, Dali 671003, China; lavroro@163.com
2 Dali Forestry and Grassland Science Research Institute, Dali 671000, China; yinjiting234@163.com
3 Institute of Remote Sensing and Geographic Information System, School of Earth and Space Sciences, Peking University, Beijing 100871, China; chenyuling92@pku.edu.cn
4 School of Information Science and Technology, Beijing Forestry University, Beijing 100083, China; wubg@bjfu.edu.cn
* Correspondence: wangjianming618@163.com

Abstract: This study aims to investigate the potential and advantages of multi-agent reinforcement learning (MARL) in forest management, offering innovative insights and methodologies for achieving sustainable management of forest ecosystems. Focusing on the *Pinus yunnanensis* secondary forests in Southwest China, we formulated the objective function and constraints based on both spatial and non-spatial structural indices of the forest stand structure (FSS). The value of the objective function (VOF) served as an indicator for assessing FSS. Leveraging the random selection method (RSM) to select harvested trees, we propose the replanting foreground index (RFI) to enhance replanting optimization. The decision-making processes involved in selection harvest optimization and replanting were modeled as actions within MARL. Through iterative trial-and-error and collaborative strategies, MARL optimized agent actions and collaboration to address the collaborative optimization problem of FSS. We conducted optimization experiments for selection felling and replanting across four circular sample plots, comparing MARL with traditional combinatorial optimization (TCO) and single-agent reinforcement learning (SARL). The findings illustrate the superior practical efficacy of MARL in collaborative optimization of FSS. Specifically, replanting optimization based on RFI outperformed the classical maximum Delaunay generator area method (MDGAM). Across different plots (P1, P2, P3, and P4), MARL consistently improved the maximum VOFs by 54.87%, 88.86%, 41.34%, and 22.55%, respectively, surpassing those of the TCO (38.81%, 70.04%, 41.23%, and 18.73%) and SARL (54.38%, 70.04%, 41.23%, and 18.73%) schemes. The RFI demonstrated superior performance in replanting optimization experiments, emphasizing the importance of considering neighboring trees' influence on growth space and replanting potential. Following selective logging and replanting adjustments, the FSS of each sample site exhibited varying degrees of improvement. MARL consistently achieved maximum VOFs across different sites, underscoring its superior performance in collaborative optimization of logging and replanting within FSS. This study presents a novel approach to optimizing FSS, contributing to the sustainable management of *Pinus yunnanensis* secondary forests in southwestern China.

Keywords: stand structure optimization; selective cutting; replanting; multi-agent reinforcement learning; co-optimization

Citation: Xuan, S.; Wang, J.; Yin, J.; Chen, Y.; Wu, B. Multi-Agent Reinforcement Learning for Stand Structure Collaborative Optimization of *Pinus yunnanensis* Secondary Forests. *Forests* **2024**, *15*, 1143. https://doi.org/10.3390/f15071143

Academic Editors: Zengxin Zhang and Guojie Wang

Received: 20 May 2024
Revised: 23 June 2024
Accepted: 27 June 2024
Published: 30 June 2024

1. Introduction

Secondary forests generally exhibit issues such as unsustainable forest stand structures (FSSs), diminished biodiversity, heightened vulnerability to forest fires, and susceptibility to natural calamities such as pest infestations, diseases, and wildfires [1–3]. The optimization and adjustment of FSS represent pivotal technical interventions in forest management and planning, offering indispensable strategies for managing secondary forests effectively [4,5]. Central to this endeavor is the meticulous determination of FSS indices, the formulation of optimization models, and the design of solution algorithms. Initially, the optimization model is crafted through the judicious selection of FSS indexes, informed by the unique characteristics of FSS. Subsequently, relevant algorithms are deployed to resolve the optimization model, thereby facilitating management optimization through a spectrum of adjustment measures including selective cutting, replanting, tending, and pruning. FSS indices encompass both spatial and non-spatial dimensions, encompassing non-spatial metrics such as diameter scales counts [6], species counts, and plant density [7], alongside spatial indices including the mingling index (M) [8,9], canopy competition index (CI) [10,11], angle index (W) [12,13], story index (S) [14], open comparison (OP) [15], and neighborhood comparison (U) [16].

Selective cutting emerges as a pivotal strategy in the optimization of FSS. This approach entails the targeted removal of trees with limited growth potential, thereby orchestrating a more conducive distribution pattern of trees, enhancing understory light conditions, and alleviating competitive pressures within the forest milieu, ultimately culminating in FSS optimization. However, the efficacy of singular selective cutting endeavors often falls short in achieving the desired optimization outcomes. Complementarily, replanting constitutes another indispensable mechanism for optimization and regulation, endeavoring to bolster the stability and biodiversity of forest ecosystems through the strategic introduction of young trees of indigenous species in judiciously chosen spatial domains. The crux of replanting optimization resides in the discernment of pivotal indicators such as location and species.

Conventional replanting methodologies typically rely on techniques like the Voronoi diagram [17] or Delaunay triangulation [18], Kriging interpolation [19], among others, for determining replanting locations. However, these methodologies often overlook the intricate interplay between replanted trees and their neighboring counterparts, resulting in rigid replanting locations and potentially exacerbating inter-tree competitive pressures. Moreover, the scholarly discourse surrounding the optimization of stand structure based on replanting strategies remains relatively sparse. Existing studies predominantly rely on statistical analyses to ascertain replanting positions and pertinent information pertaining to the replanted flora [17–19], with scant attention devoted to replanting research underpinned by intelligent algorithms.

Numerous scholars have devised optimization and adjustment models, such as multi-objective operation and spatial structure, building upon the aforementioned optimization and adjustment strategies [17,20–22]. These models typically serve to simulate and optimize individual interventions, such as selective cutting or replanting, or iteratively refine stand structure through a sequence of actions, notably selective cutting followed by replanting (note: specific references are provided for sequential adjustments). However, none of the aforementioned optimization frameworks have comprehensively considered the collaborative synergies among multiple adjustment measures during model formulation and solution. Particularly within the realm of replanting adjustments, addressing complex multidimensional information encompassing spatial coordinates of replanted trees, tree species, age, height, and related parameters poses a significant challenge. Integrating these multidimensional attributes within the model and devising solution algorithms to capture their nuances are essential for leveraging the synergistic effects of multiple measures in optimizing and controlling forest stand structure. Notably, in scenarios typified by secondary forests dominated by a single species and characterized by extensive forest cover, a holistic approach encompassing replanting, tending, and additional measures becomes imperative to enhance spatial segregation within the forest stand and optimize pertinent indices.

The optimization of FSS is a nonlinear multi-objective optimization problem [23–25]. Existing literature predominantly explores algorithms for solving FSS optimization models grounded in single measures. These encompass heuristic methodologies like Monte Carlo (MC) [26–29] and bionic algorithms such as genetic algorithms (GAs) [30–32], simulated annealing (SA) [33–35], and particle swarm optimization (PSO) [17,36]. While MC offers simplicity and strong programmability, its outcomes often lack precision due to inherent algorithmic limitations. Conversely, bionic algorithms like GAs and PSO frequently encounter challenges such as local optimization pitfalls and volatility in the values of the objective function (VOFs) during solution processes.

Reinforcement learning (RL), characterized by trial-and-error strategies, emerges as a promising intelligent algorithm owing to its conceptual simplicity, absence of a model, dynamic decision-making, and robust adaptability [37]. In a previous study, we addressed the multi-objective logging optimization problem for FSS using single-agent reinforcement learning (SARL), which excelled in single-measure multi-objective optimization. However, FSS optimization necessitates the integration of multiple measures beyond selective logging alone. Moreover, SARL is constrained by individual knowledge and experience, often requiring prolonged training periods to adapt to environments and learned strategies, thus posing challenges in balancing exploration and exploitation.

Multi-agent reinforcement learning (MARL) presents a novel approach by leveraging multiple agents to collaboratively address complex optimization problems. Each agent interacts with the environment, receiving rewards and adapting in response to the behaviors of other agents. However, practical applications of MARL encounter challenges concerning the balance between collaboration and competition among agents, as well as the intricate design of reward functions. Despite its prominence in fields like autonomous driving and smart grids, MARL remains largely unexplored in collaborative optimization of FSS.

Overall, the construction and design of FSS optimization models and algorithms necessitate a comprehensive consideration of measures such as selective cutting and replanting, effective representation of multidimensional tree characteristics during replanting regulation, and a thorough integration of neighboring tree influences on replanted tree growth. Synchronized optimization and adjustment simulations are imperative for solving the optimal configuration of FSS. While SARL offers advantages over traditional heuristic and bionic algorithms in addressing single-measure optimization models, it proves inadequate for addressing synergistic optimization models involving multiple measures. Hence, this study introduces the replanting foreground index (RFI), a logging and replanting collaborative optimization model, and a MARL to address the collaborative optimization problem of FSS based on *Pinus yunnanensis* secondary forest data in Yunnan Province, China. This elucidates the potential and advantages of the multi-agent approach in forest management, providing novel insights and methodologies for sustainable forest ecosystem management.

2. Materials and Methods

2.1. Study Areas

The forest inventories were undertaken within the geographical vicinity of Cangshan Mountain, located between longitude 99°55′–100°12′ E and latitude 25°34′–26°00′ N, situated in Yunnan Province, China (Figure 1). The surveyed regions predominantly situated on the eastern slope of Cangshan Mountain, include prominent features such as Lan Peak, Malong Peak, Foding Peak, and Zhonghe Peak. This geographical area lies within the subtropical climate belt, with an average annual temperature of around 15 °C, influenced by prevailing southwest monsoon winds [38,39]. Notably, the region experiences abundant annual precipitation exceeding 1000 mm, with distinct seasonal variations marked by pronounced dry spells interspersed with heavy rainfall. The wet season predominantly spans from May to October, accounting for approximately 84% of the total annual precipitation [40]. Predominantly, the area is characterized by red soil, contributing to its unique ecological landscape.

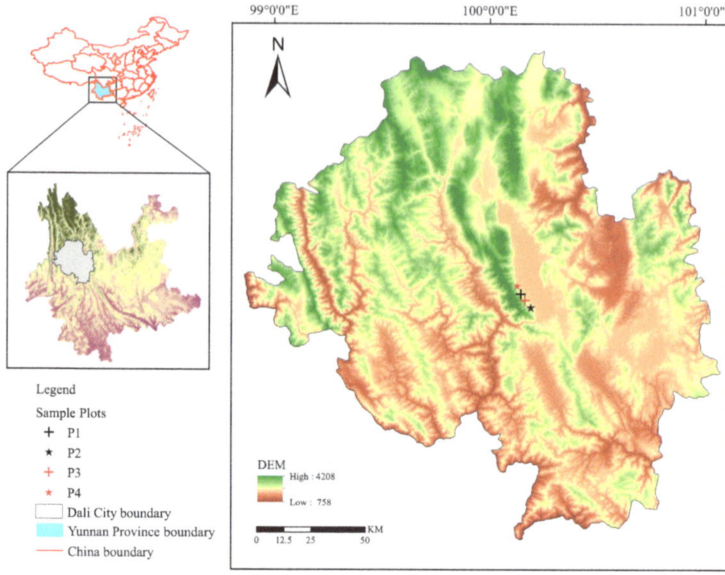

Figure 1. Description of Study Sites: Cangshan Mountain, Yunnan Province, China. P1–P4 Designated as Plot Locations.

2.2. Study Site and Data Collection

Table 1 presents key statistical data pertaining to the sample plots utilized in this study. The primary focus of the investigation centered on the dominant tree species, *Pinus yunnanensis*, spanning four circular sample plots, each strategically positioned atop distinct peaks. These circular sample plots boasted radii measuring 19 m (Lan Peak), 20 m (Malong Peak), 32 m (Foding Peak), and 35 m (Zhonghe Peak).

Between July and December 2022, comprehensive forestry operations were executed within the delineated sample plots. This phase involved meticulous measurements and recording of geographical coordinates, elevation, slope, slope direction, and plot radii. Furthermore, a systematic survey was conducted encompassing live standing trees with a diameter at breast height (DBH) equal to or exceeding 5 cm (DBH $\geq$ 5 cm) within the sample plots. Each tree underwent a thorough assessment, documenting essential forestry attributes including species, DBH, tree height (TH), crown width (CW), and crown length (CL), facilitated by specialized altimeters and distance measurer. Using a Topcon GTS-2002 autofocus total station (Topcon, Tokyo, Japan), we precisely determined the relative coordinates of each tree's base with respect to the center of the sample plots.

Table 1. Essential Details of Sample Plot Characteristics.

Sample Plots	East Long.	North Lat.	Elevation (m)	Slope (°)	Slope Dir.	Sample Plot Radius (m)	Tree Species Composition	Stand Density (trees/ha)
P1	100°08.2149″	25°41.5280″	2254	13.45	East	35	8 PY-2 PA-BA-TG	1481
P2	100°10.9639″	25°38.1518″	2271	16.15	South	32	7 PY-3 PA	1822
P3	100°09.3947″	25°39.9506″	2195	17.70	NE	20	7 PY-3 PA-QAC-VBT-GGW-BA	1830
P4	100°07.1906″	25°43.5923″	2138	5.10	NE	19	10 PY-QAC	1975

Note: NE, North-East; PY, *Pinus yunnanensis*; PA, *Pinus armandii*; QAC, *Quercus acutissima Carrut*; VBT, *Vaccinium bracteatum Thunb*; GGW, *Gaultheria griffithiana Wight*; BA, *Betula alnoide*; TG, *Ternstroemia gymnanthera*. The column 'Tree Species Composition' delineates the distribution of tree species within each plot. The numerical values preceding each species denote the relative abundance of that particular species for every 10 trees sampled within the plot.

2.3. Determination of Spatial Structure Units and Edge Correction

The Voronoi method was used to delineate spatial structure units within the forest stands [17]. This method constructs Voronoi diagrams based on the measured relative positions of individual trees, creating polygons that represent spatial structure units for each tree and its neighboring trees [41,42].

However, the spatial structure units at the edges of stands may be affected by sample boundaries, potentially leading to errors in calculating spatial structure indices. To address this issue, the study utilized the buffer zone method [21,43]. For example, for a circular plot with a diameter of a meters, the buffer zone is created by extending b meters inward from the edge of the plot towards the center. In other words, a m ring area is used as the buffer zone [44]. When calculating the spatial structure index for a unit composed of a central tree and its adjacent trees, the trees within the buffer area are only considered as adjacent trees in forming the spatial structure unit.

The determination of the buffer zone width depends on several factors, including plot size, methods for analyzing stand structure indices, and geographical location. This study carefully considered these factors and, based on existing research and experience [45], set the buffer zone width at 2 m.

2.4. Stand Structure Indexes

Quantifying stand structure is essential for optimizing FSS. In this study, the number of tree diameter classes, the diversity of tree species, cutting intensity and plant density were selected as indicators to quantify the non-spatial structure of the stand. To assess the spatial structure of the forest stand, indicators such as uniform angle index, mingling index, crown competition index, story index, and open comparison were utilized.

2.4.1. Non-Spatial Structural Indexes

(1) Tree Diameter Classes [6]

Categorizing trees into diameter classes is crucial for our analysis, as it correlates directly with stand growth. In this study, we categorized trees based on their DBH, starting from 6 cm and increasing in 2 cm increments. This systematic approach ensured consistency in the number of diameter classes throughout FSS optimization:

$$D = D_0 \tag{1}$$

D_0 denotes the count of diameter classes before selection cutting, and D signifies the count post-selection cutting.

(2) Diversity of Tree Species

During the selection cutting process, it is crucial to preserve tree species diversity to prevent unintentional extinction. We diligently ensured that the count of tree species remained unchanged throughout the process:

$$T = T_0 \tag{2}$$

T_0 signifies the initial count of tree species, and T denotes the count post-selection cutting.

(3) Cutting Intensity

The vigor of the stand after optimization relies on the cutting intensity. Ideally, the annual cutting volume should not exceed the annual growth rate of the stand. Previous research [46,47] has indicated that the ideal cutting intensity for secondary forests of *Pinus yunnanensis* should be limited to a maximum of 35%:

$$N \geq N_0(1 - 35\%) \tag{3}$$

N_0 signifies the total count of trees before selection cutting, and N represents the count after selection cutting.

(4) Plant Density (PD) [7]

PD is the key factor influencing the replanting effect. Previous studies [48] showed that the reasonable range of PD of *Pinus yunnanensis* is 1667~3333 trees/hm^2. After replanting optimization, the PD of the sample plots should be within the range of [1667, 3333]:

$$1667 \leq PD \leq 3333 \tag{4}$$

2.4.2. Spatial Structural Indexes

(1) M

The spatial segregation of tree species, represented by M, is calculated as the ratio of neighboring tree j stem count, excluding the same species as the object tree i, to the total neighboring tree stem count. It is mathematically expressed as [8,9]:

$$M_i = \frac{1}{n}\sum_{j=1}^{n} v_{ij} \tag{5}$$

M_i symbolizes the mingling index of the object tree i, and v_{ij} is a discrete variable. When the neighboring tree j is not of the same species as tree i, $v_{ij} = 1$; otherwise, $v_{ij} = 0$.

(2) CI

To measure the competitive pressures among trees, we designed a CI that employs the overlap area of canopies. The index is calculated as follows [10,11,44]:

$$CI_i = \frac{1}{Z_i} \times \sum_{j=1}^{n} AO_{ij} \times \frac{L_j}{L_i} \tag{6}$$

CI_i represents the canopy competition index for object tree i, Z_i denotes the projected canopy area of object tree i. $L_i = H_i \times CW_i \times CL_i$, where H_i, CW_i and CL_i represent the height, canopy width, and canopy length of the object tree i, respectively. $L_j = H_j \times CW_j \times CL_j$, where H_j, CW_j and CL_j denote the height, canopy width, and canopy length of the competing tree i, respectively. AO_{ij} indicates the overlap area between the canopies of object tree i and competitor tree j, with $AO_{ij} = 1$ when there is no overlap.

(3) W

The characterization of the stand's horizontal distribution pattern is represented by the index W. This index measures the proportion of angles α between the object tree i and its nearest neighbors that are smaller than a predefined standard angle α_0.It is mathematically expressed as [12,13]:

$$W_i = \frac{1}{n}\sum_{j=1}^{n} z_{ij} \tag{7}$$

W_i denotes the angle index of tree i, and z_{ij} is a discrete variable. If the angle α between trees i and j is less than α_0, $z_{ij} = 1$; otherwise, $z_{ij} = 0$.

(4) S

The vertical diversity and complexity within a stand are encapsulated by the index denoted as S, which quantifies the proportion of neighboring trees j at the same height level as the object tree i. It is calculated as follows [14]:

$$S_i = \frac{1}{n}\sum_{j=1}^{n} v_{ij} \tag{8}$$

S_i represents the story index of tree i, and v_{ij} is a binary variable. If tree i shares the same height level as tree j, $v_{ij} = 0$; otherwise, $v_{ij} = 1$.

(5) *OP*

The open comparison represents the light environment and available growing space for tall trees within the stand. This index measures the degree to which the object tree i within a spatial structure unit is overshadowed by its neighboring tree j. This index is represented as follows [15]:

$$OP_i = \frac{1}{n}\sum_{j=1}^{n} t_{ij} \tag{9}$$

OP_i is the open comparison corresponding to object tree i, t_{ij} is a discrete variable. If the horizontal distance between object tree i and neighboring tree j exceeds their height difference, $t_{ij} = 1$; contrarily, $t_{ij} = 0$.

(6) *U*

The size differentiation in the diameter of individual forest trees is described by the proportion of neighboring trees j nearest to object tree i that are larger than object tree i among the n neighboring trees. This comparison was calculated using the formula [16]:

$$U_i = \frac{1}{n}\sum_{j=1}^{n} k_{ij} \tag{10}$$

U_i is neighborhood comparison of object tree i, k_{ij} is characterized as a discrete variable. When the DBH of neighboring tree j is larger than that of object tree i, k_{ij} = 1; otherwise, k_{ij} = 0.

The aforementioned indices were calculated and analyzed using data from the stand survey with R version 4.2.0.

2.5. Methods for Selecting Trees for Cutting

The RSM (random selection method) [24,44], QVM (Q-value method) [49], and VMM (V-map method) [50] are widely employed techniques for selecting trees for cutting. RSM identifies trees for removal through rapid random sampling from the initial pool of retained trees (comprising all trees within the original stand), and ensuring selection intensity remains within bounds. QVM constructs a single-tree composite index Q_i using five spatial structural parameters (W, M, CI, OP, and S), then ranks these Q_i values within the harvesting limit to determine the trees to be harvested. Previous research [49] suggests that the probability of FSS achieving optimization is higher when trees corresponding to the top Q_i values are felled. Under ideal conditions, the average angle index of the stand ($\overline{W}$) falls within the range [0.475, 0.517], with a central value of 0.496. In VMM, the initial selection of structural units for harvesting prioritizes the nearest neighboring trees of the reference tree with the highest W value (0.496). These neighboring trees are then used to identify trees for felling, with particular focus on those exhibiting weak mixing, moderate mixing, and suppressed competition based on their M values.

In our prior study, we conducted experimental comparisons among these three selection methods, revealing the RSM as the optimal choice for optimizing stand structure when integrated with RL algorithm [44]. Consequently, the RSM has been selected as the preferred method for tree selection in the cutting optimization segment of this research.

2.6. RFI

2.6.1. Number of Replanting Trees and Species Configuration

The literature suggests that the optimal planting density for *Pinus yunnanensis* ranges from 1667 to 3333 trees per hectare. In the study, we set the upper limit of planting density for the sample plots after replanting to 3333 trees/hm^2. Considering various sample plot sizes, we rounded down to determine the upper limit of tree numbers in P1, P2, P3, and P4, resulting in 1282, 1072, 418, and 378 trees, respectively.

In mixed forests, different proportions of replanting species may lead to varying degrees of species segregation, while differences in DBH, H (tree height), CW (crown width), and CL (crown length) of replanted trees may affect the competitive relationships among neighboring trees post-replanting. The four sample plots in this study were secondary forests dominated by *Pinus yunnanensis*, with other native or companion species present, including *Pinus armandii*, *Vaccinium bracteatum*, *Quercus acutissima*, *Betula alnoides*, *Gaultheria griffithiana*, and *Ternstroemia gymnanthera*. We set the average DBH of replanted trees at 5 cm, and trees within the [5, 6) cm DBH interval were used to calculate an average H of 4.79 m, an average CW of 6.05 m, and an average CL of 1.71 m, which determined the size of replanted trees.

2.6.2. Maximum Delaunay Generator Area Method (MDGAM)

Ideal replanting locations are usually in relatively wide "forest gaps". Traditional methods for pinpointing these sites include the MDGAM [18], maximum null circle method (MNCM) [17], and Kriging method [19]. These methods assess various areas to determine the optimal replanting location rooted in maximal area calculation. The MDGAM, based on Delaunay triangulation, represents each tree as a node and the distances between neighboring trees as side lengths. This approach effectively captures both the "forest gaps" and the forest stand's distribution pattern [51]. In contrast, the MNCM and Kriging method lack the Delaunay triangulation's characteristics and cannot adequately consider the forest stand's distribution pattern, thereby presenting significant limitations in determining the replanting location. Consequently, this study relied on the generating element area of Delaunay triangulation as the primary criterion for determining replanting locations.

2.6.3. RFI

Replanted trees establish a new spatial structure within the stand alongside their neighboring trees. While the basic MDGAM primarily considers growth space and stand distribution patterns, merely positioning replanting sites within the maximal "forest gaps" overlooks the impact of neighboring trees on the growth of newly planted ones. This oversight may exacerbate competitive relationships within the stand. It becomes imperative to factor in spatial relative coordinates, tree species, tree age, tree height, and articulate these multidimensional characteristics within the replanting optimization model and solution algorithm.

Therefore, this study introduces a novel RFI to tackle these issues. Leveraging Delaunay triangulation, the RFI characterizes the relative coordinates of replanted trees and their neighbors, the mingling index, the impact of replanting species on replanting efficacy, DBH in neighborhood comparison to denote age, and the crown competition index to consider tree crown and height influences. The specific formula for the RFI is outlined as follows:

$$RFI_i = \frac{\frac{1+DAA_i}{\delta_{DAA}} \cdot \frac{1+M_i}{\delta_M} \cdot \frac{1+U_i}{\delta_U}}{\frac{1+CI_i}{\delta_{CI}}} \tag{11}$$

RFI_i represents the replanting prospect index of the tree to be replanted i, DAA_i denotes the area of the Delaunay triangulation generation element where tree i is situated, M_i signifies the mingling index of tree i, U_i denotes the neighborhood comparison of tree i, and CI_i represents the canopy competition index of tree i. Additionally, δ_{DAA}, δ_M, δ_U, and δ_{CI} denote the standard deviation of each structural parameter.

By introducing the RFI, we comprehensively consider the impacts of various factors on the growth of replant trees, including growth space, mingling index, neighborhood comparison, and canopy competition index. This integrated index enhances the accuracy and effectiveness of determining replanting locations. During replanting operations, priority is assigned to locations with higher RFIs, thereby enhancing the efficiency of stand adjustment and optimization.

2.7. Forest Stand Structure Optimization Model

FSS optimization represents a multi-objective optimization challenge [17,24,44], where the optimization model aligns with predefined objectives and constraints. This model entails selective logging and replanting strategies to enhance the overall FSS. Constructing an objective function in accordance with optimization goals is pivotal for effective FSS optimization.

2.7.1. Objective Function

Solving a multi-objective optimization problem involves finding the optimal solution considering multiple objectives while adhering to constraints [52,53]. Typically, these objectives are interconnected and constrained, making it challenging to achieve optimal solutions for each individual objective. Therefore, it's crucial to integrate and synthesize multiple sub-objectives into an overall objective function to find the optimal solution.

When optimizing FSS, a higher VOF signifies a better spatial structure for the forest stand. This study employed the concept of "multiplication and division" [54] to select and combine five spatial structure indices: angle index, mingling index, crown competition index, story index, and open comparison into the objective function for the multi-objective optimization model of FSS. This objective function was calculated using the formula:

$$\max OF = \frac{1}{N}\sum_{i=1}^{N}\frac{\frac{1+M_i}{\delta_M}\cdot\frac{1+OP_i}{\delta_{OP}}\cdot\frac{1+S_i}{\delta_S}}{\frac{1+CI_i}{\delta_{CI}}\cdot\frac{1+|W_i-0.496|}{\delta_{|W-0.496|}}} \tag{12}$$

M_i, OP_i, S_i, CI_i, and W_i denote the mingling index, open comparison, story index, canopy competition, and uniform angle index of the central tree i, respectively. Additionally, δ_M, δ_{OP}, δ_S, δ_{CI}, and δ_W represent the standard deviations of their respective structural parameters. The midpoint of the range [0.475, 0.517] is 0.496, where a smaller value of $|W_i - 0.496|$ indicates that the forest stand's horizontal distribution pattern is closer to randomness.

To ensure consistency in model evaluation and comparison, and effectively explore performance differences among different methods under optimization objectives, the same objective function was employed for all optimization models in this study.

2.7.2. Traditional Combinatorial Optimization (TCO) Model

Achieving effective optimization of FSS often requires the collaboration of multiple optimization measures, such as selective cutting and replanting strategies. The traditional logging and replanting combinatorial optimization model serves as a common approach for optimizing FSS, involving a dual optimization process. Initially, a portion of the trees within the sample plot is selectively cut according to a predefined harvesting strategy to attain a relatively optimal FSS. Subsequently, native tree species are replanted at suitable locations within the plot post-harvesting, aiming for a double optimization of the FSS.

In a previous study [44], simulated selective logging experiments on four sample plots yielded relatively optimal FSS outcomes. Building upon these findings, replanting optimization experiments were conducted, based on the improved FSS after selective cutting in each sample plot, to assess the practical application of the logging-first-then-replanting optimization concept.

The optimization model is pivotal for FSS optimization, encompassing the objective function, constraints, and other factors. The constraints of the TCO model are outlined as follows:

(1) Constraints

Following optimization and adjustments, the quality of each sub-objective should not deteriorate compared to its pre-optimization state, ensuring that the spatial structure diversity of the forest stand remains intact. This necessitates a closer-to-random horizontal distribution pattern and an enhanced mixing degree. During replanting, the plant density within the sample plot must be maintained within a reasonable range. Research findings [48,55,56] suggest that the suitable planting density for *Pinus yunnanensis* ranges between 1667~3333 trees/hm^2. The constraints of the TCO model are formulated as follows:

$$s.t.\begin{cases} \left|\overline{W_2}-0.496\right| \leq \left|\overline{W_1}-0.496\right| \\ \overline{M_2} \geq \overline{M_1} \\ 1667 \leq PD \leq 3333 \end{cases} \tag{13}$$

$\overline{W_1}$ and $\overline{M_1}$ represent the average values of the angle index and mingling index of the stand after selective logging, respectively. $\overline{W_2}$ and $\overline{M_2}$ denote the average values of each parameter after replanting, and PD indicates the plant density of the stand in the sample plot after replanting optimization.

(2) Solving Algorithm

Based on the findings of prior research [44], this study derived the FSS of each sample plot following selection cutting optimization via RL. Then, the RFI was employed to designate the replanting locations. Considering the predetermined upper limit for the number of replanting trees in each plot, tree species for replanting were allocated in equal proportions, and the dimensions of individual replanted trees were determined accordingly. This process culminated in the compilation of the replanting tree set. Various quantities of trees, ranging from 0 to the predefined upper limit, were selected for replanting from this set. Consequently, their corresponding structural indices for each forest stand were calculated. Figure 2 depicts the flowchart of the TCO process, while the algorithmic pseudocode is elaborated in Appendix A.1.

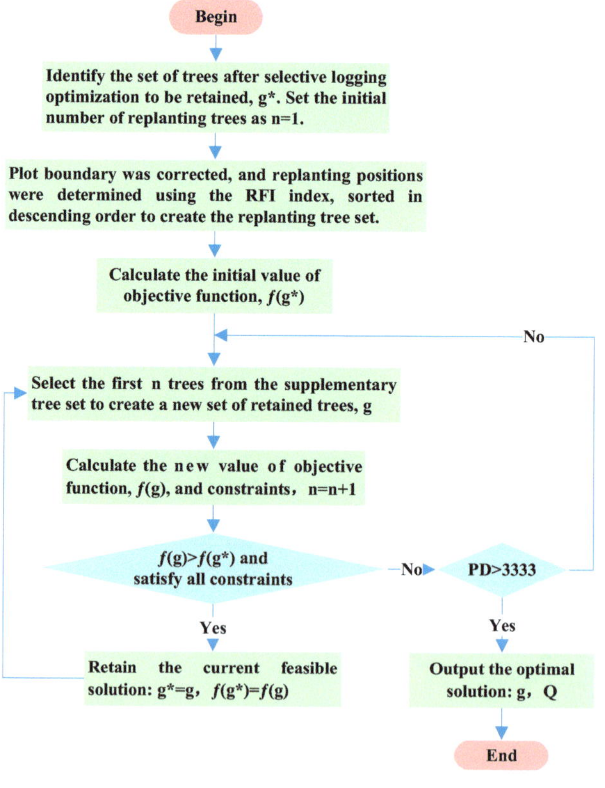

Figure 2. Flowchart of TCO algorithm.

2.7.3. Collaborative Optimization Model

The collaborative optimization model for logging and replanting aims to optimize the FSS through replanting while adhering to the objectives and constraints of selection cutting, thereby achieving dual optimization of the FSS. This model emphasizes real-time collaboration between logging and replanting, enhancing flexibility and adaptability to varying site conditions. RL presents advantages in addressing dynamic optimization problems like this, yet the performance of SARL and MARL differs significantly in collaborative optimization scenarios. SARL, constrained by individual knowledge and experience, necessitates extensive training time for complex problems, whereas MARL, leveraging collaboration among agents, encounters challenges in balancing cooperation and competition and designing intricate reward functions. In this study, the objective function of the collaborative optimization model aligns with that of the TCO model, facilitating effective comparison across different optimization approaches. The constraints of the collaborative optimization model are detailed as follows:

(1) Constraints

During the selection cutting process, it is imperative to ensure that the horizontal distribution pattern of the forest stand approaches random distribution, enhancing the mingling index, reducing stand competition, promoting vertical diversity, and increasing openness and light penetration. Regarding non-spatial structure constraints, it is essential to maintain the number of diameter classes and tree species in the stand during selective logging optimization, with the cutting intensity not exceeding 35%. Replanting optimization constraints are the same as those of the TCO model's replanting optimization segment. In the collaborative optimization model, the VOF post-selective logging F_1 must exceed the initial VOF F_0, while the VOF post-replanting F_2 must surpass F_1. A higher F_2 indicates superior structure in the cooperative optimization. In summary, the constraints of the collaborative optimization model are expressed as follows:

$$s.t.\begin{cases} \left|\overline{W_1} - 0.496 < \overline{W_0} - 0.496\right| \\ \overline{M_1} > \overline{M_0} \\ \overline{CI_1} < \overline{CI_0} \\ \overline{S_1} > \overline{S_0} \\ \overline{OP_1} > \overline{OP_0} \\ D_1 = D_0 \\ T_1 = T_0 \\ N_1 \geq N_0(1-35\%) \\ \left|\overline{W_2} - 0.496 < \overline{W_1} - 0.496\right| \\ \overline{M_2} > \overline{M_1} \\ 1667 \leq PD \leq 3333 \\ F_1 > F_0 \\ F_2 > F_1 \end{cases} \tag{14}$$

$\overline{W_0}$, $\overline{M_0}$, $\overline{CI_0}$, $\overline{S_0}$, $\overline{OP_0}$ represent the average values of the uniform angle index, mingling index, canopy competition index, story index, and open comparison, respectively, in the forest stand before the selective felling; $\overline{W_1}$, $\overline{M_1}$, $\overline{CI_1}$, $\overline{S_1}$, $\overline{OP_1}$ denote the average values of each parameter after selective cutting; D_0, T_0, N_0 indicate the number of diameter classes, tree species, and the total number of trees in the stand before selective logging, while D_1, T_1, and N_1 represent the values of each parameter after selective cutting; $\overline{W_2}$, $\overline{M_2}$ represent the values of the angle index, crown competition index, story index, and open comparison of the forest stand after replanting; PD denotes the density of forest stand plants after logging and replanting collaborative optimization.

(2) Solving Algorithm Based on SARL

This study employs the RSM (random selection method) to determine the logging trees and assesses the replanting potential using the RFI in solving the collaborative optimization problem of selective cutting and replanting of FSS using SARL. The actions of selecting logging trees and determining the number of replanting trees are translated into agent actions in RL. Agents can choose between selective and non-selective logging in the logging optimization phase, and between replanting and non-replanting in the replanting optimization phase. Four joint optimization actions are generated, including ① selective cutting and replanting, ② no selective cutting and no replanting, ③ selective cutting but no replanting, and ④ no selective cutting but replanting. Actions ③ and ④ are equivalent to single optimization actions and have been extensively discussed in previous research. This study focuses on actions ① and ②, selective logging and replanting, and non-selective logging and no replanting, as the co-optimization actions of SARL. Agents select actions successively to maximize the reward value. Figure 3 illustrates the flowchart of SARL for addressing the co-optimization problem of FSS logging and replanting, with detailed algorithmic pseudocode provided in Appendix A.2.

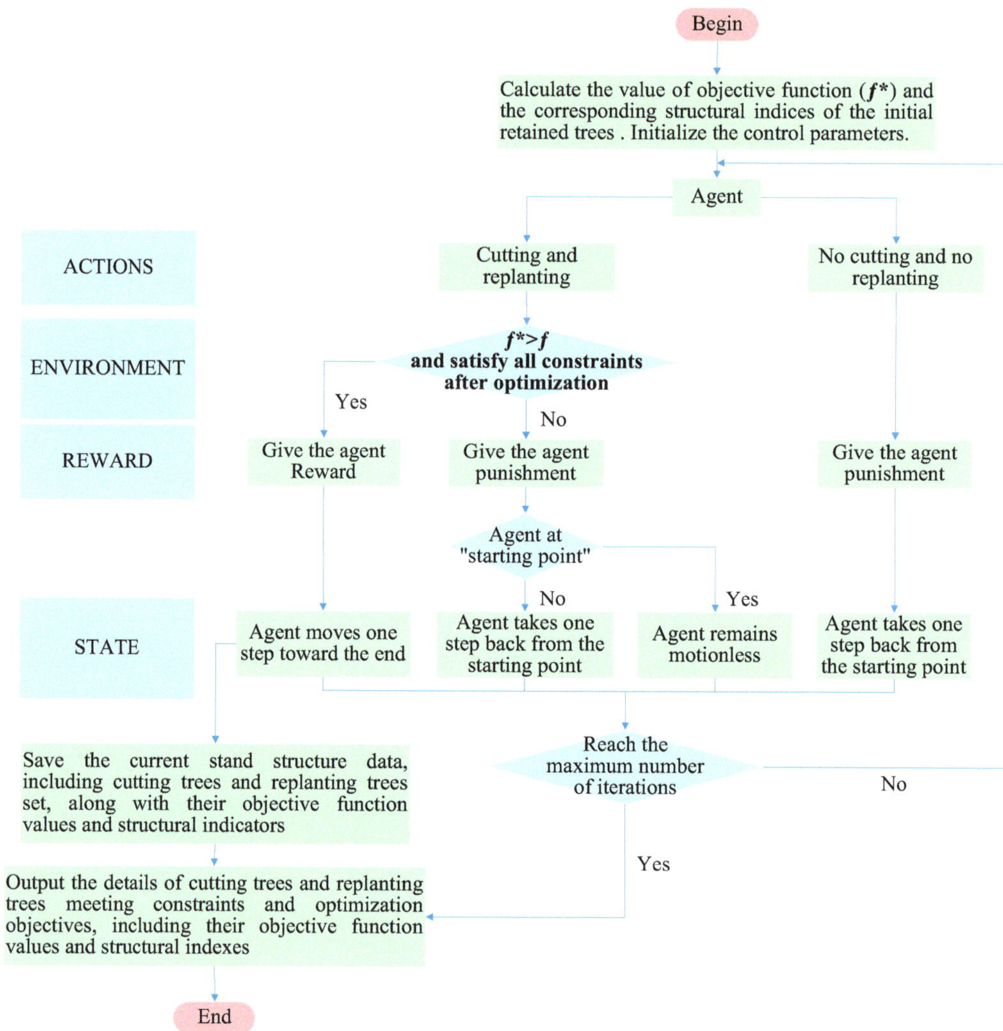

Figure 3. Flowchart of SARL algorithm.

(3) Solving Algorithm Based on MARL

Figure 4 illustrates the flowchart of MARL for addressing the collaborative optimization problem of FSS logging and replanting, while detailed algorithmic pseudocode is provided in Appendix A.3. At each iteration, Agent1 and Agent2 are positioned at the "start point" and "end point" respectively. Agent1 is responsible for logging optimization, while Agent2 achieves replanting optimization. Using trial-and-error strategy, Agent1 is trained to find the set of selective logging trees satisfying the constraints, and Agent2 is trained to determine the replanting points and the optimal number of replanting plants. The collaborative optimization strategy motivates Agent1 and Agent2 to collaborate effectively through cross-collaboration actions and ultimately obtain the optimal spatial structure of the forest stand.

Designing an effective reward function is crucial in RL to find better results efficiently. In this solving algorithm, the reward function in the selective harvesting optimization part follows the result of previous study. For replanting optimization, a method based on curve trend for sample point selection is designed, rewarding or penalizing based on the presence of extreme points. Agent2 receives rewards and penalties in the replanting process based on the rewards and penalties obtained in the first and second rounds, with specific rules outlined in Table 2. Agent1 and Agent2 collaborate to move towards each other by efficiently identifying the FSS that maximizes the objective function. The earlier meet of Agent1 and Agent2 signifies the higher efficiency of MARL in solving multi-objective forest stand structure (MOFSS) optimization.

Table 2. Reward and punishment rules in MCO.

Initial Rewards/Penalties			Subsequent Rewards/Penalties		Final Rewards/Penalties
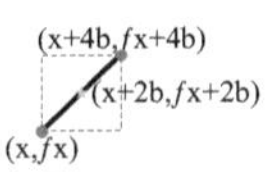	Increase	Little reward	$max(f_{x+b}, f_{x+3b}) > max(f_x, f_{x+2b}, f_{x+4b})$	Little reward	Little reward + Little reward
			Otherwise	Little reward	Little reward + Little reward
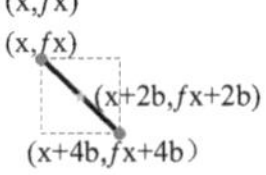	Decrease	Little reward	$max(f_{x+b}, f_{x+3b}) > max(f_x, f_{x+2b}, f_{x+4b})$	Little reward	Little reward + Little reward
			Otherwise	Little reward	Little reward + Little reward
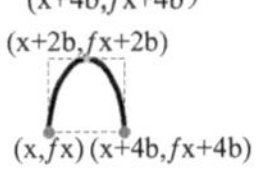	Convex	Large reward	$max(f_{x+b}, f_{x+3b}) > max(f_x, f_{x+2b}, f_{x+4b})$	Little reward	Large reward + Little reward
			Otherwise	Large reward	Large reward + Large reward
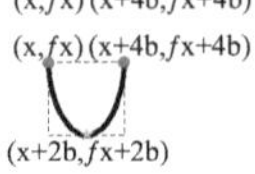	Concave	Penalty	$max(f_{x+b}, f_{x+3b}) > max(f_x, f_{x+2b}, f_{x+4b})$	Little reward	Penalty + Little reward
			Otherwise	Large penalty	Penalty + Large penalty

2.7.4. Experimental Scheme

Firstly, this study utilized an iterative algorithm to simulate replanting optimization based on the initial and optimal stands after cutting in four sample plots. The effectiveness of the RFI was compared with the MDGAM (maximum Delaunay generator area method) to validate the efficacy of RFI.

Next, to assess the applicability and effectiveness of the TCO scheme employing RFI, the stands of the four sample plots were subjected to logging optimization. The top three ranked stand structures, determined by the VOF, were selected and further optimized through replanting according to RFI. This aimed to validate the generalizability and efficacy of the TCO model.

Finally, to evaluate the performance of SARL and MARL in solving the multi-objective collaborative optimization model of FSS, SARL and MARL were employed. SARL and MARL utilized RSM for selective logging determination, while RFI guided replanting location determination, achieving collaborative optimization of forest stand structure (COFSS) (Table 3).

Table 3. Experimental schemes for FSS optimization.

Experiment			Cutting Optimization			Replanting Optimization		
Name	**Type**	**Purpose**	**Cutting Method**	**Solution Algorithm**	**Resulting Stand**	**Replanting Method**	**Solution Algorithm**	**Resulting Stand**
SR	Comparison Experiment	Compare the effectiveness of the RFI with the MDGAM	None	None	Initial stand	MDGAM RFI	Iterate	Optimal replanting stand based initial stand
OCR			RSM QVM VMM	MC PSO RL	Optimal cutting stand			Optimal replanting stand based optimal cutting stand
TCO	Optimization experiment	Validate the effectiveness and generalizability of RFI-based TCO			First three optimal cutting stands	RFI		Optimal replanting stand based first three optimal cutting stand
SCO		Evaluate the performance of SARL and MARL in solving multi-objective co-optimization models with FSSs	RSM	SARL	Optimal cutting stand based on RSM-SARL		SARL	Optimal replanting stand using RFI after the RSM-SARL optimal cutting was conducted
MCO				MARL	Optimal cutting stand based on RSM-MARL		MARL	Optimal replanting stand using RFI after the RSM-MARL optimal cutting was conducted

Note: SR: Single Replanting, OCR: Optimal Cutting–Replanting, TCO: Traditional Combinatorial Optimization, SCO: Collaborative Optimization based SARL, MCO: Collaborative Optimization based MARL

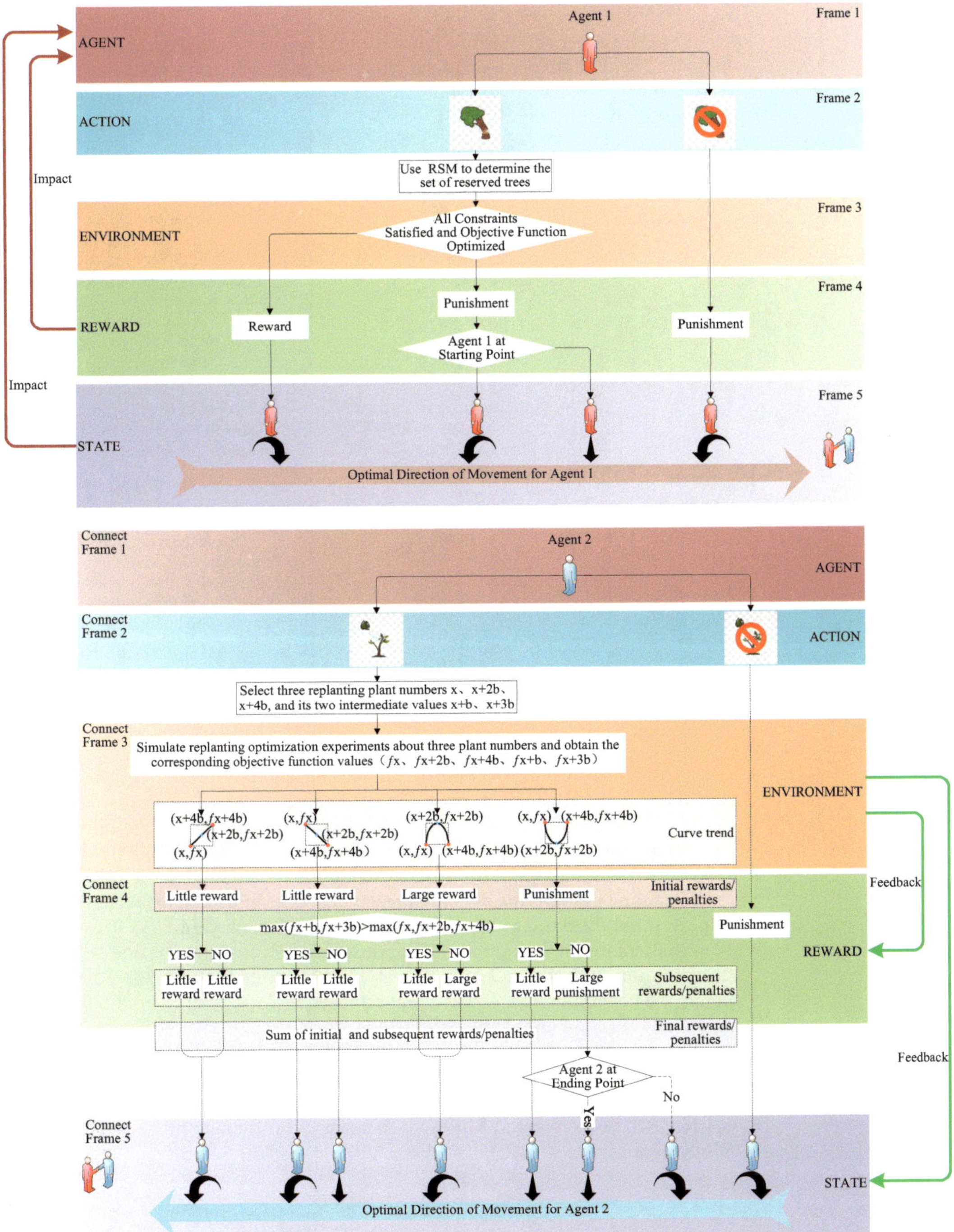

Figure 4. Flowchart of MARL algorithm.

2.7.5. Parameter Settings

The parameter settings of each solution algorithm in the experiment are shown in Table 4. The main experimental program was developed in Python version 3.10, with all algorithms implemented manually. This approach allowed for precise control and customization of the algorithmic processes without relying on pre-built libraries.

Table 4. Optimizing algorithm parameter configuration.

Algorithms	Parameters and Parameter Values	Value Meaning
Iterate	$I = 0$	The initial iteration count
	$I_{max} = 10{,}000$	Upper limit of iterations
	$U = 0$	Initial count of successive iterations without OF improvement
	$U_{max} = 500$	Upper limit of successive iterations without OF improvement
SARL	$I = 0$	The initial iteration count
	$I_{max} = 10{,}000$	Upper limit of iterations
	$state = 1$	Starting position of the agent
	$state_{max} = 100$	Farthest position of the agent
	$r1 = 150, r2 = 10, r3 = -1, r4 = -1, r5 = 1$	Reward and penalty values
RL	$I = 0$	The initial iteration count
	$I_{max} = 10{,}000$	Upper limit of iterations
	$state_1 = 1$	Starting position of the agent1
	$state_2 = 100$	Starting position of the agent2
	$state_{max1} = 100$	Farthest position of the agent1
	$state_{max2} = 1$	Farthest position of the agent2
	$r1 = 200, r2 = 20, r3 = 10, r4 = 1, r5 = -1, r6 = -50$	Reward and penalty values

3. Results

3.1. RFI

To ascertain the effectiveness of RFI, replanting optimization experiments were conducted on eight distinct FSSs. These FSSs encompassed both the optimal stands following selective logging and the initial stands. The study compared the effects of different methods for determining replanting locations in terms of the objective function and the number of iterations. The findings are summarized in Figure 5.

Regarding the VOF, the RFI-based replanting location determination method (RBRLDM) outperformed the basic MDGAM in achieving higher maximum VOFs across five FSSs (P1 initial, P2 initial, P2 optimal, P3 optimal, and P4 initial). For one FSS (P4 optimal), the maximum VOF was equal to that of MDGAM, while for two FSSs (P1 optimal and P3 initial), the maximum VOF equaled that achieved by MDGAM. Overall, RBRLDM yielded superior VOFs compared to basic MDGAM.

As to the number of iterations, RBRLDM demonstrated significantly faster convergence compared to basic MDGAM. Basic MDGAM was prone to "local optimization" pitfalls. Notably, in the replanting optimization experiments of the P3 initial stand, basic MDGAM outperformed RBRLDM in terms of both the number of iterations and the objective function.

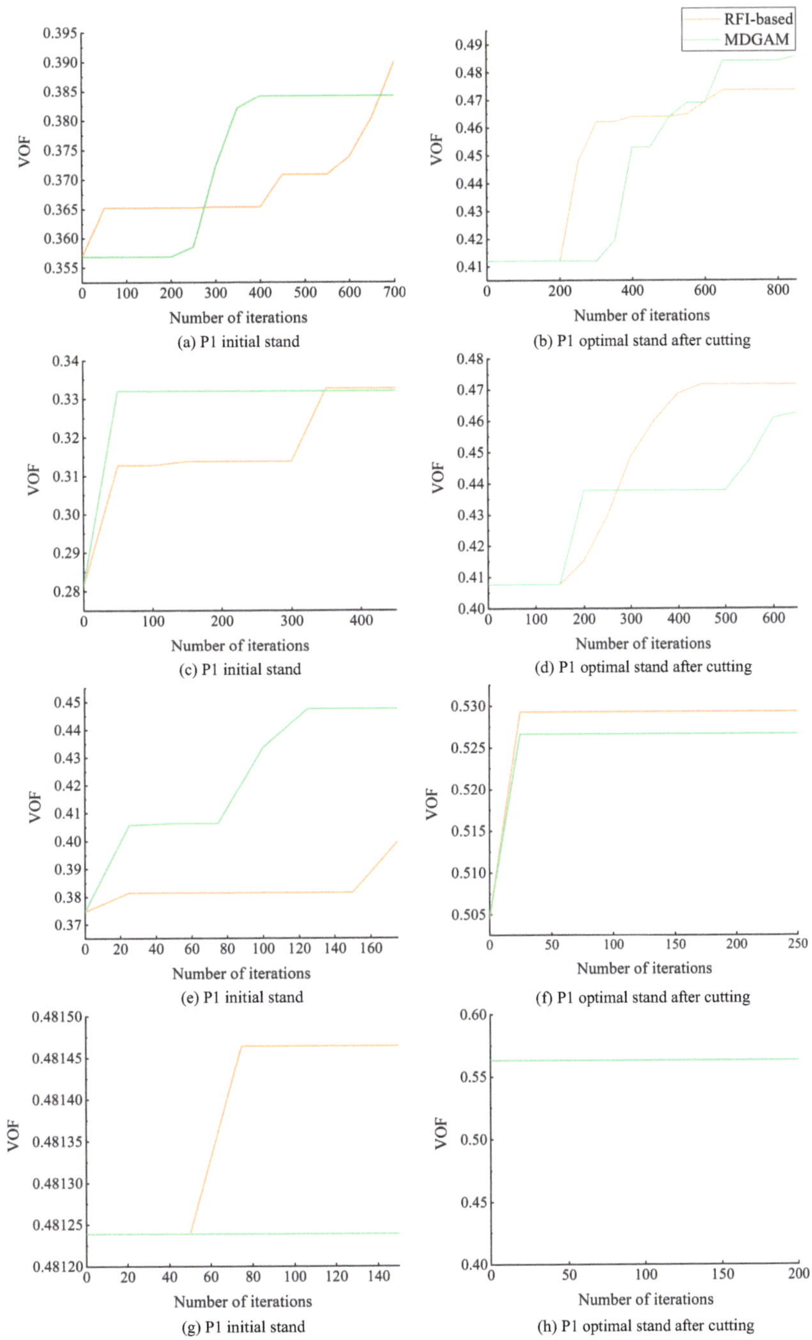

Figure 5. Results of two replanting methods.

3.2. TCO

The TCO model achieves dual optimization of cutting and replanting by optimizing replanting on the optimal FSS after selective logging. To validate its effectiveness, this study initially optimized the FSS of four sample plots through selective logging. Subsequently, the top three FSS sets were ranked in descending order according to the VOF, and replanting optimization was conducted on these sets based on RFIs.

Table 5 illustrates the results of TCO. In sample plot P1, the best replanting outcome of $FSSACA_{1st}$ (the best FSS after cutting adjustment) was inferior to $FSSACA_{2nd}$ and $FSSACA_{3rd}$ (second and third best FSS after cutting adjustment). Similar trends were observed in P2. However, in P3, replanting optimization yielded the best outcome on $FSSACA_{1st}$, achieving optimal harvesting and replanting simultaneously, thereby fulfilling the TCO model's ultimate objective. The optimization result in P4 was sub-optimal, with replanting optimization failing to achieve the best outcome on $FSSACA_{1st}$ and $FSSACA_{2nd}$, while $FSSACA_{3rd}$ achieved the desired result.

Overall, experimental results suggest that the TCO model, based on optimal FSS after selective logging, may not always attain optimal results and exhibits certain limitations.

Table 5. The results of TCO.

P1	Initial VOF	0.3515	0.3515	0.3515
	VOF after Cutting	0.4121_{1st}	0.4009_{2nd}	0.3975_{3rd}
	VOF after Replanting	0.4737_{3rd}	0.4879_{1st}	0.4829_{2nd}
P2	Initial VOF	0.2814	0.2814	0.2814
	VOF after Cutting	0.4078_{1st}	0.3983_{2nd}	0.3877_{3rd}
	VOF after Replanting	0.4718_{2nd}	0.4348_{3rd}	0.4785_{1st}
P3	Initial VOF	0.3748	0.3748	0.3748
	VOF after Cutting	0.5047_{1st}	0.4887_{2nd}	0.4729_{3rd}
	VOF after Replanting	0.5293_{1st}	0.5186_{2nd}	0.4908_{3rd}
P4	Initial VOF	0.4812	0.4812	0.4812
	VOF after Cutting	0.5635_{1st}	0.5593_{2nd}	0.5537_{3rd}
	VOF after Replanting	0.4852_{3rd}	0.5157_{2nd}	0.5713_{1st}

3.3. Collaborative Optimization (CO)

This study conducted experiments on selective cutting and replanting collaborative optimization (SCRCO) using SARL and MARL on four sample plots to compare the practical application of the CO and TCO models. As shown in Figure 6 and Table 6, both SARL and MARL under the SCRCO model can solve the multi-objective SCRCO problem of FSS, similar to the TCO model. Except for P3, SARL achieved better VOFs than TCO, with SARL improving the FSS of all four sample plots more than TCO on average. The optimization results of MARL in the four sample plots were better than those of TCO and SARL, with MARL achieving a higher average improvement in FSS than the other two methods.

Table 6. Specific VOFs for each solution algorithm.

	P1	**P2**	**P3**	**P4**	**Average**	**Lifting Amplitude**	**Average Lifting Amplitude**
Initial	0.3515	0.2814	0.3748	0.4812	0.3722		
TCO	0.4879	0.4785	0.5293	0.5713	0.5168	38.83%	
SCO	0.5426	0.5273	0.5196	0.5751	0.5412	45.38%	43.88%
MCO	0.5444	0.5315	0.5297	0.5897	0.5488	47.44%	

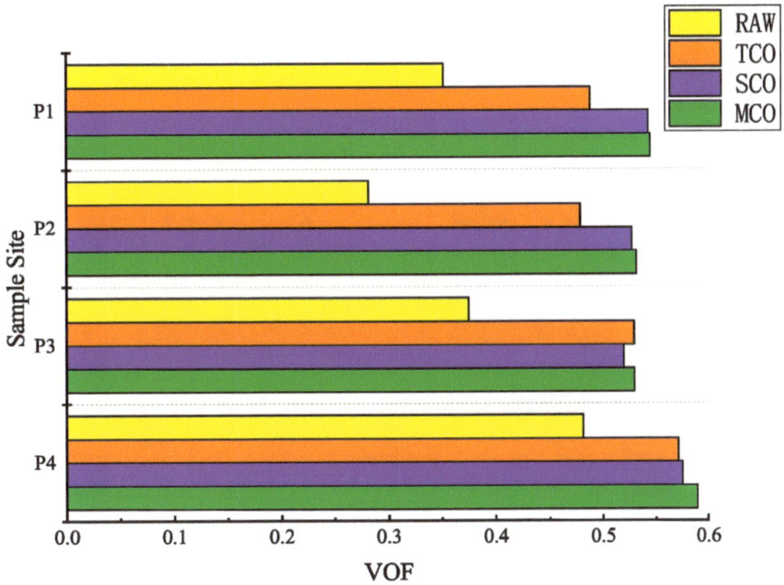

Figure 6. Best optimization results for three optimization models.

3.4. Changes in Stand Structure Parameters

A single VOF cannot comprehensively assess the actual effect of the optimization scheme. This study compared in detail the changes in stand structure indices for each sample plot under each optimization scheme to obtain a more accurate and comprehensive assessment of the optimization effect.

Figure 7 shows the optimization results of different FSS optimization schemes. The spatial structure indicators of each stand were improved to varying degrees after the combination of selective harvesting and replanting in different schemes. Overall, the gap between the average angle index of the stand and the randomly distributed angle index of 0.496 was slightly reduced, indicating a closer alignment with random distribution. The canopy competition index of the stands was reduced, suggesting alleviated pressure among trees. The mingling index of each sample site improved, particularly in P4, which likely benefited from its initially weak mixing state. Additionally, the story index and open comparison of each sample site increased, indicating an enriched vertical structure and improved light transmittance. Overall, the optimization schemes effectively improved the structure of all sample sites.

3.5. Algorithm Performance

This research compares the performance of each algorithm in terms of FSS optimization degree and convergence speed. The FSS's superiority or inferiority is measured by the VOF. As shown in Table 6, the maximum VOF of MARL (0.5444, 0.5315, 0.5297, 0.5897, respectively) was generally better than that of SARL (0.5426, 0.5273, 0.5196, 0.5751, respectively) in the cutting-replanting combination experiments in the four sample plots.

Regarding the convergence speed, MARL outperformed SARL and the TCO scheme. SARL is prone to "local optimization", and the TCO scheme requires numerous iterations in harvesting optimization, resulting in slower convergence. Overall, MARL exhibits superior convergence speed and achieves better VOFs compared to SARL and the TCO scheme. (Figures 8 and 9).

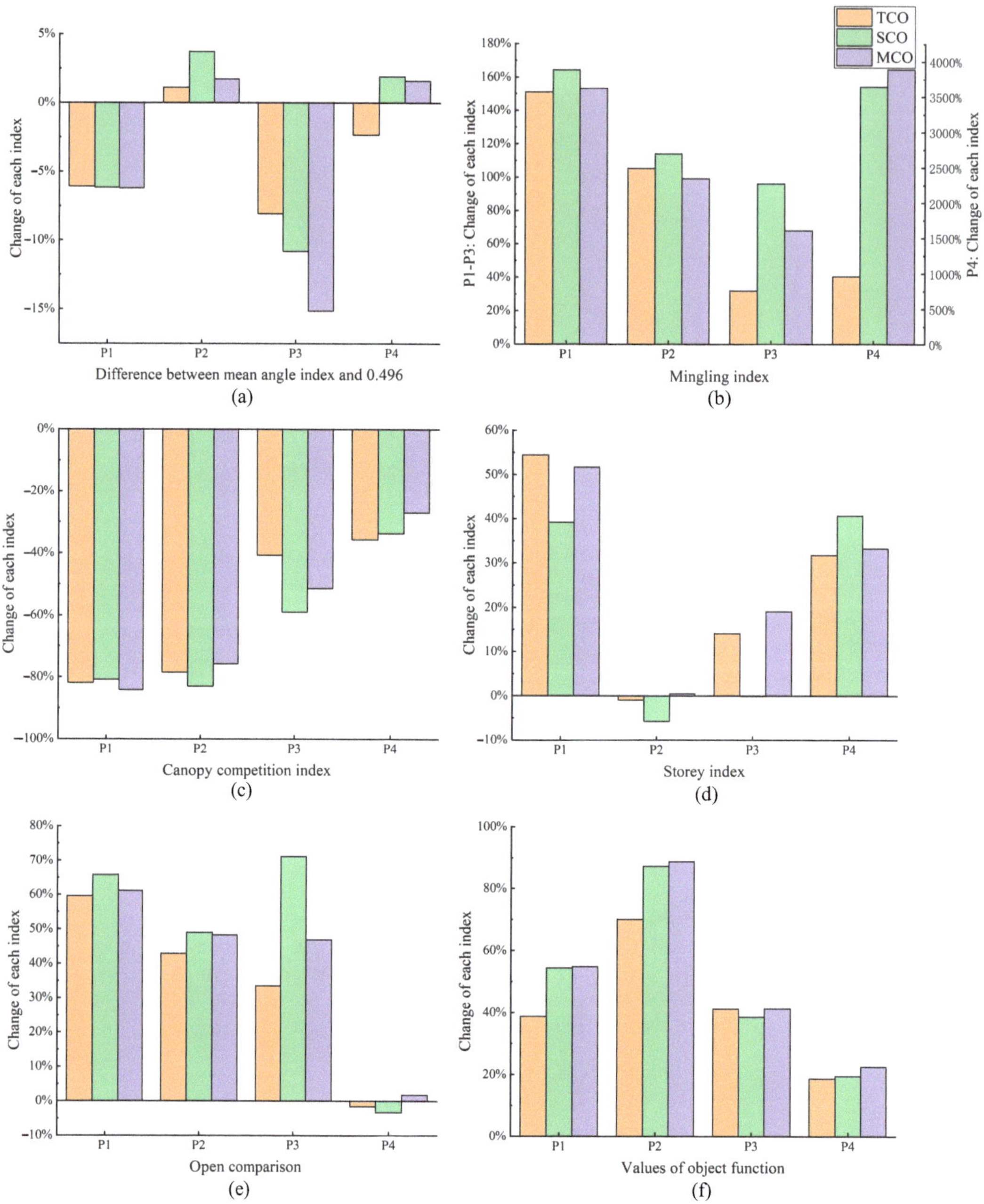

Figure 7. Parameter changes.

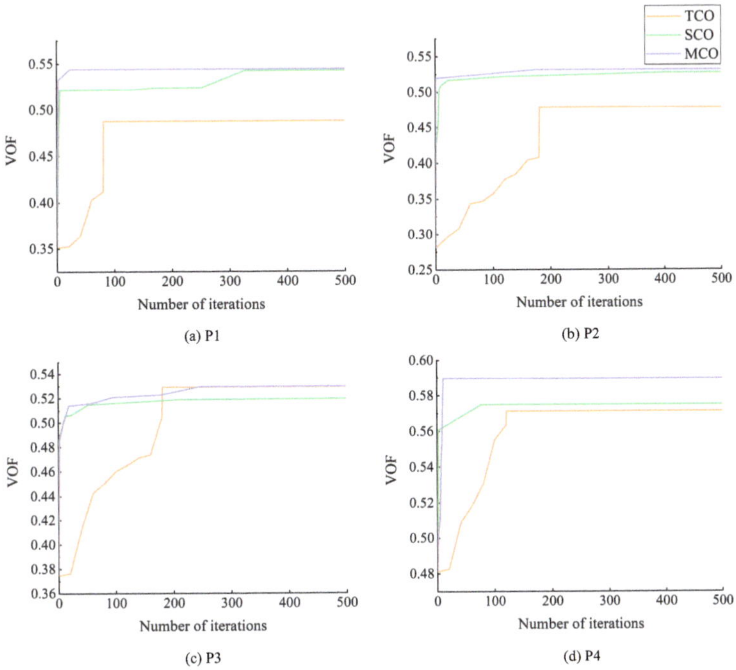

Figure 8. Number of iterations for the optimization algorithm.

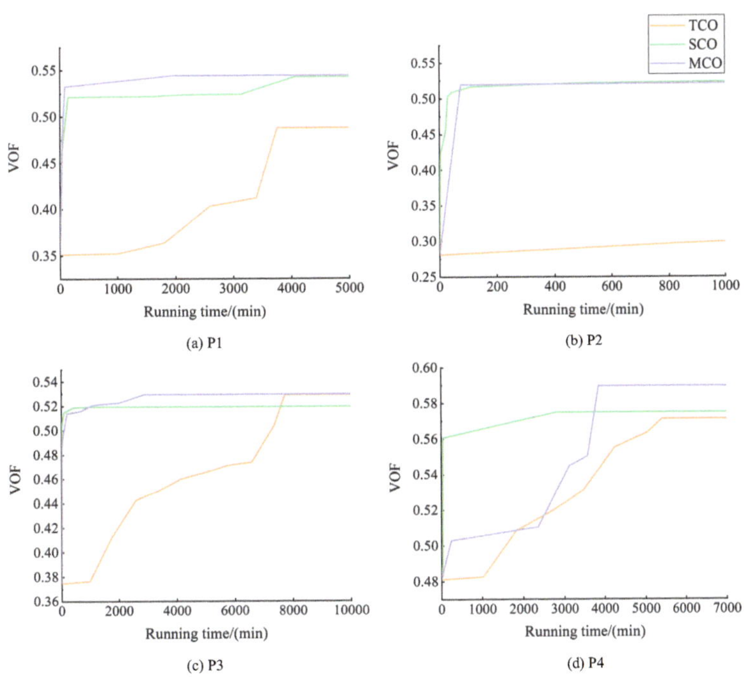

Figure 9. Running time of the optimization algorithm.

4. Discussion

Optimizing FSS is crucial for sustainable forest management, and developing a scientific and efficient quantitative model for FSS and enhancing model-solving efficiency are urgent challenges. Selective cutting and replanting are common measures for FSS optimization. However, the TCO model faces limitations such as not flexible replanting positions and difficulty in simultaneously optimizing selective cutting and replanting. To address these issues, this study proposes a new optimization scheme, aiming to introduce a multi-agent co-optimization strategy in RL to achieve the collaborative optimization of selective cutting and replanting for FSS. Simulated optimization experiments were conducted using data from four circular sample plots of *Pinus yunnanensis* plantation secondary forest on the eastern slope of Cangshan Mountain in Dali City, Yunnan Province, China, to verify the effectiveness and feasibility of this new scheme.

4.1. Superiority of RFI

The results of simulation optimization using different replanting location determination methods demonstrate that the RFI-based method can achieve better VOFs with fewer replanting iterations. This highlights the practical superiority of the RFI-based method in application schemes.

The basic MDGAM only considers the growth space of replanted trees in replanting optimization, selecting the maximum "forest gap" as the optimal location for replanting without considering other FSS indicators. In contrast, the RFI-based replanting optimization comprehensively considers various factors affecting the growth of replanted trees, including the relative coordinates of replanted trees and their neighbors, mingling degree, effects of replanting species, and tree age represented by DBH size ratio, as well as canopy competition index to account for tree crown and height effects. By incorporating multiple stand structure indices into the objective function, the RFI-based method aims to maximize VOF by improving the quality of as many stand structure indices as possible. Therefore, the RFI offers a significant advantage in replanting optimization for enhancing FSS optimization.

4.2. TCO and CO

In both P1 and P2, despite utilizing the optimal FSS adjusted after harvesting for replanting optimization experiments, the best replanting outcomes were not achieved. However, the $FSSACA_{2nd}$ and $FSSACA_{3rd}$ led to more substantial improvements in replanting optimization. This indicates that relying solely on the results from $FSSACA_{1st}$ may not always yield the best overall optimization outcomes, suggesting the need for more comprehensive optimization schemes. Conversely, in P3, replanting based on $FSSACA_{1st}$ achieved double optimization by combining selective cutting and replanting, resulting in the most optimal effect. This aligns with findings from other studies [17], emphasizing the collaborative synergy between selective logging and replanting. In P4, replanting optimization experiments based on FSS after $FSSACA_{1st}$ and $FSSACA_{2nd}$ did not yield expected results, whereas $FSSACA_{3rd}$ achieved the best replanting optimization effect. This further highlights the potential limitations of relying solely on $FSSACA_{1st}$ results, emphasizing the need for more flexible consideration of alternative optimization paths.

In simulated co-optimization experiments across the four sample plots, SARL outperformed the TCO model in terms of VOFs in all plots except P3. Moreover, the average enhancement of FSS achieved by SARL was higher than that of the TCO model. Notably, the optimization results of MARL across the sample plots were significantly superior to those of TCO and SARL. MARL exhibited a greater enhancement of FSS compared to the other methods. This underscores the exceptional performance of MARL in cooperative optimization, offering an effective approach for tackling complex FSS optimization problems.

Overall, relying solely on $FSSACA_{1st}$ results for replanting optimization may not always lead to the best outcomes in FSS cutting and replanting optimization. The TCO model may exhibit limitations in certain scenarios, necessitating the adoption of more comprehensive and flexible optimization schemes tailored to specific circumstances. SARL and MARL demonstrate significant advantages over the TCO model in solving collaborative FSS optimization problems, providing novel insights and methodologies for addressing such challenges.

4.3. Changes in Stand Structure Parameters and Algorithm Performance

After the dual optimization of selective logging and replanting, several improvements in the FSS were observed. These include the maintenance of tree diameter classes, an increase in tree species diversity, closer adherence to a random distribution pattern, improved spatial segregation of tree species, reduced competitive pressure among trees, enriched vertical structure, and enhanced understory light conditions. Overall, under the applied constraints, the FSS showed improvements compared to its pre-optimization state, validating the efficacy of RL in addressing the multi-objective logging and replanting co-optimization problem for FSS. Additionally, the study confirmed that the optimization approach of selective logging followed by replanting, as seen in the TCO model, effectively addresses such challenges, consistent with findings from previous research [17].

In terms of algorithm performance, the MARL optimization scheme consistently yielded higher maximum VOFs compared to the TCO and SARL schemes, indicating superior FSS optimization outcomes with the multi-agent approach. Furthermore, the MARL algorithm required significantly fewer iterations to converge to the optimal solution compared to SARL and TCO schemes, underscoring its faster convergence speed and enhanced efficiency. SARL exhibited a tendency towards local optima, necessitating a higher number of iterations to achieve desired results. Conversely, the TCO scheme demonstrated a slower convergence speed overall, requiring more iterations to reach the optimal solution. Regarding algorithmic time consumption, MARL performed comparably to SARL and outperformed the TCO scheme, suggesting its ability to maintain high efficiency and computational performance.

Overall, the MARL approach exhibits clear advantages over the TCO and SARL schemes in solving the optimization problem of forest stand structure through cutting and replanting combinations. Its faster convergence speed, superior FSS optimization outcomes, and competitive computational efficiency highlight its effectiveness in addressing complex optimization challenges in forestry management.

5. Conclusions

Although our previous study successfully applied SARL to the multi-objective optimization of FSS, it only considered single cutting measures and did not integrate the synergistic effects of multiple regulatory measures in the model construction and solution process. While SARL has advantages in solving single cutting optimization models, it is less effective in addressing models that require the coordination of multiple measures.

This study introduces the RFI, which comprehensively considers various factors influencing the growth of replanted trees, enhancing the accuracy and effectiveness of replanting location determination. The RFI-based method demonstrates superior performance in optimizing FSS. Moreover, this research pioneers the application of MARL to collaborative optimization of FSS cutting and replanting. By leveraging the collaborative strategy of MARL, the study successfully achieves collaborative optimization of logging and replanting in FSS. In comparison to TCO and SARL, MARL exhibits significant advantages in solving this problem. The detailed comparison of experimental results quantitatively verifies the superiority of MARL in collaborative optimization of stand structure cutting and replanting, providing strong theoretical support for addressing stand structure optimization challenges in actual forest management scenarios.

While this study successfully applied MARL to the multi-objective optimization of stand structure, the experimental results may not fully reflect real-world forest conditions due to the specific sample data and simulation settings used. Additionally, although the optimization scheme based on MARL showed better results, there is still room for performance improvement. Based on the above research results and shortcomings, future research can focus on the following aspects:

(1) To enhance the accuracy and efficiency of MARL decisions, it is essential to integrate GIS (Geographic Information System) data and tools to obtain more detailed three-dimensional spatial information and accurate canopy calculations. By combining GIS with MARL, it is possible to construct constraints or objective functions that incorporate spatial three-dimensional features. This integration ensures that MARL makes more informed and precise decisions.

(2) Improving and optimizing the algorithms and techniques used in the stand structure optimization scheme based on MARL. Enhancing its performance and efficiency in processing large-scale data and complex environments will improve its generalization ability and practical applicability.

(3) Explore and design more flexible and integrated multi-dimensional optimization strategies to balance various objectives and interests in forest management, achieving a more comprehensive and sustainable optimization of stand structure. For instance, design more efficient and appropriate methods for determining the felling and replanting trees.

Author Contributions: S.X.: Writing—original draft, Conceptualization, Data curation, Methodology, Formal analysis, Investigation. J.W.: Writing—review & editing, Visualization, Supervision, Methodology, Funding acquisition, Data curation, Conceptualization. J.Y.: Data curation, Validation, Software, Visualization. Y.C.: Validation, Software, Visualization. B.W.: Writing—review & editing, Validation, Software, Visualization. All authors have read and agreed to the published version of the manuscript.

Funding: This research was funded by Yunnan Fundamental Research Projects (grant NO.202201AT 070006), and Yunnan Postdoctoral Research Fund Projects (grant NO.ynbh20057).

Data Availability Statement: The original contributions presented in the study are included in the article. Further inquiries can be directed to the corresponding author.

Conflicts of Interest: The authors declare no conflict of interest.

Appendix A

Appendix A.1. Algorithmic Pseudocode of TCO

The Algorithm A1 as follows:

Algorithm A1: TCO for MOFSS

```
Input  :Set of retained trees after cutting g*
Output:The optimal solution g and the corresponding value of objective function f(g)
Read the set of retained trees after cutting optimization g* ;
Set the initial number of replanting trees n = 1 ;
Construct spatial structure units and correct edges of sample plots ;
Obtain replanting locations based on RFI and construct the set of replanted trees replant_set ;
Calculate the initial value of objective function f(g*) ;
while TRUE do
    Select the first n replanting trees from the replanting set and construct a new set of retained trees g ;
    Calculate the new value of objective function f(g) and judge whether it meets the constraints ;
    if f(g) > f(g*) and meet all constraints then
        Save the current feasible solution g* = g, f(g*) = f(g) ;
    end if
    else
        if Current plant density > 3333 then
            Output the optimal solution g and the corresponding value of objective function f(g)
        end if
    end if
    Increase the number of replanting plants n
end while
```

Appendix A.2. Algorithmic Pseudocode of SARL

The Algorithm A2 as follows:

Algorithm A2: SARL for MOFSS

```
Initialize states S, actions A, ε-greedy policy EPSILON, learning rate α, discount factor γ,
  maximum episodes MAXEPISODES, and Q(s, a), where s ∈ S and a ∈ A. Set initial s
  and a to 0.;
for episode = 1 to MAXEPISODES do
    Initialize s;
    if ACTION is cutting and replanting then
        Choose a from s using policy derived from Q (ε-greedy);
        Take action a, observe reward r, and next state s′;
    end if
    if S = NSTATES − 2 then
        Set S1 to 'terminal';
        Set reward R to d;
    end if
    else Selective Cutting and Replanting;
    Use R program to partition Voronoi diagram, calculate structural parameters, and
      objective values of FSS after selective cutting and replanting;
    if VOF > VOF* then
        Set S1 to S + 1;
        Set reward R to a;
        Document selective cutting tree number and values of OF after selective cutting and
          replanting;
    end if
    else if VOF = VOF* then
        Set S1 to S;
        Set reward R to b;
    end if
    else
        Set reward R to c;
        if S = 0 then
            Set S1 to S;
        end if
        else Set S1 to S − 1;
        ;
    end if
    ;
end for
else if S = NSTATES − 2 then
    Set S1 to 'terminal';
    Set reward R to d;
end if
else Set S1 to S + 1;
Set reward R to c;
;
;
Q(s, a) ← Q(s, a) + α[r + γ max_a′ Q(s′, a′) − Q(s, a)];
s ← s′;
```

Appendix A.3. Algorithmic Pseudocode of MARL

The Algorithm A3 as follows:

Algorithm A3: MARL for MOFSS

```
Initialize states S1, S2, actions A1, A2, ϵ-greedy policy EPSILON, learning rate α, discount factor γ, maximum
  episodes MAXEPISODES, and Q1(s1, a1), Q2(s2, a2), where s1 ∈ S1, a1 ∈ A1, s2 ∈ S2 and a2 ∈ A2. Set initial
  s1, a1, s2 and a2 to 0.;
for episode = 1 to MAXEPISODES do
    Initialize s1 for Agent1 and initialize s2 for Agent2;
    for Agent1 do
        if ACTION1 is cutting then
            Choose a1 from s1 using policy derived from Q1 (ϵ-greedy);
            Take action a1, observe reward r1, and next state s1';
        end if
        if S1 >= S2 then
            Set S1 to 'terminal';
            Set reward R1 to d;
        end if
        else Selective Cutting;
        Use R program to partition Voronoi diagram, calculate structural parameters, and objective values of
          FSS after selective cutting;
        if VOF1 > VOF* then
            Set S1 to S + 1;
            Set reward R1 to a;
            Document selective cutting tree number and values of OF after selective cutting;
        end if
        else if VOF1 = VOF* then
            Set S1 to S;
            Set reward R1 to b;
        end if
        ;
        Q1(s1, a1) ← Q1(s1, a1) + α[r + γ max_a1' Q1(s1', a1') − Q1(s1, a1)];
        s1 ← s1';
    end for
    for Agent2 do
        if ACTION2 is planting then
            Choose a2 from s2 using policy derived from Q2 (ϵ-greedy);
            Take action a2, observe reward r2, and next state s2';
        end if
        if S1 >= S2 then
            Set S2 to 'terminal';
            Set reward R2 to d2;
        end if
        else Replanting;
        Use R program to partition Voronoi diagram, calculate structural parameters, and objective values of
          FSS after replanting;
        if VOF_l<VOF_m<VOF_r or VOF_l>VOF_m>VOF_r then
            if max(VOF_m l, VOF_m r) > max(VOF_l, VOF_m, VOF_r) then
                Set S2 to S2 − 1;
                Set reward SmallR2_1 to a2;
            end if
            else
                Set S2 to S2 − 1;
                Set reward SmallR2_2 to a2;
            end if
        end if
        else if VOF_l<VOF_m and VOF_m>VOF_r then
            if max(VOF_m l, VOF_m r) > max(VOF_l, VOF_m, VOF_r) then
                Set S2 to S2 − 1;
                Set reward SmallR2_1 to a2;
            end if
            else
                Set S2 to S2 − 2;
                Set reward LargeR2_2 to a2;
            end if
        end if
        else if VOF_l>VOF_m and VOF_m<VOF_r then
            if max(VOF_m l, VOF_m r) > max(VOF_l, VOF_m, VOF_r) then
                Set S2 to S2 + 1;
                Set reward SmallP2 to a2;
            end if
            else
                Set S2 to S2 + 1;
                Set reward LargeP2 to a2;
            end if
        end if
        Document selective cutting tree number and values of OF after replanting;
        ;
        Q2(s2, a2) ← Q2(s2, a2) + α[r + γ max_a2' Q(s2', a2') − Q2(s2, a2)];
        s2 ← s2';
    end for
end for
```

References

1. Jiang, L.; Lu, Y.C.; Liao, S.X.; Li, K.; Li, G.Q. A study on diameteral structure of Yunnan pine forest in the plateaus of mid-Yunnan Province. *For. Res.* **2008**, *21*, 126–130.
2. Zaizhi, Z. Status and perspectives on secondary forests in tropical China. *J. Trop. For. Sci.* **2001**, *13*, 639–651.
3. Li, L.F.; Han, M.Y.; Zheng, W.; Su, J.W.; Li, W.C.; Zheng, S.H.; Gong, J.B. The causes of formation of low quality forest of *Pinus Yunnanensis* Their Classif. *J. West China For. Sci.* **2009**, *38*, 94–99.
4. Wu, J.C. Study on Structure and Adjustment of Larch and Spruce-Fir Stands. Master's Thesis, Chinese Academy of Forestry: Beijing, China, 2008.
5. Cao, Z. Research on Stand Structure Regulation Technology Based on Forest Health. Ph.D. Thesis, Beijing Forestry University: Beijing, China, 2020.
6. de Lima, R.A.F.; Batista, J.L.F.; Prado, P.I. Modeling tree diameter distributions in natural forests: An evaluation of 10 statistical models. *For. Sci.* **2015**, *61*, 320–327. [CrossRef]
7. Zeide, B. How to measure stand density. *Trees* **2005**, *19*, 1–14. [CrossRef]
8. Petritan, A.M.; Biris, I.A.; Merce, O.; Turcu, D.O.; Petritan, I.C. Structure and diversity of a natural temperate sessile oak (*Quercus petraea* L.)–European Beech (*Fagus sylvatica* L.) forest. *For. Ecol. Manag.* **2012**, *280*, 140–149. [CrossRef]
9. Bettinger, P.; Tang, M.P. Tree-level harvest optimization for structure-based forest management based on the species mingling index. *Forests* **2015**, *6*, 1121–1144. [CrossRef]
10. Bao, Q.; Liang, X.W.; Weng, G.S. The Concept of Tree Crown Overlap and its Calculating Methods of Area. *J. Northeast For. Univ.* **1995**, *23*, 103–109.
11. Wang, J.M. Study on Decision Technology of Tending Felling for Larix Principis-Rupprechtii Plantation Forest. Ph.D. Thesis, Beijing Forestry University: Beijing, China, 2017.
12. Zhang, G.Q.; Hui, G.Y.; Zhao, Z.H.; Hu, Y.B.; Wang, H.X.; Liu, W.Z.; Zang, R.G. Composition of basal area in natural forests based on the uniform angle index. *Ecol. Inform.* **2018**, *45*, 1–8. [CrossRef]
13. Wan, P.; Zhang, G.Q.; Wang, H.X.; Zhao, Z.H.; Hu, Y.B.; Zhang, G.G.; Hui, G.Y.; Liu, W.Z. Impacts of different forest management methods on the stand spatial structure of a natural Quercus aliena var. acuteserrata forest in Xiaolongshan, China. *Ecol. Informatics* **2019**, *50*, 86–94. [CrossRef]
14. Lu, Y.; Zang, H.; Wan, X.J.; Deng, Z.A.; Li, J.J. Storey structure study of Cyclobalanopsis myrsinaefolia mixed stand based on storey index. *For. Resour. Wanagement* **2012**, , 81–84. [CrossRef]
15. Joshi, C.; De Leeuw, J.; Skidmore, A.K.; Van Duren, I.C.; Van Oosten, H. Remotely sensed estimation of forest canopy density: A comparison of the performance of four methods. *Int. J. Appl. Earth Obs. Geoinf.* **2006**, *8*, 84–95. [CrossRef]
16. Zhao, Z.H.; Hui, G.Y.; Hu, Y.B.; Li, Y.F.; Wang, H.X. Method and application of stand spatial advantage degree based on the neighborhood comparison. *J. Beijing For. Univ.* **2014**, *36*, 78–82.
17. Liu, S. Research on the Analysis and Multi-Objective Intelligent Optimization of Stand Structure of Natural Secondary Forest. Ph.D. Thesis, Central South University of Forestry and Technology: Changsha, China, 2017.
18. Zhao, C.Y.; Li, J.P. Spatial location and allocation of replanting trees on pure Chinese fir plantation based on Voronoi diagram and Delaunay triangulation. *J. Cent. South Univ. For. Technol.* **2017**, *37*, 1–8.
19. Liu, J.; Xiang, J.; Yu, K.Y.; Yang, S.P.; Zhang, J.Z.; Zhao, Q.Y. Simulation of Replantation of Low-density Ecological Landscape Forest With Coupled Stand Structure. *Acta Agric. Univ. Jiangxiensis* **2018**, *40*, 1125–1133.
20. Li, J.J. Research on the Spatial Structure Optimization of Mangrove Ecosystem in Zhanjiang, Guangdong. Ph.D. Thesis, Central South University of Forestry and Technology: Changsha, China, 2010.
21. Hui, G.Y.; Hu, Y.B.; Zhao, Z.H. Research progress of structure-based forest management. *For. Res.* **2018**, *31*, 85–93.
22. Dong, L.B.; Wei, H.Y.; Liu, Z.G. Optimizing forest spatial structure with neighborhood-based indices: Four case studies from northeast China. *Forests* **2020**, *11*, 413. [CrossRef]
23. Hu, Y.B. Structure-Based Spatial Optimization Management Model for Natural Uneven-Aged Forest. Ph.D. Thesis, Chinese Academy of Forestry: Beijing, China, 2010.
24. Tang, M.P.; Tang, S.Z.; Lei, X.D.; Li, X.F. Study on Spatial Structure Optimizing Model of Stand Selection Cutting. *Sci. Silvae Sin.* **2004**, *40*, 25–31.
25. Li, J.J.; Zhang, H.R.; Liu, S.; Kuang, Z.F.; Wang, C.L.; Zang, H.; Cao, X.P. A space optimization model of water resource conservation forest in Dongting Lake based on improved PSO. *Acta Ecol. Sin.* **2013**, *33*, 4031–4040.
26. Nelson, J.; Finn, S. The influence of cut-block size and adjacency rules on harvest levels and road networks. *Can. J. For. Res.* **1991**, *21*, 595–600. [CrossRef]
27. Haight, R.G.; Travis, L.E. Wildlife conservation planning using stochastic optimization and importance sampling. *For. Sci.* **1997**, *43*, 129–139. [CrossRef]
28. Barrett, T.; Gilless, J.; Davis, L. Economic and fragmentation effects of clearcut restrictions. *For. Sci.* **1998**, *44*, 569–577. [CrossRef]
29. Boston, K.; Bettinger, P. An analysis of Monte Carlo integer programming, simulated annealing, and tabu search heuristics for solving spatial harvest scheduling problems. *For. Sci.* **1999**, *45*, 292–301. [CrossRef]
30. Bettinger, P.; Graetz, D.; Boston, K.; Sessions, J.; Chung, W. Eight heuristic planning techniques applied to three increasingly difficult wildlife planning problems. *Silva Fenn.* **2002**, *36*, 561–584. [CrossRef]

31. Zeng, H.; Pukkala, T.; Peltola, H. The use of heuristic optimization in risk management of wind damage in forest planning. *For. Ecol. Manag.* **2007**, *241*, 189–199. [CrossRef]
32. Moore, C.T.; Conroy, M.J.; Boston, K. Forest management decisions for wildlife objectives: System resolution and optimality. *Comput. Electron. Agric.* **2000**, *27*, 25–39. [CrossRef]
33. Öhman, K.; Lämås, T. Clustering of harvest activities in multi-objective long-term forest planning. *For. Ecol. Manag.* **2003**, *176*, 161–171. [CrossRef]
34. Crowe, K.A.; Nelson, J. An evaluation of the simulated annealing algorithm for solving the area-restricted harvest-scheduling model against optimal benchmarks. *Can. J. For. Res.* **2005**, *35*, 2500–2509. [CrossRef]
35. Lichtenstein, M.E.; Montgomery, C.A. Biodiversity and timber in the Coast Range of Oregon: Inside the production possibility frontier. *Land Econ.* **2003**, *79*, 56–73. [CrossRef]
36. Hu, C.Y.; Yao, H.; Yan, X.S. Multiple Particle Swarms Coevolutionary Algorithm for Dynamic Multi-Objective Optimization Problems and Its Application. *J. Comput. Res. Dev.* **2013**, *50*, 1313–1323.
37. Levine, S.; Finn, C.; Darrell, T.; Abbeel, P. End-to-end training of deep visuomotor policies. *J. Mach. Learn. Res.* **2016**, *17*, 1334–1373.
38. Dali Bai Autonomous Prefecture Cangshan Protection and Administration Bureau. *Cangshan Chronicle*; Yunnan Nationalities Publishing House: Kunming, China, 2008.
39. Yang, Z.H. Research of pH in the Er Hai plateau Lake of Western Part of Yunnan Province. *Environ. Sci. Surv.* **1997**, *16*, 22–25.
40. Yuan, R.J.; Yang, S.H.; Wang, B.R. Study on the altitudinal pattern of vegetation distribution along the eastern slope of Cangshan Mountain, Yunnan. *J. Yunnan Univ. Nat. Sci. Ed.* **2008**, *30*, 318–325.
41. Liao, J.B.; Li, Z.Q.; Quets, J.J.; Nijs, I. Effects of space partitioning in a plant species diversity model. *Ecol. Model.* **2013**, *251*, 271–278. [CrossRef]
42. Magnussen, S.; Allard, D.; Wulder, M.A. Poisson Voronoi tiling for finding clusters in spatial point patterns. *Scand. J. For. Res.* **2006**, *21*, 239–248. [CrossRef]
43. Sterner, R.W.; Ribic, C.A.; Schatz, G.E. Testing for life historical changes in spatial patterns of four tropical tree species. *J. Ecol.* **1986**, *74*, 621–633. [CrossRef]
44. Xuan, S.; Wang, J.; Chen, Y.L. Reinforcement Learning for Stand Structure Optimization of Pinus yunnanensis Secondary Forests in Southwest China. *Forests* **2023**, *14*, 2456. [CrossRef]
45. Xiang, B.W. Study on the Structure Adjustment and Optimization of Quercus Secondary Forest in Hunan Province. Master's Thesis, Central South University of Forestry and Technology: Changsha, China, 2019.
46. Su, J.W.; Li, L.F.; Zheng, W.; Yang, W.B.; Han, M.Y.; Huang, Z.M.; Xu, P.B.; Feng, Z.W. Effect of Intermediate Cutting Intensity on Growth of *Pinus yunnanensis* Plant. *J. West China For. Sci.* **2010**, *39*, 27–32.
47. Han, M.Y.; Li, L.F.; Zheng, W.; Su, J.W.; Li, W.C.; Gong, J.B.; Zheng, S.H. Effects of different intensity of thinning on the improvement of middle-aged Yunnan pine stand. *J. Cent. South Univ. For. Technol.* **2011**, *31*, 27–33.
48. Zhang, W.; Wu, X.L.; Jiang, S.N.; Liu, Y.; Weng, G.Q.; Shen, Y.C.; Yang, H.; Wang, H.C.; Liu, C.; Zhou, J.M.; et al. *Technical Regulation for Forestation*; Technical Report GB/T 15776-2023, State Administration for Market Regulation; Standardization Administration: Beijing, China, 2023.
49. Yu, Y.T. Study on Forest Structure of Different Recovery Stages and Optimization Models of Natural Mixed Spruce-Fir Secondary Forests on Selective Cutting. Ph.D. Thesis, Beijing Forestry University: Beijing, China, 2019.
50. Hao, Y.L. Study on Cutting Tree Determining Method Based on Forest Stand Spatial Structure Optimization. Ph.D. Thesis, Chinese Academy of Forestry: Beijing, China, 2012.
51. Sun, Y.; Liu, S.; Wang, S.J.; Zhao, S.B.; Li, E.P.; Luo, J.; Tian, J.X.; Cheng, F.S. Analysis and Evaluation of Forest Spatial Structure Using Weighted Delaunays Triangle Network. *J. Northeast For. Univ.* **2022**, *50*, 61–68.
52. Konak, A.; Coit, D.W.; Smith, A.E. Multi-objective optimization using genetic algorithms: A tutorial. *Reliab. Eng. Syst. Saf.* **2006**, *91*, 992–1007. [CrossRef]
53. Gunantara, N. A review of multi-objective optimization: Methods and its applications. *Cogent Eng.* **2018**, *5*, 1502242. [CrossRef]
54. Hillier, F.S.; Lieberman, G.J. *Introduction to Operations Research*; McGraw-Hill: New York, NY, USA, 2015.
55. Yang, Z.Q.; Xie, Y.P. Discussion on the Optimum Retention Density of Natural Forest of Pinus Yunnanensis. *J. Sichuan For. Sci. Technol.* **1998**, *19*, 70–72.
56. Li, Z.Z. Seedling Raising and Afforestation Techniques of Pinus Yunnanensis. *Agric. Technol. Equip.* **2022**, 104–105+108. [CrossRef]

forests

MDPI

Article

The Many Shades of the Vegetation–Climate Causality: A Multimodel Causal Appreciation

Yuhao Shao [1], Daniel Fiifi Tawia Hagan [2], Shijie Li [3], Feihong Zhou [4], Xiao Zou [5] and Pedro Cabral [6,*]

1 School of Geographical Sciences, Nanjing University of Information Science & Technology, Nanjing 210044, China; yuhaoshao@nuist.edu.cn
2 Hydro-Climate Extremes Laboratory, Ghent University, 9000 Ghent, Belgium
3 Department of Civil and Environmental Engineering, University of Florence, 50139 Firenze, Italy; 15850732387@163.com
4 Collaborative Innovation Center on Forecast and Evaluation of Meteorological Disasters, Nanjing University of Information Science and Technology, Nanjing 210044, China
5 Tai'an Meteorological Bureau, Tai'an 271000, China
6 School of Remote Sensing and Geomatics Engineering, Nanjing University of Information Science & Technology, Nanjing 210044, China
* Correspondence: cabral@nuist.edu.cn

Abstract: The causal relationship between vegetation and temperature serves as a driving factor for global warming in the climate system. However, causal relationships are typically characterized by complex facets, particularly within natural systems, necessitating the ongoing development of robust approaches capable of addressing the challenges inherent in causality analysis. Various causality approaches offer distinct perspectives on understanding causal structures, even when experiments are meticulously designed with a specific target. Here, we use the complex vegetation–climate interaction to demonstrate some of the many facets of causality analysis by applying three different causality frameworks including (i) the kernel Granger causality (KGC), a nonlinear extension of the Granger causality (GC), to understand the nonlinearity in the vegetation–climate causal relationship; (ii) the Peter and Clark momentary conditional independence (PCMCI), which combines the Peter and Clark (PC) algorithm with the momentary conditional independence (MCI) approach to distinguish the feedback and coupling signs in vegetation–climate interaction; and (iii) the Liang–Kleeman information flow (L-K IF), a rigorously formulated causality formalism based on the Liang–Kleeman information flow theory, to reveal the causal influence of vegetation on the evolution of temperature variability. The results attempt to capture a fuller understanding of the causal interaction of leaf area index (LAI) on air temperature (T) during 1981–2018, revealing the characteristics and differences in distinct climatic tipping point regions, particularly in terms of nonlinearity, feedback signals, and variability sources. This study demonstrates that realizing a more holistic causal structure of complex problems like the vegetation–climate interaction benefits from the combined use of multiple models that shed light on different aspects of its causal structure, thus revealing novel insights that are missing when we rely on one single approach. This prompts the need to move toward a multimodel causality analysis that could reduce biases and limitations in causal interpretations.

Keywords: vegetation–atmosphere interactions; kernel Granger causality; Peter and Clark momentary conditional independence; Liang–Kleeman information flow; nonlinear land–atmosphere coupling; positive and negative feedback

Citation: Shao, Y.; Hagan, D.F.T.; Li, S.; Zhou, F.; Zou, X.; Cabral, P. The Many Shades of the Vegetation–Climate Causality: A Multimodel Causal Appreciation. *Forests* **2024**, *15*, 1430. https://doi.org/10.3390/f15081430

Academic Editor: Gerhard Wieser

Received: 14 July 2024
Revised: 31 July 2024
Accepted: 1 August 2024
Published: 14 August 2024

1. Introduction

Vegetation serves as a vital life support system for human societies, interconnecting soil, atmosphere, and water; however, its interactions within the climate system are very complex [1]. On one hand, global warming profoundly impacts vegetation ecosystems through temporal and spatial changes in climate elements such as water and energy, along side the CO_2 fertilization effect [2], which contains interactions and potential processes among various influencing factors, contributing to the complexity of the mechanisms by which climate affects vegetation. On the other hand, vegetation ecosystems modulate the climate system through biophysical and chemical processes such as growth, greening, senescence, transpiration, and photosynthesis, thereby amplifying or mitigating signals of climate [3], where the response of vegetation to the climate system is relatively weak and difficult to capture. The interaction between vegetation and climate is thus marked by a complex, bidirectional feedback, with a notable emphasis on the influence of vegetation on temperature dynamics [4]. The predominant effect of climate change on vegetation ecosystems is well documented [5], whereas the reciprocal influence of vegetation on climate change remains underexplored and presents analytical challenges [6]. Methodologically, correlation and regression analysis methods are commonly employed within ecology and geographical sciences to assess the impacts of climate change on vegetation [7–10], but these methods often falter in clarifying the bidirectional "driving-response" relationship between vegetation and climate, leading to debates and controversies [11]. Climate research, on the other hand, leans toward numerical simulations to decipher vegetation's feedback to the climate system [12,13], which depend heavily on model parameterization schemes, creating variability in analytical outcomes and obstructing consensus [14].

In light of the aforementioned challenges, causality inferences can either quantitatively and/or qualitatively through interventions unravel cause–effect relationships between various subsystems of the climate [15–17]. However, a faithful causal inference is hindered by the complexity of cause–effect relationships, which are characterized by linear and nonlinear relationships, different feedback forms, time- and frequency-dependent components, and sufficient asymmetry needed to untangle the causes from the effects [18]. A good overview of these challenges is presented in [19]. Different causal frameworks have been developed over the years to tackle the different challenges mentioned above and beyond, and while most of them have seen successful applications, each approach is often blind to its limitations. This may indicate the need to move toward a multimodel approach where different approaches with different strengths and limitations are used in tandem to unravel a more comprehensive causal structure.

A classical qualitative causality test approach is the Granger causality (GC) [20], which is a statistical hypothesis test for determining whether one time series is useful in forecasting another. By definition, a variable x (Granger) causes another variable y if the knowledge of y improves the autoregressive forecast of x. Since its development, GC has found a wide application in the fields of finance [21–24], brain science [25–27], and climate science [28–30]. For instance, Friston et al. [31] pioneered the detection and estimation of directed connections in neural networks through GC. Kovács et al. [32] employed GC to reveal the underlying causal relationships between environmental drivers and global vegetation attributes, highlighting soil moisture and the availability of accumulated precipitation as critical drivers of vegetation dynamics. Attanasio et al. [33] utilized GC to investigate the relationship between anthropogenic CO_2 emissions and global temperature, uncovering statistical evidence that the increase in CO_2 content leads to a rise in global temperature. Dhamala et al. [34] proposed a time-frequency-based GC to explore multiscale relationship between various variables. However, the assumption of linearity in these earlier forms restricts the scope of Granger causality applications to linear cases. Yet, complex nonlinear feedback relationships often exist in the climate datasets including the vegetation–temperature interaction, prompting the need to develop new methods for inferring nonlinear causal relationships. In that endeavor, Marinazzo et al. [35] applied the theory of reproducing kernel Hilbert spaces to Granger causality analysis to generalize a nonlinear form of GC. This approach was applied to understand the linear and nonlinear characteristics between different variables. Bueso et al. [36] extended kernel Granger causality (KGC) to include timescale applications and applied it to the interactions between El Niño–Southern Oscillation and soil moisture (ENSO-SM).

Land–atmosphere interactions are also characterized by positive and negative feedback signs, which are important for trend analysis in climate signs [37,38]. In general, statistical correlations are often used to represent positive and negative feedback [8]. However, although correlation is a necessary prerequisite for causality, it does not imply causation [39]. Thus, a causal inference model based on a Bayesian network was proposed, by first establishing a Markov graph with correlation to determine the links between network nodes and then through a series of intervention and statistical tests, extract the causal structure from full-time graph, realizing the leap from correlation to causal inference [16]. On this basis, the Peter and Clark momentary conditional independence (PCMCI) method combines the Peter and Clark (PC) algorithm [40] with the momentary conditional independence (MCI) approach to obtain causal graphs [19,41–43]. PCMCI has been widely used in many geological and climatological studies [44,45]. For example, Krich et al. [46] utilized PCMCI methods to estimate causal networks in biosphere–atmosphere interactions. Capua et al. [47] used PCMCI to study tropical and mid-latitude teleconnections interacting with Indian summer monsoon rainfall. Qu et al. [48] also employed PCMCI methods to detect the lagged and contingent relationships between wildfire burn area and drought patterns alongside vegetation conditions.

Beyond the above methods, information theoretic (IT) approaches also provide ways to assess causality between time series quantitatively [49]. Schreiber [50] derived an IT measure, the transfer entropy (TE), an asymetric conditional variation of the entropy-based statistics known as mutual information, which has been applied to distinguish drivers from responses within interactions [51,52]. Additionally, IT methods based on entropy, in the Shannon sense [53], provide a way to assess causal histories in systems [54]. Ruddell used mutual information the transfer entropy to assess the cause–effect relationships within a network of ecohydrological variables.

Liang et al. [49] proposed a rigorously formulated causality formalism, from first principles, based on the Liang–Kleeman information flow theory (L-K IF) where causality is quantitatively realized. This has been verified in well-known theoretical models such as the Rossler and Lorenz systems [55], as well its maiden real-world application to reveal the causal relationship between El Nino phenomenon and the Indian Ocean dipole (IOD). Since then, the L-K IF causality has been applied to several areas of research interest in climate science [56], finance [57], and psychology [58], among others, to unravel cause–effect relationships. For instance, Stips et al. [59] used this method to study the causal relationship between greenhouse gas emissions and global warming and found that the increase in the CO_2 concentration in the past 150 years led to the rise of global surface temperature. Hagan et al. [18] applied Kalman filtering and wavelet analysis techniques to extend the Liang–Kleeman information flow method to the time-frequency domain and analyzed the bidirectional mutual feed mechanism between soil moisture and air temperature. Tao et al. [60] used the information flow method to quantitatively assess the effects of climate warming and Atlantic and Pacific decadal oscillations on global precipitation and their regional differences. Docquier et al. [61] applied the Liang–Kleeman information flow method to identify the potential causal drivers of the Arctic sea ice. They found that recent and future changes in Arctic sea ice are primarily driven by air and sea surface temperatures, as well as ocean heat transport, which in turn significantly influence temperature and ocean heat transport dynamics. More recently, Zhou et al. [62] proposed a time-dependent form of the formalism to the multivariate form of the LK causality and applied it to reveal novel findings of time-varying causal structures between soil moisture, vapor pressure deficit, and gross primary productivity.

Understanding natural systems, particularly the climate, is challenging due to the complexity of the systems and the limitations in causal inquiry methods and observations [1]. In scientific research, identifying reliable and relatively certain patterns often relies on extracting and summarizing objective facts [19]. Furthermore, historical observational data provide authentic information and evidence, while causality analysis methods offer rigorous and reliable data analysis and exploration tools [19]. Like many complex natural interactions, the holistic causal structure of the vegetation–climate interaction may not be sufficiently realized with only one approach, since each method comes with its unique strengths and limitations as shown in Figure 1.

Figure 1. The many shades of the vegetation–climate causality. That is, the vegetation–temperature interaction mechanism will be revealed from different causal aspects, where LAI is the leaf area index; T is the temperature; and KGC (kernel Granger causality), PCMCI (Peter and Clark momentary conditional independence), and L-K IF (Liang–Kleeman information flow) are three different causal analysis methods.

In conducting our exploration of the causal relationships between vegetation and climate, we recognize that different causal analysis methods illuminate these causal relationships from varied aspects [63], as shown in Figure 1. This is because causality analysis of natural systems is a very complex problem that has many facets to it. As a result, different causality methods may only provide skewed perspectives of the full causal structure such as the nature of the temporal delays [64,65], degrees of nonlinearity [29], and feedback signs [4,46], among others. Consequently, relying on a single approach limits interpretation of the cause–effect relationship to a specific framework, making it blind to other potentially useful interpretations. This limitation stems from the fact that each method offers a unique lens, focusing on specific aspects of the causal relationship, and thus provides a partial view that falls short of encapsulating the entire complexity of the vegetation–temperature coupling (Figure 1). This study provides a blue–print on how to capture a fuller interpretation of the vegetation–climate coupling by harnessing a multimethod analysis approach.

Nonlinearity is a critical characteristic of the climate system. Therefore, we first conducted a causal analysis of the vegetation–climate system coupling based on the degree of nonlinearity as a function of location. Using the KGC, we aimed to understand global regions of nonlinearity in their causal relationship where knowledge of vegetation improves air temperature prediction. Next, the PCMCI was used to investigate the coupling signs, alongside providing potential temporal delays between the driver and response. Finally, Liang–Kleeman information flow (L-K IF) was applied to highlight regions of uncertainty and predictability in air temperature due to information flow from vegetation changes.

The remaining sections are structured as follows: Section 2 introduces the data and methods applied in this study. Section 3 presents the experimental results. Section 4 discusses and analyzes the experimental findings. Section 5 presents the conclusions.

2. Data and Methods

2.1. Data

2.1.1. CRU TS

For the temperature variable, datasets of global mean air temperature (T) monthly observations at meteorological stations across the world's land areas, acquired from the Climatic Research Unit (CRU), version TS v4.2.0 [66], were used. CRU TS v4.2.0 Mean Temperature is a high-resolution monthly grid dataset interpolated from 5583 observation stations. It covers the average temperature station data over the last 35 years and was interpolated into $0.5° \times 0.5°$ (longitude × latitude) covering the global land surface. We took the monthly mean temperature from 1981 to 2018 as the temperature variable.

2.1.2. GLASS LAI

For vegetation variables, we utilized LAI data from the Global Land Surface Satellite (GLASS) dataset [67,68]. GLASS LAI employs a generalized regression neural network to train LAI data from satellite observations and AVHRR band data from high-resolution radiometers. The trained model was then applied to estimate the LAI on a global scale. The product closely aligns with mainstream leaf area products, has a spatial resolution of 0.05 degrees and a temporal resolution of 8 days, and spanned the period from 1981 to 2018. In this study, we resampled the spatial resolution of these data to $0.5° \times 0.5°$ (longitude × latitude) to ensure consistency in spatiotemporal resolution with CRU data.

2.1.3. GLEAM ET

Evapotranspiration is very important in land-atmosphere interactions [69]. GLEAM (Global Land Evaporation Amsterdam Model) is a suit of specially designed algorithms aimed at accurately estimating global land evaporation by assimilating soil moisture and vegetation optical depth observations [70,71]. The model divides land evaporation into several key components including transpiration, canopy interception evaporation, bare soil evaporation, ice and snow sublimation and open water evaporation. Here, we employed the transpiration from GLEAM V3.6a to represent evapotranspiration from 1981 to 2018, with a monthly temporal resolution, resampled from a spatial resolution of $0.25° \times 0.25°$ to $0.5° \times 0.5°$ (longitude × latitude), consistent with the vegetation and temperature data specifications.

2.2. Methods

We utilized a series of causality-based analysis methods to construct the global vegetation–temperature causal structure as comprehensively as possible, including KGC, PCMCI, and L-K IF. The main features and differences of these methods are presented in Table 1.

Table 1. Description of the causal methods used in this paper.

Methods	Kernel Granger Causality	Peter and Clark Momentary Conditional Independence	Liang–Kleeman Information Flow
Abbreviation	KGC	PCMCI	L-K IF
Type of method	Qualitative causality	Qualitative causality	Quantitative causality
Theoretical basis	Granger Causality, spectral representation, kernel function	Conditional independence test, structure causality graph	Liang–Kleeman information flow
Use of time delays	Not by default	Always	Not by default
Use of iterative conditioning	No	Yes	No
Sign meaning [1]	No negative value	Positive value: Increases in drivers result in an increase in the target. Negative value: Increases in drivers result in an increase in the target.	Positive value: The driver functions to increase the variability of the target, thereby making it more uncertain. Negative value: The driver functions to reduce variability in the target.
Key references	Marinazzo et al. (2007) [35]	Runge et al. (2019) [19]	Liang (2014) [49]

[1] The sign meaning property of PCMCI and L-K IF is reproduced from [72], with permission from Nonlinear Processes in Geophysics, 2024.

The methodological workflow (Figure 2) comprised three steps and is described in the following subsections.

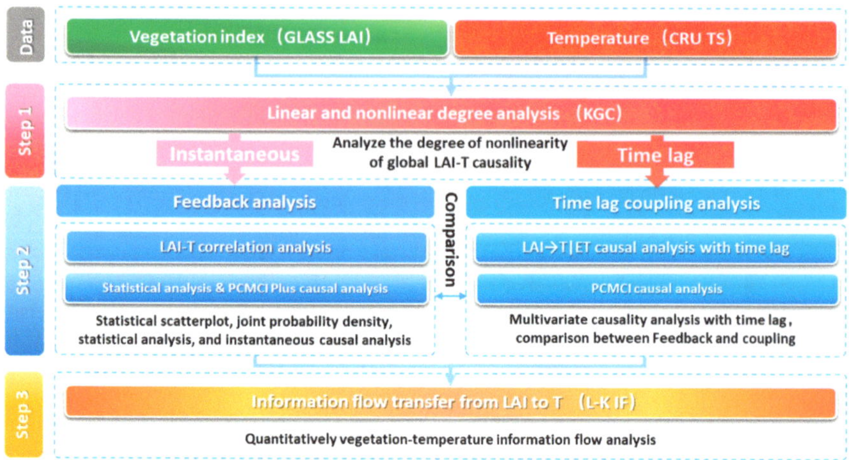

Figure 2. The research outline of this study, where the GLASS LAI is the leaf area index from the Global Land Surface Satellite (GLASS) dataset; CRU TS represents the temperature data from the Climatic Research Unit (CRU) dataset; ET represents the evapotranspiration data from the Global Land Evaporation Amsterdam Model (GLEAM) dataset; and KGC (kernel Granger causality), PCMCI (Peter and Clark momentary conditional independence), and L-K IF (Liang–Kleeman information flow) are three different causality analysis methods.

2.2.1. Kernel Granger Causality (KGC)

In GC, a set of stationary time series $\{x(t)\}_{t=1,.,N+m}$ can be commonly modeled as [20]

$$X_n = \sum_{i=1}^{m} A_i' X_{n-i} + \sum_{i=1}^{m} B_i Y_{n-i} + E_n' \tag{1}$$

where Y represents another time series that can be similarly defined as X. The matrices A_i' and B_i represent the autoregression coefficients, while E_n' denotes white noise [73]. The parameter m corresponds to the order of the autoregressive model, typically determined through selection criteria such as the Bayesian information criterion (BIC) [74]. A standard least squares optimization method was utilized for estimating model coefficients. Instead of measuring the strength of the causal interaction based on the basic concept of GC, kernel Granger causality (KGC) employs kernel function to project the original linear indivisible time series to reproducing Hilbert spaces (H). For each $\alpha \in \{1, \ldots, m\}$, the samples of the αth component of X form a vector $\mathbf{u}_\alpha \in \mathfrak{R}^N$. Calling $\mathbf{X}$ the $m \times N$ matrix having vectors $\mathbf{u}_\alpha$ as rows, H coincides with the range of the $N \times N$ matrix $\mathbf{K} = \mathbf{X}^{\mathrm{T}}\mathbf{X}$, redefining the Granger causality index as

$$\delta_{GC} = \frac{\| P^{\perp} u \|^2}{1 - \tilde{\mathbf{x}}^{\mathrm{T}} \tilde{\mathbf{x}}} \tag{2}$$

where $\tilde{\mathbf{x}}$ is the projection of an organized vector $\theta = (x_{1+m}, \ldots, x_{N+m})^{\mathrm{T}}$ on Hilbert space $H \subseteq \mathfrak{R}^N$. $\mathbf{P}$ represents the projector on the space H. Observing $H^{\perp}$ as the range of the matrix $\tilde{\mathbf{K}} = \mathbf{K}' - \mathbf{PK}' - \mathbf{K}'\mathbf{P} + \mathbf{PK}'\mathbf{P}$, the natural choice of the orthonormal basis in $H^{\perp}$ is the set of the eigenvectors of $\tilde{\mathbf{K}}$. Hence, $\| \mathbf{P}^{\perp}\mathbf{x} \|^2 = \sum_{i=1}^{m} c_i^2$, where c_i is the Pearson's coefficient of x and eigenvectors t_i. With a hand of Bonferroni test, a filtered linear Granger causality index can be drawn as

$$\delta_F(Y \to X) = \frac{\sum_{i'} c_{i'}^2}{1 - \overset{\sim}{\mathbf{x}}^T \overset{\sim}{\mathbf{x}}} \tag{3}$$

Through spectral representation $K(X, X') = \sum_a \lambda_a \phi_a(X)\phi_a(X')$, where ϕ_a are eigenfunctions of the kernel k, it is observed that $\overset{\sim}{\mathbf{x}}$, in the feature space spanned by $\sqrt{\lambda_a}\Psi_a$, coincides with the nonlinear regression of x in the original variables. Consider the space H spanned by zero-mean vectors u_φ, where φ represents eigenvectors of the Gram matrix [70], with elements $K_{ij} = k(X_i, X_j)$. The Gram matrix $K\prime$ is evaluated using both X and Y to predict α, where elements of K are defined as $K'_{ij} = k(Z_i, Z_j)$. The regression values now form a vector that is equal to the projection of α, representing the range of $K\prime$. In subsequent analysis, we consider the inhomogeneous polynomial (IP) of integer order P as our choice to meet our task. The IP kernel [23,24] of integer order p is defined as $K_p(X, X') = \left(1 + X^T X'\right)^P$. Along similar lines as described for the linear case, we constructed the kernel Granger causality, taking into account only eigenvectors that passed the Bonferroni test:

$$\delta_F^K = \sum_{i'} c_{i'}^2 \tag{4}$$

The IP kernel of integer order p is expressed as $X\prime = X \otimes X^T$, and X^T is orthonormal to X (orthogonal unit vectors to $X \otimes X^T$) up to order p and p = 1; consider only the linear regression, where $p \geq 2$ suggests that the information transfer mechanism is nonlinear. The corresponding KGC formula can be expressed as

$$KGC_{x \to y} = \sum_t \left(P^{\perp} y(t)\right)^2 / \sum_t (y(t) - \overline{y}(t))^2 \tag{5}$$

2.2.2. Peter and Clark Momentary Conditional Independence (PCMCI)

PCMCI combines the Peter and Clark (PC) [40] algorithm momentary conditional independence (MCI) approach. Consider an underlying time-dependent system $X_t = \left(X_t^1, \ldots, X_t^N\right)$ with

$$X_t^j = f_j\left(\mathcal{P}\left(X_t^j\right), \eta_t^j\right) \tag{6}$$

where f_i signifies a potential nonlinear functional dependency, and η_t^j denotes the mutually independent dynamic noise components; the nodes within a time series graph symbolize the variables X_t^j at various time lags. Here, $\mathcal{P}\left(X_t^j\right) \subset X_t^- = (X_{t-1}, X_{t-2}, \ldots)$ is defined to represent the causal progenitors, or "parents", of the variable X_t^j, identified from the historical data across all N variables.

For each variable X_t^j, the PC algorithm begins with the establishment of initial parent candidates $\mathcal{P}\left(X_t^j\right) \subset X_t^- = (X_{t-1}, X_{t-2}, \ldots)$. Subsequently, the PC algorithm undertakes nonconditional independence evaluations, excising $X_{t-\tau}^i$ if the null hypothesis $X_{t-\tau}^i \perp\!\!\!\perp X_t^j$ stands unrefuted at the significance threshold α_{PC}. By means of an a-value assessment, causal links in PC algorithms can be defined as

$$p\left(X_{t-\tau}^i \to X_t^j\right) = \underset{\{\mathcal{S}\}}{max} p\left(X_{t-\tau}^i \perp\!\!\!\perp X_t^j \mid \mathcal{S}\right) \tag{7}$$

The aggregated p-value of a causal link is determined as the maximum of all values resulting from conditional independence tests conducted on various condition sets $\mathcal{S}$, as illustrated in Equation (4). Then, the momentary conditional independence (MCI) test is used to test whether $X^i_{t-\tau} \rightarrow X^j_t$ with

$$X^i_{t-\tau} \perp\!\!\!\perp / X^j_t \mid \hat{\mathcal{P}}\left(X^j_t\right) \setminus \left\{X^i_{t-\tau}\right\}, \hat{\mathcal{P}}\left(X^i_{t-\tau}\right) \tag{8}$$

There are various MCI methods provided by PCMCI, including partial correlation (ParCorr), a nonlinear two-step conditional independence test, and two fully nonparametric tests: (i) a test based on Gaussian process regression and a distance correlation (GPDC) for the detection of additive nonlinear causality and (ii) another test based on conditional mutual information (CMI) for the detection of multiplicative nonlinear causality.

To study the positive and negative feedback relationship between LAI and T, we specifically focused on the linear independence test known as ParCorr.

The ParCorr conditional independence test relies on partial correlations and a t-test, assuming the following model:

$$X^i = \mathcal{S}\beta_{X^i} + \epsilon_{X^i}, X^j = \mathcal{S}\beta_{X^j} + \epsilon_{X^j} \tag{9}$$

where β represents coefficients and ϵ denotes Gaussian noise. This model leads to the subsequent residuals

$$r^{X^i} = X^i - \mathcal{S}\hat{\beta}_{X^i}, r^{X^j} = X^j - \mathcal{S}\hat{\beta}_{X^j}, \tag{10}$$

with X^i being the estimated value. ParCorr eliminates the influence of $\mathcal{S}$ on X^i and X^j through ordinary least-squares regression, then testing the independence of the residuals by the Pearson correlation with a t-test. The independence test yields a p-value and a test statistic value I, which represents the correlation coefficient in the case of ParCorr.

2.2.3. Liang–Kleeman Information Flow (L-K IF)

The causality in Liang–Kleeman Information Flow is defined as the time rate of the flow of information transferred from variable X_2 (e.g., leaf area index (LAI)) to X_1 (e.g., temperature (T)) when the system guides a state forward, which can be formulated as:

$$\frac{dX_1}{dt} = F(x,t) = \frac{dX_1^*}{dt} + T_{2\rightarrow 1} \tag{11}$$

where X_1 and X_1^* are n-dimensional vectors, $F = (F_1, F_2, \ldots, F_n)^T$ is the vector field. $T_{2\rightarrow 1}$ is the information flow (IF) rate from X_2 to X_1 demonstrated by Liang in [75]. The time rate of change in X_1 is precisely equivalent to the mathematical expectation of the divergence of the vector field. Specifically, for the case of bivariate analysis in this study ($n = 2$), the information flow (IF) rate is:

$$\begin{aligned} T_{2\rightarrow 1} &= \tfrac{dX_1}{dt} - \tfrac{dX_1^*}{dt} \\ &= -E\left(\tfrac{F_1}{\sigma_1}\tfrac{\partial \rho_1}{\partial x_1}\right) - E\left(\tfrac{\partial F_1}{\partial x_1}\right) \\ &= E\left[\tfrac{1}{\sigma_1}\tfrac{\partial (F_1 \rho_1)}{\partial x_1}\right] \end{aligned} \tag{12}$$

where E denotes the mathematical expectation and σ_1 the marginal probability density of X_1. It is important to note that the IF rate is asymmetric ($T_{2\rightarrow 1} \neq T_{1\rightarrow 2}$), distinguishing it from correlation.

In the real cases where the underlying system dynamics are unknown, Equation (2) is estimated using maximum likelihood estimation (MLE). Given two time series X_1 and X_2, the MLE of $T_{2\rightarrow 1} = 0$, as described by Liang in [49], is:

$$\hat{T}_{2\rightarrow 1} = \frac{C_{11}C_{12}C_{2,d1} - C_{12}^2 C_{1,d1}}{C_{22}Co_{11} - C_{12}^2}, \tag{13}$$

where C_{ij} represents the sample covariance between X_i an X_j, and $C_{i,dj}$ is the covariance between X_i and a forward differenced series derived from X_j, i.e., $\dot{X}_{j,n} = \left(X_{j,n+k} - X_{j,n}\right)/(k\Delta t)$, with $k \geq 1$ as an integer and Δt as the time step. Although the Euler forward difference method has lower accuracy, it is often indispensable because it inherently assumes that present conditions can affect future outcomes, but not vice versa.

A nonzero $T_{2\to1}$ indicates that X_2 causes X_1. Conversely, X_2 is not causal if $T_{2\to1} = 0$, aligning with the principle of nil causality, which is a core theorem in the IF theory, has been empirically tested and verified [55]. As shown in equation (13), if C_{12}=0, then $T_{2\to1} = 0$, but the reverse is not true. This highlights the concept that, in a linear sense, causation implies correlation, but correlation does not imply causation [49]. This holds even when the total information flow is zero [76], conclusively addressing the long-standing debate on causation versus correlation. The Fisher information matrix was utilized for significance testing. Its inverse provides a covariance matrix, establishing a significance level at 5%. A significant result indicates that X_2 exerts a causal influence on X_1 [49].

3. Results

3.1. Nonlinearity of the Causal Relationship between Vegetation and Temperature

The global distribution of the KGC results from LAI to T (LAI→T) from 1981 to 2018 in Figure 3 shows the spatial pattern of vegetation–temperature coupling, presenting differences at various degrees of nonlinearity ((a) P = 1, linear, (b) P = 3, nonlinear, (c) P = 5, strong nonlinear; the rest of the results are provided in Figure S1). Generally, the KGC result of LAI→T is stronger in the mid and low latitudes (intertropical regions) and weaker in the high latitudes. However, stronger signals appear in the high latitudes with increasing nonlinearity (*p* value). For example, relatively strong KGC values also occur in Eurasia and northern North America as shown in Figure 3b,c.

The linear KGC (P = 1) of the LAI→T distribution shown in Figure 3a predominantly highlights high-value regions in intertropical regions which coincide regions of strong land–atmosphere coupling. These areas were characterized by moderate vegetation cover where changes in LAI could impact changes in T.

In Figure 3b,c, the nonlinear KGC from LAI to T presents a more extensive range of regions with strong LAI responses to temperature, encompassing the intertropical regions, temperate grasslands and parts of high-latitude regions. Notable examples include the central grasslands of North America, temperate forests and grasslands in the Eurasian continent, the Sahel; and southern Africa and southeastern Australia in the Southern Hemisphere. Conversely, regions with weak nonlinear responses of LAI to temperature included extreme arid or cold regions, such as the deserts of Central Asia, and polar regions like Antarctica and Greenland. In these areas, the response of LAI to temperature was minimal, reflected in lower nonlinear GC values. Comparing Figure 3a,c into comparison, it is apparent that the regions of strong LAI→T are characterized by nonlinear processes, which reflects the complexity of the coupling.

For the boreal forest in northern Eurasia, linear KGC analysis at P = 1 (Figure 3a) revealed that the causality strength from LAI to T was comparatively weak. This potentially indicates that changes in T are not directly linked to changes in LAI, and is modulated by other mediating factors which may include precipitation, soil moisture and VPD variability. As a result, P = 5 (Figure 3c) shows stronger signals in the region indicating that the coupling is a potentially complex relationship, and hence, a lower predictive power of T anomalies with LAI.

The box plots in Figure 3d generally show increasing KGC values with increasing P values for LAI→T considering both median and mean distributions. Similar results are also obtained for T→LAI (Figure S1f). These results highlight the need to move beyond linear methods to adequately analyse this interaction. While linear methods may provide some useful information for the nature of the global distribution of the coupling, it may miss out on other key regions which limits their use for regional studies. Fundamentally, these KGC results indicates that accurately predicting T in almost all global regions would require considering more climate factors and processes beyond vegetation since they are mostly nonlinearly linked.

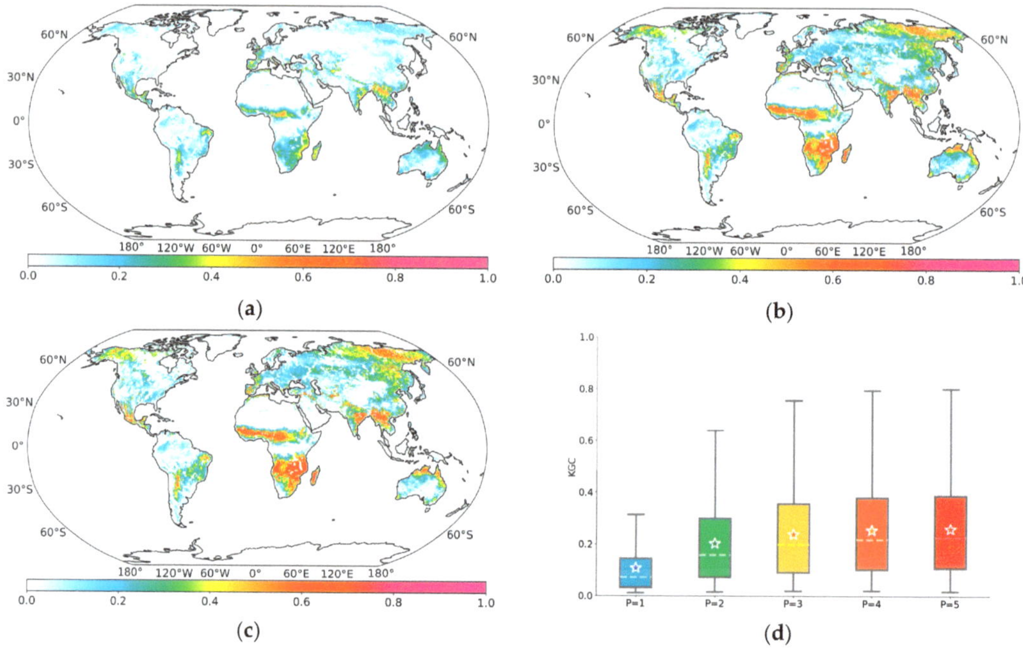

Figure 3. The kernel Granger causality (KGC) results from leaf area index (LAI) to temperature (T), indicated as LAI→T, are shown in (**a**–**c**), representing P equal to (**a**) 1, (**b**) 3, and (**c**) 5. Spatial results for P equal to 4, 5 are shown in the Supplementary Materials Figure S1. The statistical boxplots of KGC results for LAI→T across different degrees of nonlinearity are shown in (**d**), indicated by the parameter P, which ranges from 1 to 5. In (**d**), the star symbol in the middle of each boxplot represents the median value, the dashed line indicates the mean value, and outliers are not displayed. Results are computed at a 5% statistical significance.

3.2. Feedback/Coupling Signs and Timescales between Vegetation and Temperature

In this section, we compute both the Pearson correlation and PCMCI to represent the linear and nonlinear correlation-based forms respectively. Figure 4 presents the linear Pearson correlation results for the interaction (Figure 4a) and PCMCI results (Figure 4b) for LAI→T globally from 1981 to 2018. We note that Figure 4a, may more rightly represent the feedback relationship while Figure 4b would represent the coupling of LAI to T. Furthermore, the analysis in Figure 4b considered evapotranspiration (ET) as a conditional variable. Positive and negative signals are distinguished by warm red and cool blue colors respectively. White regions are statistically insignificant locations. As shown in Figure 4a, positive feedback signals were predominantly observed in the northern high latitudes, such as Siberia and northern Canada, and the mid-latitude regions, including parts of Europe and North America. Conversely, negative feedback was primarily seen in tropical regions like parts of Brazil, the Sahel and India.

In Figure 4a, in the northern high latitudes and mid-latitude regions, the positive influence of LAI on T suggests that increased vegetation cover led to increases in temperatures, which could be attributed to processes such as reduced albedo, where dense vegetation absorbed more sunlight, leading to increased surface temperatures. Additionally, evapotranspiration from vegetation can contribute to local humidity. During the growing season, increased LAI enhanced photosynthetic activity, leading to higher carbon sequestration and surface cooling through evapotranspiration. This cooling effect can further enhance vegetation growth, creating positive feedback. The coupling observed in these regions, as shown in Figure 4b, may indicate reinforcement of this mechanism, where the delay in response highlighted the time taken for vegetation to significantly influence the local climate. These above potentially illustrate that in the northern high latitudes and mid-latitude regions, energy control was the dominant factor, while water control was relatively weaker. Therefore, the strength of the causal relationship reflected in the water cycle was not as strong as that in the energy cycle. We also note that the spatial distribution of the PCMCI results in agree more with that of the KGC results for P = 5, indicating that the PCMCI does capture the nonlinearity within the interaction.

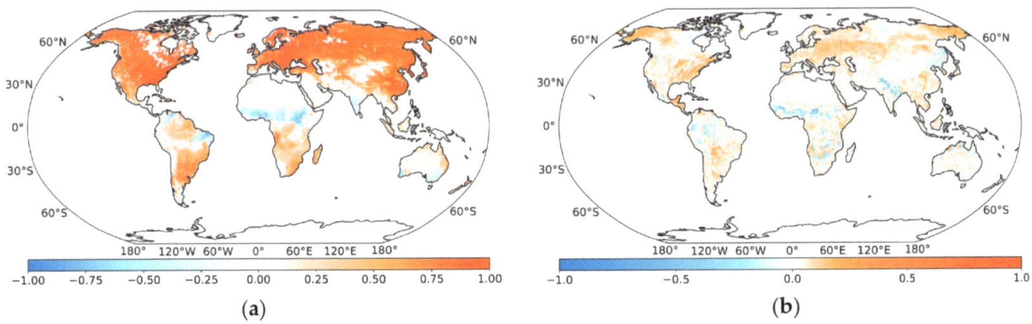

Figure 4. The global distribution of the Pearson correlation between LAI and T is shown in (**a**), while the influence of LAI on T with a one-month time lag considering the influence of ET, is shown in (**b**), both of which are computed at a statistical significance level of 1%, where warm colors (red-orange) indicate positive values, while cool colors (blue) indicate negative values.

Conversely, in tropical regions such as Brazil and the Sahelian region, the negative influence of LAI on T was observed. In these areas, increased vegetation cover was associated with lower temperatures. This negative relationship could be explained by the cooling effects of evapotranspiration, where the process of water vapor release from plants led to heat absorption from the surrounding air, thus cooling the surface. Furthermore, dense vegetation could enhance cloud formation, which in turn reflected solar radiation and reduced surface temperatures. These results add a layer of helpful interpretation to the KGC results in Figure 3 which only shows more of a binary causal–noncausal result.

To further analyse the coupling in detail, we selected the northern forest, Central Asian monsoon zone, Sahel region, and Amazon tropical rainforest as representative areas for further analysis. The analysis of four distinct regions—boreal forest (60–65° N, 90–95° E), East Asian monsoon region (26–31° N, 110–115° E), Sahel (5–10° N, 30–35° E), and Amazon rainforest (0–10° S, 55–65° W)—revealed significant insights into the relationship between LAI and temperature T. Across these regions, scatter plots, joint probability density functions (PDFs), and causal structure consistently showed a positive correlation between LAI and T, although with varying degrees of strength and temporal dynamics influenced by regional climatic conditions.

Overall, the plots demonstrate that the interaction is nonlinear, although the degree of nonlinearity varies from region to region. These validate the KGC results. In the boreal forest, Figure 5a shows a robust correlation of 0.88 between LAI and T at a lag of 1 month for LAI→T. Both Figure 5a,b suggest that LAI→T in the region is not a direct or strictly linear relationship. The causal structure (Figure 5c) emphasized a bidirectional causality portrayed as a positive feedback between vegetation and climate in the region.

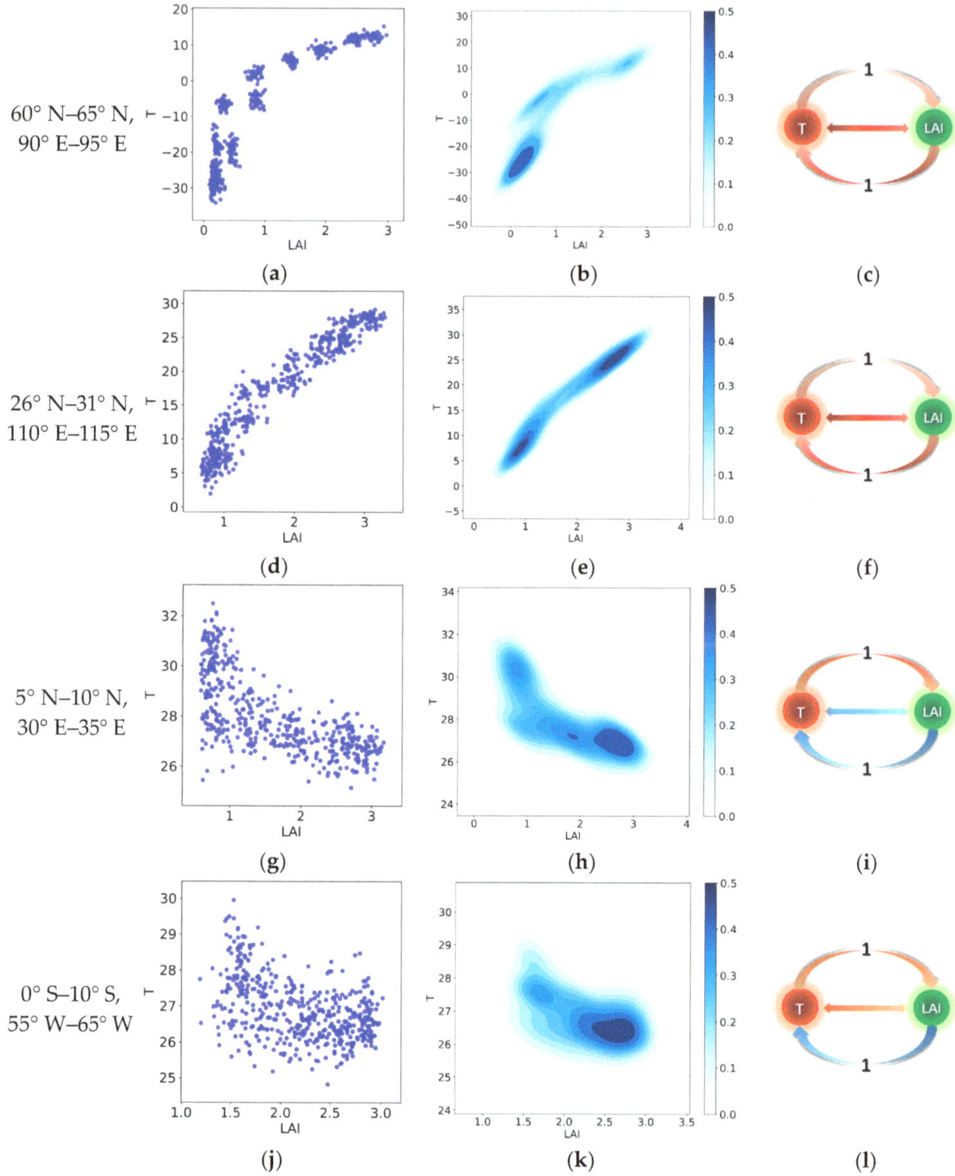

Figure 5. Time series statistics and causal analysis results for selected typical regions. The first column represents the scatter plot between LAI (x-axis) and T (y-axis) of each region, and the second column represents contour plots of the kernel densities of the scatter plot for (**a–c**) the boreal forest (60–65° N, 90–95° E), (**d–f**) East Asian monsoon region (26–31° N, 110–115° E), (**g–i**) Sahel (5–10° N, 30–35° E), and (**j–l**) Amazon rainforest (0–10° S, 55–65° W). The third column, (**c**,**f**,**i**,**l**) shows the causal structure of LAI and T in these regions. The unidirectional curved arrows represent the causal relationship with a delay of 1 calculated with PCMCI, and the bidirectional straight arrows represent the results calculated by PCMCI Plus with no time delay. The colors of the arrows are blue for negative causality and red for positive causality.

Similarly, in the East Asian monsoon region, the scatter plots (Figure 5d,e) show a clear positive trend, indicating an even more direct LAI→T relationship at a 1 month lag.

Figure 5e shows that this more direct relationship (correlation = 0.96) is likely at all vegetation and temperature scenarios in the region. The causal structure (Figure 5f) here also shows a positive vegetation–climate feedback. In the Sahel, the scatter plots from Figure 5g,h the lagged correlation between LAI and T is −0.69, and the joint PDFs (Figure 5h) demonstrate a multimodal distribution, suggesting complex interactions indicating that multiple environmental factors influences the interaction. The causal structure (Figure 5i) indicated a positive coupling of from T to LAI, and a negative coupling of LAI on regional temperature patterns.

Over the selected Amazon rainforest region, the scatter plots (Figure 5j,k) also highlighted the complexity of vegetation–climate interactions in this region (correlation = −0.43). The PDFs (Figure 5k) also showed a multimodal distribution between vegetation and climate in the region. Like the Sahel, the causal structure (Figure 5l) here also shows positive T→LAI and negative LAI→T. These plots provide a detailed view of the statistical relationship between vegetation and climate which shows the complex relationship between them.

3.3. Information Flow between Vegetation and Temperature

The Liang–Kleeman information flow here was to measure the information flow between LAI and T. Notably, similar to the PCMCI results, the Liang–Kleeman information flow (L-K IF) exhibited both positive and negative values, although their interpretations differ, as indicated in Table 1. L–K IF entropy values qualitative, measured in nats per unit time. As noted in Table 1, for PCMCI, a positive value means that the increasing or decreasing changes in the cause variable results in increases or decreases in the effect, while a negative values denote the contrary. However, in L-K IF, a positive IF indicates that one variable is a source of uncertainty for the other, such that, changes in the causal variable results in increased amplitude or variability of the other variable. Negative IF, on the other hand, suggests that changes in the causal variable reduces the amplitude of the other variable, making the causal variable a source of equilibrium and consequently, predictability. Figure 6 shows the L-K IF from LAI to T from 1981 to 2018 across the globe. As noted above, that the signs of L-K IF, based on entropy, should not be interpreted exactly as that of PCMCI, which is fundamentally based on correlation analysis. Positive values in Figure 6 suggest that vegetation was a source of uncertainty for temperature. Thus, changes in vegetation could result in anomalous temperature events such as heat waves or significant cooling of the temperature in the region. In other words, changes in vegetation could amplify temperature variability in the positive IF regions [77]. Negative L-K IF values suggested that vegetation functions to reduce the variability of temperature, keeping it within a range of equilibrium and predictability as "surprises" or anomalous events in the temperature variability would be less probable (henceforth referred to as the stabilizing effect). Overall, positive values were found in regions of strong land–atmosphere coupling [78,79]. Furthermore, these are also water–limiting regions where changes in soil moisture could potentially drive atmospheric conditions. On the other hand, the blue regions were found in energy-limiting regions across the globe where net radiation drives atmospheric conditions like temperature and precipitation by its control on evaporation [80–82]. In Figure 6, northern Eurasia depicts a stabilizing effect of LAI on T, as evidenced by the prevalence of negative L-K IF from LAI to T. This suggests that changes in vegetation in this region contributed to the stabilization of the temperature regime, a phenomenon potentially modulated by evapotranspiration mechanisms (see Figure S4). In northern North America, the L-K IF maps similarly denoted a modest stabilizing role of vegetation on the thermal environment, as indicated by softer blue shades in the LAI→T visual representation. In the Amazon basin, the LAI→T map presented a notably dynamic interaction, with pronounced red-orange gradients which showed that LAI is a source of uncertainty for regional climate system.

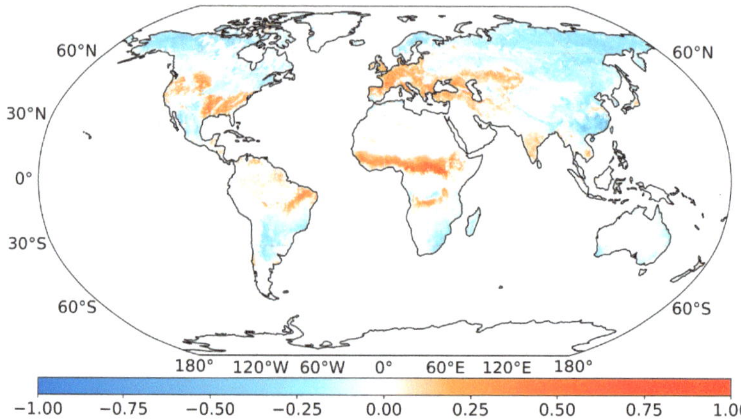

Figure 6. The global Information flow from LAI to T. Red colours indicate positive IF rates and blue colours indicate negative IF rates. All results are computed at a 5% statistical significance. White regions are statistically insignificant regions or masked out due to the absence of vegetation.

4. Discussion

In this study, we used three different causality frameworks to study complex vegetation–climate interactions, demonstrating some of the many facets of causality analysis. Previous studies were based on correlation and regression analysis methods [7–9], which remain inherently uncertain. In fact, Li et al. [3] cautioned that it was difficult to conclude that "vegetation greening in northern Eurasia will lead to temperature increase" based on remote sensing data and regression statistical methods. Thus, this study attempts to look into this problem by utilizing causality analysis methods to capture the essential characteristics potentially eliminating spurious causalities based from observational records.

In our study, we took advantage of KGC, which can derive the degree of nonlinearity within an interaction reflecting the different degrees of the vegetation-temperature coupling complexity as a function of location across the globe. The increasing nonlinearity with higher degrees of the polynomial kernel (P, see Figure 3) indicated that the influence of LAI on T involved multifaceted interactions, possibly including other ecological factors and feedback processes. Additionally, our study identified regions of strong nonlinear causal signals consistent with earlier studies like Schwingshackl et al. [83], who found these areas to be characterized by strong soil moisture–evapotranspiration coupling. Combining these findings, we concluded that the hydrological cycle and processes in transitional zones significantly influenced the causal relationship between vegetation and climate, often in a complex, nonlinear manner. The subtleties of these interactions challenge the notion of a straightforward, predictive relationship between vegetation and climate, pointing instead to a system characterized by complex, emergent properties.

Meanwhile, we focused on elucidating the positive and negative feedback and coupling features of the interaction PCMCI, finding that vegetation's impact on climate exhibited positive feedback and coupling in mainly in the mid- to high-latitude regions, while showing negative feedback in lower mid-to low latitude tropical regions. These results are consistent with recognized global patterns of the feedback between vegetation and air temperature found by Forzieri et al. in [4]. Furthermore, our results showed different feedback signals of vegetation on climate under energy and moisture controls, which is very important in global warming discussions. These findings are also consistent with Krich et al. [46] who also relied on PCMCI. In this study, the combined use of PCMCI and KGC results, provide a detailed analysis of climate transition zones, offering insights into climate change and associated critical scenarios. In regions with strong nonlinearity in land–atmosphere interactions, PCMCI methods extract information on positive and negative feedbacks, along with corresponding time delays, addressing limitations of linear theories and methods that neglect time delays. This detailed analysis, particularly in critical ecological–climatic transition zones globally, enhances our understanding and provides a basis for studying climate change and associated critical scenarios. The L-K IF approach complemented these findings by capturing the dynamics of information flow within the LAI–T coupling. By examining the information flow, this method provided insight into how changes in LAI might induce variability in T, thus indicating regions where changes in LAI could lead to extreme temperature events across the globe, which we find are predominantly in water–limited regions.

In addition to natural factors, human activities also significantly impacted vegetation–climate interactions. However, we note that our study did not take anthropogenic factors into account for the purposes of simplifying the problem. This will be for future studies. The human factor has a great impact on the vegetation on the globe. Real-world scenarios involve complex indirect effects, such as global warming leading to human or animal migration and changes in water use and subsequently affecting urbanization, forest fragmentation, and thus vegetation and ecology [84]. These processes contain important causal information to more adequately understand the causal structure between vegetation and climate. For example, in Figure 4a,b of the study results, there were a few causal signals in the Amazon region that were contrary to the overall regional situation, such as in the southeast of the Amazon in Figure 4a and the scattered positive and negative signals in Figure 4b. These atypical opposite signals may be influenced by human activities, such as the severe human-induced deforestation in the southeastern Amazon found by Ometto et al. in [85] and the impact of road construction on forest fragmentation mentioned by das Neves et al. in [86]. The indirect effects or subprocesses caused by these human factors may be inferred in this paper, but ought to be done with caution.

5. Conclusions

This study systematically studied the causality between vegetation and climate from 1981 to 2018 from observations through different causal characterstics that stem from the added value of different approaches. From these various perspectives, we elucidated the complex impact of vegetation on temperature across the globe. We advocate for an integrated approach to address the climatic and ecological challenges of our time, attaching importance to vegetation dynamics in the climate system. Integrating multiple causality methods, we observed that while KGC highlighted the complexity and nonlinearity in the LAI–T relationship, PCMCI provided a nuanced understanding of the positive or negative feedback, as well as the coupling with time lag, and L-K IF shed light on the system's equilibrium. The divergence in the conclusions drawn by each method illuminates the multifaceted nature of the LAI–T interaction. The KGC results suggested that the relationship between observed LAI and T is far more complex than a simple cause–effect, depending on the location over the globe. The PCMCI analysis, indicated that the LAI–T relationship was characterized by positive or negative influences. Finally, the L-K IF results on system equilibrium implied that vegetation could be both an agent of uncertainty or equilibrium depending on the location under consideration.

Overall, this study highlights the need to go beyond the absolute conclusions of a single method since each method has its strengths and limitations in providing a targeted perspective on the nature of the causal structure of any system under investigation. Thus, there is the need to employ a multimodel analysis that could bring us closer to obtaining more holistic views of the causal structure of natural systems.

Supplementary Materials: The following supporting information can be downloaded at https://www.mdpi.com/article/10.3390/f15081430/s1: Figure S1: The kernel Granger causality (KGC) results between LAI and T, showing KGC results from T to LAI in panels (a) to (e) and from LAI to T in panels (g) to (k), corresponding to parameter p values of 1 to 5. Panel (f) and (l) display statistical boxplots of KGC results for T→LAI and LAI→T relationships across different degrees of nonlinearity indicated by parameter P ranging from 1 to 5. Figure S2: The global distribution of the causal effects from LAI to T based on the PCMCI method, which shows the influence of LAI to T with a time lag of one month. Figure S3. The global Liang–Kleeman Information flow (L-K IF) from T to LAI.

Author Contributions: Conceptualization, Y.S., D.F.T.H., and P.C.; data curation, Y.S., S.L., and X.Z.; formal analysis, Y.S., X.Z., and P.C.; investigation, Y.S., S.L., and X.Z.; methodology, Y.S., D.F.T.H., and F.Z.; project administration, F.Z., X.Z., and P.C.; resources, Y.S., X.Z., and P.C.; software, Y.S. and X.Z.; supervision, S.L., F.Z., and P.C.; validation, D.F.T.H., S.L., and P.C.; visualization, Y.S.; writing—original draft, Y.S.; writing—review and editing, Y.S., D.F.T.H., and P.C. All authors have read and agreed to the published version of the manuscript.

Funding: This research received no funding.

Data Availability Statement: The data presented in this study are available on request from the corresponding author.

Acknowledgments: We are grateful to Daniele Marinazzo, X. San Liang, Samuel Asher Bhatti and David Docquier for their suggestions that helped to shape the motivation and structure of the study.

Conflicts of Interest: The authors declare no conflicts of interest.

References

1. Cui, J.; Piao, S.; Huntingford, C.; Wang, X.; Lian, X.; Chevuturi, A.; Turner, A.; Kooperman, G. Vegetation forcing modulates global land monsoon and water resources in a CO_2-enriched climate. *Nat. Commun.* **2020**, *11*, 5184. [CrossRef] [PubMed]
2. Piao, S.; Wang, X.; Park, T.; Chen, C.; Lian, X.; He, Y.; Bjerke, J.W.; Chen, A.; Ciais, P.; Tømmervik, H.; et al. Characteristics, drivers and feedbacks of global greening. *Nat. Rev. Earth Environ.* **2019**, *1*, 14–27. [CrossRef]
3. Li, Y.; Piao, S.; Li, L.Z.X.; Chen, A.; Wang, X.; Ciais, P.; Huang, L.; Lian, X.; Peng, S.; Zeng, Z.; et al. Divergent hydrological response to large-scale afforestation and vegetation greening in China. *Sci. Adv.* **2018**, *4*, eaar4182. [CrossRef] [PubMed]
4. Forzieri, G.; Alkama, R.; Miralles, D.; Cescatti, A. Satellites reveal contrasting responses of regional climate to the widespread greening of Earth. *Science* **2017**, *356*, 1180–1184. [CrossRef] [PubMed]
5. Sun, G.Q.; Li, L.; Li, J.; Liu, C.; Wu, Y.-P.; Gao, S.; Wang, Z.; Feng, G.-L. Impacts of climate change on vegetation pattern: Mathematical modeling and data analysis. *Phys. Life Rev.* **2022**, *43*, 239–270. [CrossRef] [PubMed]
6. Mao, R.; Jinxi, S.; Bin, T.; Xu, W.; Feihe, K.; Sun, H.; Yuxin, L. Vegetation variation regulates soil moisture sensitivity to climate change on the Loess Plateau. *J. Hydrol.* **2022**, *617*, 128763. [CrossRef]
7. Jiang, M.; He, Y.; Song, C.; Pan, Y.; Qiu, T.; Tian, S. Disaggregating climatic and anthropogenic influences on vegetation changes in Beijing-Tianjin-Hebei region of China. *Sci. Total Environ.* **2021**, *786*, 147574. [CrossRef]
8. Zhang, P.; Cai, Y.; Yang, W.; Yi, Y.; Yang, Z.; Fu, Q. Contributions of climatic and anthropogenic drivers to vegetation dynamics indicated by NDVI in a large dam-reservoir-river system. *J. Clean Prod.* **2020**, *256*, 120477. [CrossRef]
9. Zhao, J.; Huang, S.; Huang, Q.; Wang, H.; Leng, G.; Fang, W. Time-lagged response of vegetation dynamics to climatic and teleconnection factors. *Catena* **2020**, *189*, 104474. [CrossRef]
10. Zhou, Z.; Ding, Y.; Shi, H.; Cai, H.; Fu, Q.; Liu, S.; Li, T. Analysis and prediction of vegetation dynamic changes in China: Past, present and future. *Ecol. Indic.* **2020**, *117*, 106642. [CrossRef]
11. Li, Y.; Zeng, Z.; Huang, L.; Lian, X.; Piao, S. Comment on "Satellites reveal contrasting responses of regional climate to the widespread greening of Earth". *Science* **2018**, *360*, eaap7950. [CrossRef] [PubMed]
12. Kerr, Y.H.; Wigneron, J.P. Vegetation models and observations A review. In *Passive Microwave Remote Sensing of Land—Atmosphere Interactions*; CRC Press: Boca Raton, FL, USA, 2023; pp. 317–344.
13. Lian, X.; Jeong, S.; Park, C.E.; Xu, H.; Li, L.Z.X.; Wang, T.; Gentine, P.; Peñuelas, J.; Piao, S. Biophysical impacts of northern vegetation changes on seasonal warming patterns. *Nat. Commun.* **2022**, *13*, 3925. [CrossRef] [PubMed]
14. Gutman, G.; Skakun, S.; Gitelson, A. Revisiting the use of red and near-infrared reflectances in vegetation studies and numerical climate models. *Sci. Remote Sens.* **2021**, *4*, 100025. [CrossRef]
15. Pearl, J.; Mackenzie, D. *The Book of Why: The New Science of Cause and Effect*; Basic Books: New York, USA, 2018.
16. Pearl, J. *Causality*; Cambridge University Press: Cambridge, UK, 2009.
17. Imbens, G.W.; Rubin, D.B. *Causal Inference in Statistics, Social, and Biomedical Sciences*; Cambridge University Press: Cambridge, UK, 2015.
18. Hagan, D.; Wang, G.; Liang, X.; Dolman, H. A Time-Varying Causality Formalism Based on the Liang–Kleeman Information Flow for Analyzing Directed Interactions in Nonstationary Climate Systems. *J. Clim.* **2019**, *32*, 7521–7537. [CrossRef]
19. Runge, J.; Bathiany, S.; Bollt, E.; Camps-Valls, G.; Coumou, D.; Deyle, E.; Glymour, C.; Kretschmer, M.; Mahecha, M.D.; Muñoz-Marí, J.; et al. Inferring causation from time series in Earth system sciences. *Nat. Commun.* **2019**, *10*, 2553. [CrossRef] [PubMed]
20. Granger, C.W.J. Investigating Causal Relations by Econometric Models and Cross-spectral Methods. *Econometrica* **1969**, *37*, 424. [CrossRef]

1. Hong, Y.; Liu, Y.; Wang, S. Granger causality in risk and detection of extreme risk spillover between financial markets. *J. Econom.* **2009**, *150*, 271–287. [CrossRef]
22. Ajayi, R.A.; Friedman, J.; Mehdian, S.M. On the relationship between stock returns and exchange rates: Tests of Granger causality. *Glob. Financ. J.* **1998**, *9*, 241–251. [CrossRef]
23. Shojaie, A.; Fox, E.B. Granger causality: A review and recent advances. *Annu. Rev. Stat. Its Appl.* **2022**, *9*, 289–319. [CrossRef]
24. Kar, M.; Nazlıoğlu, Ş.; Ağır, H. Financial development and economic growth nexus in the MENA countries: Bootstrap panel granger causality analysis. *Econ. Model.* **2011**, *28*, 685–693. [CrossRef]
25. Seth, A.K.; Barrett, A.B.; Barnett, L. Granger causality analysis in neuroscience and neuroimaging. *J. Neurosci.* **2015**, *35*, 3293–3297. [CrossRef] [PubMed]
26. Stokes, P.A.; Purdon, P.L. A study of problems encountered in Granger causality analysis from a neuroscience perspective. *Proc. Natl. Acad. Sci. USA* **2017**, *114*, E7063–E7072. [CrossRef] [PubMed]
27. Deshpande, G.; LaConte, S.; James, G.A.; Peltier, S.; Hu, X. Multivariate Granger causality analysis of fMRI data. *Hum. Brain Mapp.* **2009**, *30*, 1361–1373. [CrossRef]
28. McGraw, M.C.; Barnes, E.A. Memory matters: A case for Granger causality in climate variability studies. *J. Clim.* **2018**, *31*, 3289–3300. [CrossRef]
29. Papagiannopoulou, C.; Miralles, D.G.; Decubber, S.; Demuzere, M.; Verhoest, N.E.; Dorigo, W.A.; Waegeman, W. A non-linear Granger-causality framework to investigate climate–vegetation dynamics. *Geosci. Model Dev.* **2017**, *10*, 1945–1960. [CrossRef]
30. Smirnov, D.A.; Mokhov, I.I. From Granger causality to long-term causality: Application to climatic data. *Phys. Rev. E Stat. Nonlinear Soft Matter Phys.* **2009**, *80*, 016208. [CrossRef] [PubMed]
31. Friston, K.J.; Moran, R.; Seth, A. Analysing connectivity with Granger causality and dynamic causal modelling. *Curr. Opin. Neurobiol.* **2013**, *23*, 172–178. [CrossRef]
32. Kovács, D.D.; Amin, E.; Berger, K.; Reyes-Muñoz, P.; Verrelst, J. Untangling the Causal Links between Satellite Vegetation Products and Environmental Drivers on a Global Scale by the Granger Causality Method. *Remote Sens.* **2023**, *15*, 4956. [CrossRef]
33. Attanasio, A.; Pasini, A.; Triacca, U. A contribution to attribution of recent global warming by out-of-sample Granger causality analysis. *Atmos. Sci. Lett.* **2012**, *13*, 67–72. [CrossRef]
34. Dhamala, M.; Rangarajan, G.; Ding, M. Estimating Granger causality from Fourier and wavelet transforms of time series data. *Phys. Rev. Lett.* **2008**, *100*, 018701. [CrossRef]
35. Marinazzo, D.; Pellicoro, M.; Stramaglia, S. Kernel method for nonlinear granger causality. *Phys. Rev. Lett.* **2007**, *100*, 144103. [CrossRef] [PubMed]
36. Bueso, D.; Piles, M.; Camps-Valls, G. Explicit Granger causality in kernel Hilbert spaces. *Phys. Rev. E* **2020**, *102*, 062201. [CrossRef] [PubMed]
37. Menon, S.; Denman, K.L.; Brasseur, G.; Chidthaisong, A.; Ciais, P.; Cox, P.M.; Dickinson, R.E.; Hauglustaine, D.; Heinze, C.; Holland, E. *Couplings between Changes in the Climate System and Biogeochemistry*; Lawrence Berkeley National Lab. (LBNL): Berkeley, CA, USA, 2007.
38. Rial, J.; Pielke, R.; Beniston, M.; Claussen, M.; Canadell, J.; Cox, P.; Held, H.; Noblet-Ducoudré, N.d.; Prinn, R.; Reynolds, J.; et al. Nonlinearities, Feedbacks and Critical Thresholds within the Earth's Climate System. *Clim. Chang.* **2004**, *65*, 11–38. [CrossRef]
39. Kennaway, R. When causation does not imply correlation. In *The Interdisciplinary Handbook of Perceptual Control Theory*; Elsevier: Amsterdam, The Netherlands, 2015; pp. 49–72.
40. Spirtes, P.; Glymour, C.; Scheines, R. *Causation, Prediction, and Search*; MIT Press: Cambridge, MA, USA, 2001.
41. Runge, J.; Nowack, P.; Kretschmer, M.; Flaxman, S.; Sejdinovic, D. Detecting and quantifying causal associations in large nonlinear time series datasets. *Sci. Adv.* **2017**, *5*, eaau4996. [CrossRef] [PubMed]
42. Runge, J. Causal network reconstruction from time series: From theoretical assumptions to practical estimation. *Chaos Interdiscip. J. Nonlinear Sci.* **2018**, *28*, 075310. [CrossRef] [PubMed]
43. Gerhardus, A.; Runge, J. High-recall causal discovery for autocorrelated time series with latent confounders. *Adv. Neural Inf. Process. Syst.* **2020**, *33*, 12615–12625.
44. Kretschmer, M.; Coumou, D.; Donges, J.; Runge, J. Using Causal Effect Networks to Analyze Different Arctic Drivers of Midlatitude Winter Circulation. *J. Clim.* **2016**, *29*, 4069–4081. [CrossRef]
45. Dubey, N.; Ghosh, S. The relative role of soil moisture and vapor pressure deficit in affecting the Indian vegetation productivity. *Environ. Res. Lett.* **2023**, *18*, 064012. [CrossRef]
46. Krich, C.; Runge, J.; Miralles, D.; Migliavacca, M.; Pérez-Priego, Ó.; El-Madany, T.; Carrara, A.; Mahecha, M. Estimating causal networks in biosphere–atmosphere interaction with the PCMCI approach. *Biogeosciences* **2020**, *17*, 1033–1061. [CrossRef]
47. Capua, G.d.; Kretschmer, M.; Donner, R.; Hurk, B.v.d.; Vellore, R.; Krishnan, R.; Coumou, D. Tropical and mid-latitude teleconnections interacting with the Indian summer monsoon rainfall: A theory-guided causal effect network approach. *Earth Syst. Dynam.* **2019**, *11*, 17–34. [CrossRef]
48. Qu, Y.; Montzka, C.; Vereecken, H. Causation Discovery of Weather and Vegetation Condition on Global Wildfire Using the PCMCI Approach. In Proceedings of the IGARSS 2021—2021 IEEE International Geoscience and Remote Sensing Symposium, Brussels, Belgium, 11–16 July 2021; pp. 8644–8647. [CrossRef]
49. Liang, X.S. Unraveling the cause-effect relation between time series. *Phys. Rev. E* **2014**, *90*, 052150. [CrossRef]
50. Schreiber, T. Measuring information transfer. *Phys. Rev. Lett.* **2000**, *85*, 461. [CrossRef] [PubMed]

51. Ruddell, B.; Kumar, P. Ecohydrologic process networks: 1. Identification. *Water Resour. Res.* **2009**, *45*, W03419. [CrossRef]
52. Goodwell, A.; Jiang, P.; Ruddell, B.; Kumar, P. Debates—Does Information Theory Provide a New Paradigm for Earth Science? Causality, Interaction, and Feedback. *Water Resour. Res.* **2020**, *56*, e2019WR024940. [CrossRef]
53. Shannon, C.E. A mathematical theory of communication. *Bell Syst. Tech. J.* **1948**, *27*, 379–423. [CrossRef]
54. Deacon, T.W. Shannon–Boltzmann–Darwin: Redefining information (Part I). *Cogn. Semiot.* **2007**, *1*, 123–148. [CrossRef]
55. Liang, X. Information flow and causality as rigorous notions ab initio. *Phys. Rev. E* **2015**, *94*, 052201. [CrossRef] [PubMed]
56. Liang, X.S.; Xu, F.; Rong, Y.; Zhang, R.; Tang, X.; Zhang, F. El Niño Modoki can be mostly predicted more than 10 years ahead of time. *Sci. Rep.* **2021**, *11*, 17860. [CrossRef] [PubMed]
57. Cincinelli, P.; Pellini, E.; Urga, G. Systemic risk in the Chinese financial system: A panel Granger causality analysis. *Int. Rev. Financ. Anal.* **2022**, *82*, 102179. [CrossRef]
58. Cong, J.; Zhuang, W.; Liu, Y.; Yin, S.; Jia, H.; Yi, C.; Chen, K.; Xue, K.; Li, F.; Yao, D. Altered default mode network causal connectivity patterns in autism spectrum disorder revealed by Liang information flow analysis. *Hum. Brain Mapp.* **2023**, *44*, 2279–2293. [CrossRef]
59. Stips, A.; Macias, D.; Coughlan, C.; Garcia-Gorriz, E.; Liang, X.S. On the causal structure between CO_2 and global temperature. *Sci. Rep.* **2016**, *6*, 21691. [CrossRef] [PubMed]
60. Tao, L.; Liang, X.S.; Cai, L.; Zhao, J.; Zhang, M. Relative contributions of global warming, AMO and IPO to the land precipitation variabilities since 1930s. *Clim. Dyn.* **2021**, *56*, 2225–2243. [CrossRef]
61. Docquier, D.; Vannitsem, S.; Ragone, F.; Wyser, K.; Liang, X. Causal Links Between Arctic Sea Ice and Its Potential Drivers Based on the Rate of Information Transfer. *Geophys. Res. Lett.* **2021**, *49*, e2021GL095892. [CrossRef]
62. Zhou, F.; Hagan, D.F.T.; Wang, G.; Liang, X.S.; Li, S.; Shao, Y.; Yeboah, E.; Wei, X. Estimating Time-Dependent Structures in a Multivariate Causality for Land–Atmosphere Interactions. *J. Clim.* **2024**, *37*, 1853–1876. [CrossRef]
63. Krakovská, A.; Jakubík, J.; Chvosteková, M.; Coufal, D.; Jajcay, N.; Paluš, M. Comparison of six methods for the detection of causality in a bivariate time series. *Phys. Rev. E* **2018**, *97*, 042207. [CrossRef] [PubMed]
64. Ding, Y.; Li, Z.; Peng, S. Global analysis of time-lag and -accumulation effects of climate on vegetation growth. *Int. J. Appl. Earth Obs. Geoinf.* **2020**, *92*, 102179. [CrossRef]
65. Wu, D.; Zhao, X.; Liang, S.; Zhou, T.; Huang, K.; Tang, B.; Zhao, W. Time-lag effects of global vegetation responses to climate change. *Glob. Change Biol.* **2015**, *21*, 3520–3531. [CrossRef]
66. Harris, I.; Osborn, T.; Jones, P.; Lister, D. Version 4 of the CRU TS monthly high-resolution gridded multivariate climate dataset. *Sci. Data* **2020**, *7*, 109. [CrossRef]
67. Liang, S.; Cheng, J.; Jia, K.; Jiang, B.; Liu, Q.; Xiao, Z.; Yao, Y.; Yuan, W.; Zhang, X.; Zhao, X.; et al. The Global Land Surface Satellite (GLASS) Product Suite. *Bull. Am. Meteorol. Soc.* **2020**, *102*, E323–E337. [CrossRef]
68. Liang, S.; Zhao, X.; Liu, S.; Yuan, W.; Cheng, X.; Xiao, Z.; Zhang, X.; Liu, Q.; Cheng, J.; Tang, H.; et al. A long-term Global LAnd Surface Satellite (GLASS) data-set for environmental studies. *Int. J. Digit. Earth* **2013**, *6*, 33–35. [CrossRef]
69. Shan, Y.; Wei, J.; Zan, B. Improving Estimates of Land–Atmosphere Coupling Through a Novel Framework of Land Aridity Classification. *Geophys. Res. Lett.* **2024**, *51*, e2023GL106598. [CrossRef]
70. Miralles, D.; Holmes, T.; Jeu, R.D.; Gash, J.; Meesters, A.; Dolman, A. Global land-surface evaporation estimated from satellite-based observations. *Hydrol. Earth Syst. Sci.* **2010**, *15*, 453–469. [CrossRef]
71. Martens, B.; Miralles, D.; Lievens, H.; Schalie, R.V.D.; Jeu, R.; Fernández-Prieto, D.; Beck, H.; Dorigo, W.; Verhoest, N. GLEAM v3: Satellite-based land evaporation and root-zone soil moisture. *Geosci. Model Dev.* **2016**, *10*, 1903–1925. [CrossRef]
72. Docquier, D.; Capua, G.d.; Donner, R.V.; Pires, C.A.L.; Simon, A.; Vannitsem, S. A comparison of two causal methods in the context of climate analyses. *Nonlinear Process. Geophys.* **2024**, *31*, 115–136. [CrossRef]
73. Granger, C. Long memory relationships and the aggregation of dynamic models. *J. Econom.* **1980**, *14*, 227–238. [CrossRef]
74. Neath, A.A.; Cavanaugh, J.E. The Bayesian information criterion: Background, derivation, and applications. *Wiley Interdiscip. Rev. Comput. Stat.* **2012**, *4*, 199–203. [CrossRef]
75. Liang, X. The Liang-Kleeman Information Flow: Theory and Applications. *Entropy* **2013**, *15*, 327–360. [CrossRef]
76. Liang, X. Normalized Multivariate Time Series Causality Analysis and Causal Graph Reconstruction. *Entropy* **2021**, *23*, 679. [CrossRef] [PubMed]
77. Hagan, D.; Dolman, H.; Wang, G.; Sian, K.T.C.L.K.; Yang, K.; Ullah, W.; Shen, R. Contrasting ecosystem constraints on seasonal terrestrial CO_2 and mean surface air temperature causality projections by the end of the 21st century. *Environ. Res. Lett.* **2022**, *17*, 124019. [CrossRef]
78. Koster, R.; Dirmeyer, P.; Guo, Z.; Bonan, G.; Chan, E.; Cox, P.; Gordon, C.T.; Kanae, S.; Kowalczyk, E.; Lawrence, D.; et al. Regions of Strong Coupling Between Soil Moisture and Precipitation. *Science* **2004**, *305*, 1138–1140. [CrossRef]
79. Miralles, D.G.; Jimenez, C.; Jung, M.; Michel, D.; Ershadi, A.; McCabe, M.F.; Hirschi, M.; Martens, B.; Dolman, A.; Fisher, J.B.; et al. The WACMOS-ET project—Part 2: Evaluation of global terrestrial evaporation data sets. *Hydrol. Earth Syst. Sci.* **2015**, *20*, 823–842. [CrossRef]
80. Koster, R.; Guo, Z.; Dirmeyer, P.; Bonan, G.; Chan, E.; Cox, P.; Davies, H.; Gordon, C.T.; Kanae, S.; Kowalczyk, E.; et al. GLACE: The Global Land-Atmosphere Coupling Experiment. Part I: Overview. *J. Hydrometeorol.* **2006**, *7*, 590–610. [CrossRef]
81. Dirmeyer, P.A. The terrestrial segment of soil moisture–climate coupling. *Geophys. Res. Lett.* **2011**, *38*, L16702. [CrossRef]

82. Hsu, H.; Dirmeyer, P.A. Soil moisture-evaporation coupling shifts into new gears under increasing CO_2. *Nat. Commun.* **2023**, *14*, 1162. [CrossRef] [PubMed]
83. Schwingshackl, C.; Hirschi, M.; Seneviratne, S. Quantifying Spatiotemporal Variations of Soil Moisture Control on Surface Energy Balance and Near-Surface Air Temperature. *J. Clim.* **2017**, *30*, 7105–7124. [CrossRef]
84. Cotton, W.R.; Pielke, R.A., Sr. *Human Impacts on Weather and Climate*; Cambridge University Press: Cambridge, UK, 1995.
85. Ometto, J.; Aguiar, A.; Martinelli, L. Amazon deforestation in Brazil: Effects, drivers and challenges. *Carbon Manag.* **2011**, *2*, 575–585. [CrossRef]
86. das Neves, P.B.T.; Blanco, C.J.C.; Duarte, A.A.A.M.; das Neves, F.B.S.; das Neves, I.B.S.; dos Santos, M.H.d.P. Amazon rainforest deforestation influenced by clandestine and regular roadway network. *Land Use Policy* **2021**, *108*, 105510. [CrossRef]

MDPI AG
Grosspeteranlage 5
4052 Basel
Switzerland
Tel.: +41 61 683 77 34

Forests Editorial Office
E-mail: forests@mdpi.com
www.mdpi.com/journal/forests

www.ingramcontent.com/pod-product-compliance
Lightning Source LLC
Chambersburg PA
CBHW041407210926
51850CB00013B/3204
* 9 7 8 3 7 2 5 8 3 4 5 5 6 *